Developing Intelligent Agents for Distributed Systems

Developing Intelligent Agents for Distributed Systems

Exploring Architecture, Technologies, and Applications

Michael Knapik
Jay Johnson

McGraw-Hill

New York San Francisco Washington, D.C. Auckland Bogotá
Caracas Lisbon London Madrid Mexico City Milan
Montreal New Delhi San Juan Singapore
Sydney Tokyo Toronto

Library of Congress Cataloging-in-Publication Data

Knapik, Michael.
    Developing intelligent agents for distributed systems : exploring architecture, technologies, and applications / Michael Knapik, Jay Johnson.
       p.   cm.
    Includes bibliographical references and index.
    ISBN 0-07-035011-6
     1. Intelligent agents (Computer software) 2. Electronic data processing—Distributed processing. 3. Computer software—Development. I. Johnson, Jay. II. Title.
    QA76.76.I58K58 1998
    006.3—dc21   97-24364
                                                                                CIP

McGraw-Hill
A Division of The McGraw-Hill Companies

Copyright © 1998 by The McGraw-Hill Companies, Inc. All rights reserved. Printed in the United States of America. Except as permitted under the United States Copyright Act of 1976, no part of this publication may be reproduced or distributed in any form or by any means, or stored in a data base or retrieval system, without the prior written permission of the publisher.

1 2 3 4 5 6 7 8 9 0 FGR/FGR 9 0 2 1 0 9 8 7

ISBN 0-07-035011-6

The sponsoring editor for this book was Steven M. Elliot, the editing supervisor was Paul R. Sobel, and the production supervisor was Sherri Souffrance. It was set in Century Schoolbook by Estelita F. Green of McGraw-Hill's Professional Book Group composition unit.

Printed and bound by Quebecor Fairfield.

McGraw-Hill books are available at special quantity discounts to use as premiums and sales promotions, or for use in corporate training programs. For more information, please write to the Director of Special Sales, McGraw-Hill, 11 West 19th Street, New York, NY 10011. Or contact your local bookstore.

Information contained in this work has been obtained by The McGraw-Hill Companies, Inc. ("McGraw-Hill") from sources believed to be reliable. However, neither McGraw-Hill nor its authors guarantees the accuracy or completeness of any information published herein and neither McGraw-Hill nor its authors shall be responsible for any errors, omissions, or damages arising out of use of this information. This work is published with the understanding that McGraw-Hill and its authors are supplying information, but are not attempting to render engineering or other professional services. If such services are required, the assistance of an appropriate professional should be sought.

 This book is printed on recycled, acid-free paper containing a minimum of 50% recycled, de-inked fiber.

Contents

Foreword   xiii
Preface   xvii
Trademarks   xix

Section 1 Introduction   1

1.1 Intelligent Agents—Some definitions   2
1.2 What This Book Is About   6
1.3 Section Summaries   7
1.4 Audience and Structure of This Book   9

Section 2 From Artificial Intelligence Comes Intelligent Agents   11

2.1 Introduction   11
2.2 40 Years of Classical AI   12
2.3 Hierarchy: Bridging the Gap between Natural and Artificial Computational Worlds   17
    2.3.1 Consciousness at the top of the hierarchy   19
    2.3.2 Hierarchy and IA systems   20
2.4 The Connectionist Revolution: Human Brain as Inspiration   21
2.5 Agents of the Mind   23
2.6 Agents of the Computer—AI Begets IAs   26
2.7 AI and Agents—the Specifics   29
    2.7.1 Solving problems with search methods   30
    2.7.2 Knowledge-based agents and agents that can reason   30
    2.7.3 Goal-driven agents and agents that can plan   31
    2.7.4 Agents that can reason under uncertainty   32
    2.7.5 Agents that learn   33
    2.7.6 Agents and communications   34

Section 3 Converging Technologies that Facilitate and Enable Agents   37

3.1 Introduction   37
3.2 Expert Systems and Knowledge Bases   39
    3.2.1 "Classical" expert systems   39
    3.2.2 Agents with common sense: Lenat's universal expert system   43
        3.2.2.1 The Cyc knowledge base   44
        3.2.2.2 CycL—the Cyc representation language   44
        3.2.2.3 Inferencing in Cyc   44

		3.2.2.4	Cyc Interface tools	45
		3.2.2.5	Applications of Cyc	46
		3.2.2.6	Distributed IAs using Cyc	48
		3.2.2.7	Accessing Cyc	51
3.3	Object Orientation—An Overview			52
	3.3.1	OO concepts and terminology		53
		3.3.1.1	What is an object?	53
		3.3.1.2	Abstration	53
		3.3.1.3	The states of the art	54
		3.3.1.4	Attributes	55
		3.3.1.5	Methods in the madness	55
		3.3.1.6	Encapsulation	57
		3.3.1.7	Inheritance	57
		3.3.1.8	Polymorphism	58
		3.3.1.9	Persistence	60
	3.3.2	OO analysis and design		60
		3.3.2.1	Static model	61
		3.3.2.2	Dynamic model	61
		3.3.2.3	Making a model work	62
		3.3.2.4	Detailed design and implementation	63
		3.3.2.5	Tasks and concurrency	63
		3.3.2.6	Network and interobject communication considerations	64
		3.3.2.7	Dynamic binding	65
		3.3.2.8	Classic programming	65
		3.3.2.9	OO is not the holy grail or silver bullet	67
		3.3.2.10	OO and team-based development	67
		3.3.2.11	OO metrics	68
		3.3.2.12	OO and maintenance	70
	3.3.3	OO agents		71
		3.3.3.1	OO agents are reusable	71
		3.3.3.2	Reduced agent-development costs	72
		3.3.3.3	OO development enables flexible agent structuring	73
		3.3.3.4	OO agents are maintainable	73
		3.3.3.5	OO agents are extensible	74
		3.3.3.6	OO agent systems are understandable	74
		3.3.3.7	OO agent development supports interconnected hierarchies of agents and domains	74
		3.3.3.8	OO agents and intrinsic system knowledge	75
		3.3.3.9	OO agents plus OO domains enables modeling and simulations	76
		3.3.3.10	What is an object-oriented agent?	76
		3.3.3.11	What are OO agents made of?	76
		3.3.3.12	Agent attributes	77
		3.3.3.13	Agent operations	77
		3.3.3.14	Encapsulation—the agent boundary	77
		3.3.3.15	Agent classes and agent instantiation	78
		3.3.3.16	Agent classes—the root agent class	79
		3.3.3.17	Domain-specific agent classes	80
	3.3.4	Relationships among agents		81
		3.3.4.1	Relationship defined	81
		3.3.4.2	The composition relationship	82
		3.3.4.3	Relationships can be modeled as connector objects	85
		3.3.4.4	Dynamic relationships—scenarios revisited	85
	3.3.5	OO agent architecture benefits from uniformity of representaton		88
	3.3.6	Agent development by extension and refinement		88
		3.3.6.1	Reuse by specialization and generalization	88
		3.3.6.2	Reuse by composition and encapsulation	89
		3.3.6.3	Reuse by instantiation	90

3.4 Intelligent Agents via Soft Computing ... 90
3.4.1 Fuzzy systems and fuzzy logic ... 91
3.4.1.1 Introduction ... 92
3.4.1.2 Need for fuzzy reasoning techniques in IA systems ... 92
3.4.1.3 An overview of fuzzy technologies ... 94
3.4.1.4 Applications of fuzzy systems technologies ... 97
3.4.1.5 Making fuzzy agents ... 99
3.4.2 Evolutionary computing ... 101
3.4.2.1 Introduction ... 102
3.4.2.2 Need for evolutionary computing techniques in IA systems ... 103
3.4.2.3 How developers and agents can make better agents: incorporating evolutionary computing technology into agent systems ... 105
3.4.3 Artificial neural networks—computers that learn while they compute ... 000
3.4.3.1 Introduction: the brain as model ... 107
3.4.3.2 Need for neural networks ... 108
3.4.3.3 Overview of artificial neural networks ... 110
3.4.3.4 Applications of neural systems technologies ... 112
3.4.3.5 Making brainy agents ... 112

Section 4 Agent-Enabling Infrastructures ... 115

4.1 Introduction ... 115
4.1.1 A word about interoperability "standards" ... 119
4.1.2 A word about client/server ... 120
4.1.3 Evaluating infrastructures for agents ... 121
4.1.3.1 Interoperability infrastructures—agent communications and agent application models ... 121
4.1.3.2 Criteria for choosing an infrastructure ... 123
4.2 OpenDoc ... 125
4.3 Object Linking and Embedding and ActiveX ... 130
4.3.1 OLE—an overview ... 130
4.3.1.1 OLE common object model ... 131
4.3.1.2 Structure of OLE ... 132
4.3.2 Active X ... 133
4.3.2.1 ActiveX technologies ... 134
4.3.2.2 ActiveX and Java ... 135
4.3.3 Developing agents based on OLE and ActiveX ... 135
4.4 The Common Object Request Broker Architecture ... 137
4.4.1 Interoperability between ORBs and agents over the Internet ... 141
4.5 The Distributed Computing Environment ... 143
4.6 Other Infrastructures ... 146
4.6.1 Networked objects ... 146
4.6.2 Portable distributed objects ... 147
4.6.3 Publish/Subscribe and AppleEvents ... 147
4.6.4 Operating systems ... 148
4.6.5 A potpourri of agent environments ... 149
4.6.5.1 Database management systems ... 149
4.6.5.2 Collaborative computing and groupware ... 149
4.6.5.3 Expert systems environments and agents ... 150

Section 5 Agent Architectures ... 151

5.1 Introduction ... 151
5.1.1 Architecture and infrastructure—a fine line ... 152
5.1.2 What makes a good agent architecture? ... 153
5.1.3 Analyzing an agent architecture ... 154

viii   Contents

5.2 Spectrum of Architectural Complexity — 155
    5.2.1 Simple, single agent equals simple architecture — 155
    5.2.2 A few simply interacting agents equal moderately complex architecture — 156
    5.2.3 Many complex agents plus complex interaction equals complex architecture — 156
5.3 Reiken's M Architecture—A Complex Architecture of Integrated, Diversified Agents — 157
    5.3.1 Integrating diversified agents—the issues — 157
    5.3.2 Basic structure of integration — 157
    5.3.3 Application of M—the virtual meeting room — 160
5.4 Genesereth's Architecture—Architectural Concepts Emphasizing Interoperability — 161
    5.4.1 The importance of a metaprotocol — 161
    5.4.2 Language issues — 163
    5.4.3 Communications issues — 164
    5.4.4 The agent communications language — 165
        5.4.4.1 ACL and agent behavior — 167
    5.4.5 An interoperability facilitator — 167
5.5 Leveraging Existing Intelligence — 169
5.6 Rosenschein's Approach: An Architecture for Agent Negotiations — 170
    5.6.1 Negotiation protocols—the issues — 171
    5.6.2 Negotiation protocols from game theory — 172
    5.6.3 Quest for an agent salutation protocol — 172
    5.6.4 Agent negotiations—considering the domain — 176
    5.6.5 Agent negotiation protocols—utility maximization — 179
5.7 Kautz's Architectural Approach—Bottom-Up Prototyping and Interation — 179
5.8 Kuo-Cho Lee's ITX Architecture for Agent Control — 181
5.9 Edmond's Collaborating Agents—A Federation Architecture — 184
5.10 Agent Architectures and Emergence — 187
5.11 Sugawara's Architectural Concepts—Dealing with Change — 188

Section 6 Agent-Design Considerations — 193

6.1 Introduction — 193
6.2 Designing Agents—The Big Picture — 194
    6.2.1 Agent requirements analysis—general aspects — 194
    6.2.2 Agent requirements analysis—specific aspects — 195
    6.2.3 General design considerations — 196
6.3 Agents, Platforms, and Environments: Where Agents Fit In — 200
    6.3.1 Execution environments — 200
        6.3.1.1 Requirements for agent execution environments — 200
        6.3.1.2 Multithreaded, multitasking operating systems — 202
    6.3.2 Distributed computing paradigms — 203
        6.3.2.1 The client/server paradigm — 204
        6.3.2.2 Agent mobility as extension of the client/server paradigm — 205
        6.3.2.3 When the Internet/Intranet is the computer—agent collaboration and the impending peer-to-peer paradigm — 207
    6.3.3 Considerations for a common agent platform — 210
6.4 Agents and Humans — 212
    6.4.1 Agents and humans working together — 212
    6.4.2 Anthropomorphic considerations — 215
    6.4.3 Considering the agent's audience—the user — 216
    6.4.4 Agent-user interaction—trust, competence, and learning — 217
    6.4.5 Helping humans work: designing work-flow agents — 220
6.5 Incorporating Agent Capabilities in Shrink-Wrap Software — 220
6.6 Topics on Intelligence — 221
    6.6.1 Simple agents versus intelligent agents — 221
    6.6.2 Designing in intelligence and autonomy — 222

			6.6.2.1 Goals and plans	222
			6.6.2.2 Etzioni's Softbot example of goals and plans	224
			6.6.2.3 An agent's environment and planning models	225
6.7	Agent Components			228
6.8	Classifying Agents Based on Degree of Mobility			230
	6.8.1	Fixed or stationary agents		231
	6.8.2	Mobile or itinerant agents		232
		6.8.2.1 Mobile agent paradigm		233
		6.8.2.2 Mobile agent concepts		236
		6.8.2.3 Puttings things together		243
		6.8.2.4 Mobile agent technology		243
		6.8.2.5 Mobile agents—conclusion		246
6.9	Authentication, Exceptions, and Security			248
	6.9.1	Handling Exceptions		249
	6.9.2	Security considerations—general		250
	6.9.3	Security considerations—authentication and digital signatures		251
6.10	Agents Programmed or Configured by the End User			251
6.11	Agent Communications			255
	6.11.1	Knowledge query and manipulation language		255
	6.11.2	Knowledge Interchange Format (KIF)		255
	6.11.3	Agent communications example		255

Section 7 Developing Intelligent Agents Now — 259

7.1	Introduction			259
7.2	Building Simple Agents			260
	7.2.1	Agents (almost) without programmers		260
	7.2.2	Power-user environments for building agents		261
7.3	Serious Agent Tool Kits and Development Environments—Introduction			265
	7.3.1	Smalltalk agents		265
		7.3.1.1 Smalltalk development environments		267
		7.3.1.2 Basic components of a Smalltalk-based agent		268
		7.3.1.3 Smalltalk agents by example—an agent auction		269
		7.3.1.4 Specializing Smalltalk agents for particular tasks		270
	7.3.2	Distributed Smalltalk		274
		7.3.2.1 Interactive environment		274
		7.3.2.2 Open architecture		275
		7.3.2.3 Standards-based object services—Distributed Smalltalk frameworks		275
		7.3.2.4 How Distributed Smalltalk enables agent interaction		277
7.4	Java			279
	7.4.1	Agent applictions in Java		279
	7.4.2	Java as an OO language		280
	7.4.3	Java tools		281
	7.4.4	Applets and servelets as agents		282
	7.4.5	Agile agents in Java		283
	7.4.6	The Java agent template		285
		7.4.6.1 Using the Java agent template		285
		7.4.6.2 The Java agent template application programming interface		286
	7.4.7	Agent-support facilities in Java		289
7.5	Telescript: The Complete Mobile Agent Environment			290
	7.5.1	Mobile agent technology		290
		7.5.1.1 Language		290
		7.5.1.2 Engine		292
		7.5.1.3 Protocols		292
	7.5.2	Programming mobile agents—the Telescript object model		293

			7.5.2.1	Object structure	293
			7.5.2.2	Object classification	294
			7.5.2.3	Object manipulation	295
		7.5.3	Programming a place		295
			7.5.3.1	The catalog entry class	295
			7.5.3.2	The warehouse class	297
			7.5.3.3	The price reduction class	299
		7.5.4	Programming an agent		299
			7.5.4.1	Shopper class	299
			7.5.4.2	Product unavailable class	302
		7.5.5	Using mobile agents—monitoring changing conditions		302
			7.5.5.1	The user experience	303
		7.5.6	Using mobile agents—doing time-consuming legwork		305
			7.5.6.1	The user experience	305
		7.5.7	Mobile agents—using services in combination		307
			7.5.7.1	The user experience	307
		7.5.8	Telescript in the real world		310
	7.6	Agent Development Environments and Security			311
	7.7	Modern Object-Oriented Languages and the Agent Paradigm			312
		7.7.1	The coming Java wave		312
		7.7.2	Java, Telescript, and Smalltalk		314
		7.7.3	Joining forces?		314
		7.7.4	Summary		315

Section 8 Agent Applications   317

8.1	Network Agents		317
	8.1.1	LAN management agents	318
	8.1.2	NetWare management agent	318
	8.1.3	NetWare hub services agent	318
	8.1.4	NetWare LANalyzer agent	319
	8.1.5	Network software distribution agents	319
	8.1.6	Automatic access and connection agents	320
8.2	Database Agents		320
	8.2.1	Data integrity checking agent	320
	8.2.2	Constraint agents	320
	8.2.3	Database report-distribution agents	320
	8.2.4	A distributed database backup agent	321
8.3	Communications Managemnet Agents		321
8.4	Search Agents		322
8.5	Avatars and Cyberspace		322
8.6	Assistants and Work-Flow-Automation Agents		323
8.7	Financial Agents		324
8.8	Filtering Agents		324
	8.8.1	NewT—personalization of Usenet news	325
8.9	Agents as Researchers and Reporters		325
8.10	Telephony Agents		326
8.11	Commerce—Deal-Making Agents in a Worldwide Marketplace		327
	8.11.1	Bits of commerce—agents finding products	328
	8.11.2	Bits of commerce—agents finding people	329
8.12	Advertising agents		330
8.13	E-Mail Agents		330
8.14	Database Access via the Web		331
8.15	Agents in Industrial Automation and Control Domains		331
8.16	Governmental Agents		332

8.17	Medical Agents	333
8.18	Military Agents	334
8.19	Computer-Aided Design Helpers	334
8.20	JAT-Based Agents for Interactive, Collaborative, Concurrent Design and Engineering	334
8.21	Technical Assistance Agent	336
8.22	Decision Support Agents	336
8.23	"Bots"	337
8.24	Author's Assistant	337
8.25	Anthropomorphic Agents: Firefly	338
8.26	Big Brother Agents	338
8.27	Agents as Meeting Facilitators	339

Section 9 Agent Futures — 341

9.1	Introduction	341
9.2	The Future of Network Management	343
9.3	No Surfing	344
9.4	Commercial Agents	344
9.5	Net Searching and Information Mining	346
9.6	Agents to Infiltrate Applications	346
9.7	Military Agents	347
9.8	Database Agents	347
9.9	"Big supplier" Is Watching You	347
9.10	Trust	348
9.11	Information Agents and Cooperative Information Systems	348
9.12	Future Agent Builders	349
9.13	Social Issues Pertaining to Agent Technology	350
9.14	Replacing Humans with Software	351
9.15	The Global Desktop	353
9.16	Agents at Home	353
	9.16.1 A day in the life of an agent-enhanced human	354
9.17	Agents: The Dark Side	354
	9.17.1 Future agent security	355
	9.17.2 Privacy	355
	9.17.3 You are your agent(s) (at least in cyberspace)	356
9.18	Inventing the Future of Agents at MIT: Work at Software Agents Group, MIT Media Laboratory	357
	9.18.1 Modeling intelligent autonomous agents	357
	9.18.2 Computational model of emotion for autonomous agents	358
	9.18.3 Software agents	358
	9.18.4 Agents that reduce information overload	358
	9.18.5 Amalthaea—a multi-agent system that discovers, monitors, and filters information resources	359
	9.18.6 Yenta—matchmaking agents	359
	9.18.7 Remembrance agents	359
	9.18.8 Using simulated evolution to create adaptive systems	360
	9.18.9 Anthropomorphizing software agents	360
	9.18.10 Browsing large information spaces—emergent structure from collective action	360
	9.18.11 Kasbah—an agent marketplace for buying and selling goods	361
	9.18.12 ALIVE—artificial life interactive video environment	361
	9.18.13 Modeling synthetic characters for games and interactive storytelling	361
9.19	Miscellaneous Agent-Related Projects	362

9.19.1	Intelligent browsing agents	362
9.19.2	Persona project	362
9.19.3	On-line cooperating agent architecture	362
9.19.4	Agents and ontologies	362
9.19.5	Agent architecture	363
9.19.6	Operating system support for agents	363
9.19.7	Internet search agent	363
9.19.8	ARPA Intelligent Integration of Information (I"3) project	363
9.19.9	Guardian: a prototype intelligent agent for monitoring intenstive-care and other medical patients	363
9.19.10	Intelligent, ethical agents	364
9.19.11	Mail agent	364
9.19.12	Knowledge-based agents	364
9.19.13	Agent collaboration languages	364
9.19.14	Sulla—a user agent for the web	364
9.20	The Future of AI equals The Future of IAs	365
9.21	Summary	367

Acronyms 369
Bibliography 375
Index 381

Foreword

The Internet in general and the Web in particular are the most significant recent developments in data communication—arguably in all of computing. An old idea takes on new significance in the world that the Web creates.

Personal Computers and Public Networks

Consider for a moment the personal computer. The PC put application software closer to the user than had the mainframe or minicomputer. In so doing, it altered two critical parameters of computing, performance and control:

- The increased performance of the PC made possible the graphical user interface. This in turn made computers accessible to ordinary people by making an application's functionality visible on screen (in menus).
- The increased control that the PC gave to the individual user made possible today's mass market for computer software. The user could select and purchase software for his or her own computer.
- Consider now the browser, the PC application that gives users access to the World Wide Web. In particular, consider it in terms of two critical parameters of networking analogous to the parameters of computing the PC affected:
- The user extracts value from the Web by interacting with it. The browser demands time-consuming *interaction* even for tasks that lend themselves to *automation*. This limits the user's productivity—his performance.
- The Web site operator provides not just information, but also the one application through which the information is accessed. On his own computer, the user runs only the browser. The result is a loss in user control.

The Web's value lies in the goods and services it offers, that they're not limited to those any one on-line service provider can assemble, and that they're universally accessible despite this diversity. The importance of the browser is that it places these vast resources at the fingertips of ordinary people. The browser's drawback is that it makes those fingertips an absolute necessity.

The Problem with Browers

An example illustrates the problem. Suppose I'm interested in a certain author—Umberto Eco, perhaps. If he publishes a new paper on semiotics or a new novel, I want a copy. If he schedules a lecture at a nearby university, I want to know about it as soon as possible; I'll go if my schedule allows.

It's quite possible that all the information I need to keep abreast of Eco's activities is to be found on the Web today. It's likely that every publisher that might be an outlet for Eco's work has a home page. It's even more likely that Stanford, UC Berkeley, San Francisco State, etc. have home pages. Let's assume that these publishers and universities provide, through their home pages, news of new books and upcoming events, respectively.

How do I proceed? I make a list of publishers and universities. Then, say, monthly, I go to their home pages in search of Eco. Although all of the home pages are laid out differently, after a visit or two I become familiar with the terrain. The real problem—the reason, in fact, that I abandon the entire endeavor after a month or two—is that it requires too much of my time.

The problem is clear. The Web offered me information in principle but not in practice. I needed an application that would track the activities of well-known authors and report those activities to me, say, by email, but I had only a browser. I wanted automated access to the Web, but I had to use it interactively. I'd have been happy to interact with the Web once—to tell it of my interest in Eco—but not once a month. I call this "the browser problem."

The Promise of Agents

Intelligent agents arose in artificial intelligence, but have arisen anew in computer networking because they promise to solve the browser problem. Rather than keep track of Eco myself, I can delegate that task to an agent.

An *agent* is software that can carry out information-related tasks without ongoing human supervision. Above all else, an agent is self-reliant. An agent may exhibit expert knowledge of a particular field; act with a user's authority so that it can access the information it needs; communicate with the user to get occasional advice or to report results; and survive crashes and restarts of its host computer so that it can execute long enough to complete its work. One agent, furthermore, may be able to meet and interact with another. This important ability enables groups, or societies, of agents to collaborate.

The transposition of the agent concept from artificial intelligence to computer networking introduces one important new possibility: running an agent on a network server, rather than in the user's PC. The agent I instruct to keep tabs on Eco, for example, is much better off on a server than in my PC. Once it has its assignment, all it needs—publishers, universities, even my mailbox—it finds in the Internet. It should run there. The choice of venue is not a matter of taste. If it ran in my homePC, with my one voice telephone line as its means of communication, the agent would inconvenience me by tying up my line. If it ran in my wirelessPDA, it would cost me a lot of money.

Note that to do its job, my Eco agent must draw upon the resources of many Web sites, not just one. If the agent had to execute on a single server, it would have to access most or all of those resources remotely. Alternatively, if during the course of its execution it could move from server to server at will, it could access the resources of each server locally. In many applications, such *mobile* agents are more efficient and less expensive than *stationary* agents.

The conventional *standalone* applications most familiar to us (for example, work processors) are quite at home in the PC. That's where they find the resources they require—keyboard, mouse, screen, disk. Many of the new *agent-based* applications will be at home in the network for the same reason. The resources they need—directories, databases, mailboxes, communication media, and goods and services of all kinds—are out there in the network.

The AI community has produced many agent technologies since it conceived of agents years ago. In the last several years, the networking community has produced agent technologies for networks. At a September 1996 workshop on mobile agents at Dartmouth College, no less than ten such technologies were represented. The number of stationary agent technologies is even greater.

In an October 1995 presentation, John Ousterhout, who leads the Tcl effort at Sun, called intelligent agents "the new software high ground". Consider the present book your guide to that new region of the computing landscape.

James E. White
General Magic, Inc.

Preface

When we started this book almost 2 years ago, agents technologies were just beginning to come into the limelight in certain circles, both in and out of computer science proper. During the ensuing 2 years, agents have, pardon the expression, been coming out of the woodwork; not only do the major computer science and engineering literature contain more in-depth and more frequent articles about agents, but the mainstream press, such as the Wall Street Journal and local daily newspapers are carrying articles about them. This does not mean, however, that intelligent agent technologies have matured into mainstream, usable commercial realities except in the most trivial applications. We hope this book, among others recently appearing on agents, will contribute to a proliferation of intelligent agents.

In addition, we wrote this book during a series of standards upheavals and technology transitions; in other words, we were (and still are) on the "bleeding edge". Certain technologies that we included here may now be only important technological milestones on the roadway to subsequent standards and technologies.

For example, although comprising a relatively small portion of this book, we present a fair amount of detail on Telescript and OpenDoc. Both of these agent-enabling technologies were poised to takeoff during the 1995 to early 1997 period. However, both of these technologies have struggled in the marketplace recently and, indeed, certain OpenDoc technologies are now being transitioned and integrated into a Java-based approach (called JavaBeans) to some of the problems OpenDoc addressed.

What is important to remember though, is that the concepts inherent in both OpenDoc and Telescript, along with all the other technologies presented in this book, whether existing within current, viable commercial products, or only in academic laboratories, all contribute to the total space of ideas that will make computers more intelligent and easier to use. What we hope to have accomplished here, was to present enough of these concepts, from whatever the source, to help you understand where intelligent agents came from and, most importantly, where they may be headed.

We are sorry if anything in this book is a misconstrual of anyone's position, especially as to what is important and what is not in this field; in a survey-type work such as this, we took a broad stroke and sometimes missed the subtleties. Additionally, we apologize for missing anyone's work in the agent field—but in this length-limited medium, content decisions had to be made; In the 2 years since beginning this work, agent-related information has exploded, and would currently fill several CDs! Check out the bibliography and web pointers and do some research—you'll have fun!

Finally, we express our sincerest thanks to the many people who supported us in this project. This also includes the many researchers in the field with whom we personally conversed, authors of articles and books, all those who reviewed this work, especially our literary agents (real live people at Waterside Productions, Inc.), our families and associates, and, finally, all those folks at McGraw-Hill who got this thing published.

The URLs for websites given in this book are known to be accurate at press time; however, as with everything technical, things change—so does the web. If you have trouble reaching a reference site, try doing a search (e.g., via AltaVista) based on the context in which the site is mentioned.

Michael Knapik
Jay Johnson

Trademarks

The following is a list of the trademark names found throughout this book: Merriam-Webster, IBM's Glendale, IBM, OpenDoc, Microsoft's Common Object Management, OLE, Sun Microsystems Java language, servelets, IBM/Lotus Notes, Smalltalk, VisualBasic, C++, Telescript, DCE, CORBA, Common-Lisp, Prolog, Distributed Smalltalk, Intel's Pentium MMX, OPS-5, Clips, OWL, Conceptual Graphs, ACL, KQML, KIF, DENDRAL, CADAUDEUS, MACSYMA, MYCIN, DEC's XCON, WEBLS, RETE match, HTML, MCC, KQML, GeoAgent, PolAgent, EcoAgent, Cycorp, Inc., SPARC SunOS 4.1, Alpha AXP OSF/1, Macintosh System 7, UNIX, Rational Software Corp., UML, AT&T, ACM Sigsoft Software Engineering, VisualWorks, VisualWave, C, FORTRAN, COBOL, Pascal, ParcPlace\Digitalk Apple's Toolbox/Shared Library, Honeywell Inc, Blue Circle Cement, Sendai Subway Train, Sony PalmTop, Modico Inc., SireneF, WindowPC, ZeTec GmbH, MathWorks Inc., Matlab, Exsys Inc., AXCELIS Inc., Excel, Evolver API, FlexTool GA, Agentware, NeuroGenesis, TCP/IP, Apple's Publish/Subscribe, AppleEvents, X-Window, Orbix+Isis, Component Ingration Labs, Fresco, Xerox, Iona Technologies, Orbix 2.0, Mac OS 7.5, Mail/News, ActiveX, X/Open Company LTD, NT 4.0., Word, Active Group, Computer Associates International Inc., Digital Equipment Corp., Hewlett-Packard Corp., NCR Corp., The Powersoft Division of Sybase Inc., SAP AG, Microsfot App Wizard, NeXT, OPENSTEP, WebObjects, IONA Orbix, Siemens Nixdorf, Tandem, Bull, Gradient Technologies, Pyramid, SGI, Stratus and Transarc, NEO, Magic CapOS, Netscape Navigator, InternetWare, MacDraw, LonWorks, MagicCap-based PDAs, HotJava, Aglets, Marimba Inc., Castanet, ACL, Telescript, Cross Route, Software Inc., Lucent Technologies, Inc., Limbo, Sun Microsystems, Knowledge Query and Manipulation Language, and KIF.

Developing Intelligent Agents for Distributed Systems

Section

1

Introduction

If nothing else, prehistoric cave dwellers were self-sufficient. They had little need for extensive communications or information. As long as they knew where the nearest woolly mammoth was, they were in pretty good shape. Similarly, programmers in the early decades of the computer revolution were content in their relative isolation and had little interest in anything beyond the nearest mammoth mainframe.

Humans have progressed from self-sufficiency to complex interdependence. While we may long for the days of rugged individualism in the face of modern society's problems, the advantages of civilization are undeniable. In the modern world, specialized raw material providers supply the producers of specialized material goods, services, and information, who, in turn, supply consumers, all via vast communication and transportation grids. This allows each provider or consumer to use tightly focused expertise to analyze and combine elements in order to produce or consume ever more advanced products.

Just as humans have gone beyond a solo existence, the days of isolated software are quickly fading away as the Internet, the World Wide Web (WWW), and intranets spread across the computing landscape. Modern software developers must fit the systems they build into a complex information grid consisting of servers and clients connected by a plethora of local- and wide-area networks. However, networking is a double-edged sword. Like the connections among humans within a community, interlocking software, spread across multiple platforms, provides tremendous power; but at the same time it threatens to create potentially disastrous complexity. As in many successful human interactions, the keys to reducing the complexity of the information grid are task decomposition, hierarchical structuring, and delegation.

One of the most promising ideas for both unleashing the power of distributed systems and reducing their complexity is agent technology, especially "intelligent" agents. We believe intelligent agents will become pervasive within our computer and communication systems: creeping into user interfaces (such as via "wizards" and Microsoft's "Bob"), helping to map information into knowledge, underpinning electronic commerce, and acting as proxies for people in all sorts of situations. There are many ideas and technologies, discussed in this book, that will help usher in the age of the intelligent agent. But first, let's get a clear understanding of what we mean by intelligent agent (IA).

1.1 Intelligent Agents—Some Definitions

There are many definitions of agents. By reading and understanding several, from different viewpoints which usually relate to the definers' specialties, you can gain an understanding of the important aspects of "agenthood." First, for the layperson, three definitions found in the dictionary are:

> **agent** (ay-jent) *n.*
> ...
> 1. something that produces or is capable of producing an effect: an active or efficient cause
> 2. one who acts for or in the place of another by authority from him...
> 3. a means or instrument by which a guiding intelligence achieves a result.
> ... (Merriam-Webster, 1996)

IAs have a more expanded meaning according to the point of view of the computer science community, especially from the artificial intelligence (AI) researchers and practitioners. We present a few of them next.

> An intelligent agent is considered to be a computer surrogate for a person or process that fulfills a stated need or activity. The surrogate entity provides decision-making capabilities that are similar to the described intentions of a human. This surrogate can be given enough of the persona of a user or the gist of a process to perform a clearly defined or delimited task. An intelligent agent can operate within the confines of a general or precisely represented need and within the boundaries of a given information space. (King, 1995)

Franklin and Graesser at the Institute for Intelligent Systems, University of Memphis, define an agent, and, indeed, base a whole taxonomy of agents on the principal notion of autonomy: "An autonomous agent is a system situated within and a part of an environment that senses that environment and acts on it, over time, in pursuit of its own agenda and so as to affect what it senses in the future." (Franklin and Graesser, 1996).

Russell and Norvig define agents in three ways, ordered by increasing sophistication, or intelligence:

1. In a generic, but powerful behavioral sense: "An agent is anything that can be viewed as perceiving its environment through sensors and acting upon that environment through effectors." (Norvig and Russell, 1994)

2. With the addition of rationality: "For each possible percept sequence, an ideal rational agent should do whatever is expected to maximize its performance measure, on the basis of the evidence provided by the percepts' sequence and whatever built-in knowledge the agent has." (Norvig and Russell, 1994)

3. A rational agent takes the next step toward intelligent behavior if it can act autonomously. In other words: "An agent is autonomous to the extent that its actions and choices depend on its own experience, rather than on knowledge of the environment that has been built-in by the designer." (Norvig and Russell, 1994).

A more expansive definition of IAs is given by Atkinson et al. at IBM's Glendale Programming Lab.*

> Intelligent agents are software entities that carry out some set of operations on behalf of a user or another program with some degree of independence or autonomy, and in so doing, employ some knowledge or representation of the user's goals or desires. Intelligent agents can then be described in terms of a space defined by these two dimensions of agency and intelligence.
>
> Agency is the degree of autonomy and authority vested in the agent, and can be measured, at least qualitatively, by the nature of the interaction between the agent and other entities in the system. At a minimum, an agent must run asynchronously. The degree of agency is enhanced if an agent represents a user in some way. This is one of the key values of agents. A more advanced agent can interact with other entities such as data, applications, or services. Further advanced agents collaborate and negotiate with other agents.
>
> Intelligence is the degree of reasoning and learned behavior: the agent's ability to accept the user's statement of goals and carry out the task delegated to it. At a minimum, there can be some statement of preferences, perhaps in the form of rules, with an inference engine or some other mechanism to act on these preferences. Higher levels of intelligence include a user model or some other form of understanding and reasoning about what a user wants done. Further out on the intelligence scale are systems that learn and adapt to their environment, both in terms of the user's objectives, and in terms of the resources available to the agent. Such a system might, like a human assistant, discover new relationships, connections, or concepts independently from the human user, and exploit these in anticipating and satisfying user needs. (Atkinson et al., 1995)

A generic operational definition might be couched as follows:

- *Autonomy:* Agents operate without the direct intervention of humans or others, and have some kind of control over their internal state.
- *Social ability:* Agents interact with other agents (and possibly humans) via some kind of agent communication language.
- *Reactivity:* Agents perceive their environment and respond in a timely fashion to changes that occur in it.

---

*http//www.raleigh.ibm.com/iag/iahome.html

- *Proactivity:* Agents do not simply act in response to their environment, they are able to exhibit goal-directed behavior by taking the initiative. (Majewski, 1996).

Agents can also be defined operationally in terms of the domains in which they provide their services, including:

- Searching for information
- Filtering data
- Monitoring conditions and alerting users when alarm or set point conditions are detected
- Performing actions on behalf of a user; for example, a travel "agent" would arrange a vacation (buying tickets, reserving a hotel room, renting an automobile, providing information on points of interest, etc.)
- Providing access and transactional security
- Providing context-sensitive help, on-line tutoring
- Ameliorating network and system complexity, network management
- Being virtual actors and actresses in movies
- Optimizing system use via goal-directed techniques

Once successfully delegated to perform tasks such as those listed, agents can be powerful tools for managing the complexity in today's computer systems, particularly within distributed computing environments such as client–server and peer-to-peer systems.

As noted by Atkinson et al., software agents span a spectrum of "intelligence." Most agents used in the past and still in use today are "dumb," in that they operate using narrowly defined sets of brittle rules, and have little or no autonomy or flexibility. For example, a typical agent might sort E-mail based on the name of the sender. At the other extreme, the more advanced types of agents now beginning to emerge are called "intelligent." They can exhibit behaviors commonly associated with "intelligent" entities, namely, humans. More to the point, IAs can exhibit these behaviors because they usually incorporate various AI software technologies. We'll examine some of these AI technologies and how they contribute to the creation of IAs later in this book.

As computer technologies advance, the advent of ever more complex distributed, heterogeneous systems and networks means that developing software using traditional programming methods has become more difficult, and incorporating or using agents within distributed-systems applications has become more attractive. For example, users attempting to sift manually through hundreds of thousands of pages on the Internet's WWW, together with millions more news groups and other dedicated information sources, are easily (and understandably) frustrated. However, as agents emerge as locators, analyzers, and sifters of information, users will find their facility essential, rather than just a nicety.

When it comes to handling and making sense of the information flood, agents can be a curse as well as a blessing. While your agents are trying to make sense of piles of data, other agents are generating and transmitting an ever-increasing number of more complex messages that are sent to you, a potential consumer. For example, much of the mail you receive may be automatically generated messages. Many are advertisements masquerading as "important" information, that is, junk mail. A junk mailer may find out your electronic address and your interests by tracking the activities of your agents. Indeed, some of the static E-mail you used to get may now be replaced by active agents that attempt to interrogate your machine about various aspects of your system(s) or your interactions within cyberspace. This scenario raises questions about security and privacy, which we'll discuss in more detail later in this book.

Even as we travel inexorably toward hooking every computer into global networks, major roadblocks loom ahead. Agents are emerging as essentially mobile pieces of software that can execute autonomously on any accommodating network node. Today computers on a network can share data blindly, but there is little common understanding of shared information. What we need is a common infrastructure and an agent architecture to support the intelligent sharing of data and processing resources across networks—a mechanism to exchange messages, facilitate the mobility of executable programs or components, and a common language for those components (or agents) to exchange knowledge and collaborate on tasks and goals.

Certain standards committees are (seemingly always) in the process of wringing out standards that address many of these requirements. For example, the common object request broker architecture (CORBA) is an evolving standard that addresses the manner in which objects interact. [Agents, in our view, are—or should be!—more than likely designed and produced using object-oriented (O-O) concepts and development environments—but not necessarily so.] Two other examples of de facto interoperability standards that are also used within commercial software are OpenDoc and Microsoft's distributed common object management (DCOM) technologies, such as ActiveX and OLE. In addition, there are knowledge-sharing standards such as the knowledge interchange format (KIF) and the Knowledge Query and Manipulation Language (KQML), which facilitate agent collaboration.

Certain other commercially available systems all represent examples of technologies directly conducive to the creation of IA systems, including:

- General Magic's Tabriz agentware development environment and Tabriz agent tools, based on the Telescript language, include an environment for developing agent-based Internet solutions and services. Magic Cap is an agent-focused operating environment and Graphical User Interface (GUI) was created for handheld computers.

- Distributed Smalltalk from ParcPlace-Digitalk enables the creation of CORBA-compliant agents, which are executed within distributed virtual machines.

- Sun Microsystems' Java language, applets, servelets, and their virtual-machine-based execution environments herald a major shift in how information and computational functionality are designed and delivered.
- IBM/Lotus Notes collaborative groupware environment has an agent creation and execution capability.

More information about these and other standards and development environments is presented later in this book.

1.2 What This Book Is About

This book is a "big picture" presentation of several important technologies, techniques, ideas, and mechanisms that enable the creation of IAs and their supporting infrastructures and architectures. The principal aim of this book is to help software developers understand IAs, how to approach their design and development, and how to use them to make systems smarter. We explore the ways agents are being used to simplify the world, and how they may be used in the future. We explain (in relatively nontechnical terms) how IAs work, as well as some of their history. We include case studies of applying IA technology within certain domains, as well as information about current and future standards and environments that software engineers should follow. Some code examples are provided in the languages Telescript, Smalltalk, Java, and VisualBasic.

We realize that there are hundreds, if not thousands, of sources on AI. In fact, one book we would like to mention right now is: *AI: A Modern Approach* (AIAMA), also known as the "intelligent agent book." Here Norvig and Russell profess that the raison d'être of AI is to make IAs! Their book focuses on explaining AI technologies in terms of their efficacy and application to the creation of IAs. AIAMA contains a plethora of information on AI; we heartily recommend it. It can serve as a perfect one-volume reference on AI; an adjunct, if you will, to this book for those interested in the details of several AI technologies—especially if you want to incorporate those techniques in your IAs.*

In contrast, this book's approach, in addition to providing a high-level review of selected AI technologies, is also intended to cover many non-AI-related materials from academia and commercial sources on agent-related work. For example, we discuss several languages (Smalltalk, C++, Java, Telescript), distributed computing technologies (such as, CORBA), and several IA architectures. Our goal is to collect and summarize in one place as much information pertinent to agents as possible, and to sort it out for you. *Before* you commit to any technology, language, or software design presented in this book, do the necessary research to find out the latest information in any of these areas. Together with some examples of some agents and discussions for applications of agents in several domains, this book gives you, again, the big picture.

We would also like to state what not to expect from this book. This book is *not:*

---

*http://www.cs.berkeley.edu/7Erussell/aima.html

- A programming manual
- A particular agent (or other!) language reference manual
- A "cookbook" approach to building particular types of agents
- A detailed presentation of a particular architecture, infrastructure, or operating system (OS)
- An endorsement of any particular "way" of making IAs
- A dogmatic definition of what intelligence is, or assertions of exactly what attributes an agent must possess, or by what mathematical formalisms an agent would be designated *intelligent*
- A detailed treatise on the application of AI techniques in any particular domain

Let us now preview what you can expect to find in subsequent sections of this book.

1.3 Section Summaries

This book's contents are arranged as follows:

Section 2, "From Artificial Intelligence Come Intelligent Agents," explains how IAs came to be, and the technological roots that underpin them. We explore the conceptual milestones of AI and discuss how these concepts are related to and, indeed, led to the development of IAs. Topics include speculation about the brain and mind, and how minds can be thought of as hierarchies of IAs executing autonomous or semiautonomous tasks. We discuss how these concepts of mind and brain affect the way we build intelligent computing machinery. To complement and greatly expand on this section's contents, AIAMA is highly recommended.

Section 3, "Converging Technologies that Facilitate and Enable Agents," provides an overview of major technologies contributing to and facilitating the development and use of IAs. Among these are selected AI technologies such as expert- and knowledge-based systems, fuzzy logic, genetic algorithms, neural networks, and object-oriented development. One specific knowledge-based system that we cover in more detail is Cyc, "a formalized representation of a vast quantity of fundamental human knowledge: facts, rules of thumb, and heuristics for reasoning about the objects and events of everyday life." (Cycorp's WWW page: http://www.cyc.com/tech.html#kb.) As you will see, Cyc has great potential to become a source of the I in many an IA!

Section 4, "Agent-Enabling Infrastructures," examines the various interagent communications infrastructures within (or on) which you can build your agent architecture. Agents depend on well-behaved, structured systems to interoperate. Summaries of interobject communications standards such as distributed commuting envrionment (DCE, CORBA, OLE/ActiveX, and OpenDoc are presented, along with comments about their suitability as a foundation for agent development.

Section 5, "Agent Architectures," presents several ways in which agents can be put together to form a working, collaborative system. Certain architectures include discussions of communications protocols and special agent communication languages. From this section we expect you to gain insight into how to organize your own agents into a collaborative, intelligent system.

Section 6, "Agent-Design Considerations," is a collection of ideas, constraints, and tips you should keep in mind as you start your agent-design process.

Section 7, "Developing Intelligent Agents Now," provides examples of specific languages and development environments that can be used to create and support agents. Topics range from open-ended languages such as Smalltalk and Java to more agent-specific environments such as Telescript, to development environments like IBM/Lotus Notes, where agents can be embedded into an otherwise non-agent-oriented application. Examples of code snippets, scripts, configurations, and object definitions used to do specific IA-like functions are included.

Section 8, "Agent Applications," presents several applications that can use agent technology to perform *specific* tasks that enhance the capabilities of the application, including the following:

- Programmer's assistants and wizards
- E-mail/FAX management agents
- Knowledge search and retrieval agents
- Network management agents
- Financial agents
- Filter agents
- Network-based software distribution agents
- Deal-making agents in a worldwide marketplace
- Agents and databases
- Agents in telephony
- Industrial automation and control agents
- Work-flow agents
- Patient-monitoring agents
- Author's assistant
- Agent-based collaborative engineering

Section 9, "Agent Futures," is a look into where intelligent agent technologies are headed. We speculate on the sophistication and proliferation of agents to come, and the social consequences of the increasing use of IAs. We forecast that as agents pervade different types of cyberspaces, they will become increasingly autonomous. We discuss what this means in terms of control (who has it), privacy, and effects on human–machine interaction. We discuss how agent-to-agent negotiations will affect companies and whole economies. We then survey current research that can lead to many exciting agent-based systems in the future.

1.4 Audience and Structure of This Book

The structure of this book follows from the way we, as software engineers, thought about making IAs. With the application, domain, and user needs always in mind, the following scenario seems reasonable:

1. Gain an understanding of what the "I" in IAs means via a basic understanding of AI (Sections 1 and 2).
2. Decide what kind of intelligence your IAs need, and, at a high level, how to incorporate intelligence within them (Section 3).
3. Investigate those technologies we refer to as infrastructures; substantial implementations that will save you from reinventing the distributed-object-computing wheel (Section 4).
4. Proceed with the agent architecture design, or reuse an agent architecture suitable to the application (Section 5).
5. Review some specific agent-design tips and techniques (Section 6).
6. Choose your language and development environment, with special emphasis on those that are "agent-ready" (Section 7).
7. If you still don't know what to do with an IA, read some more about their potential application and what the future portends (Sections 8 and 9).

With that somewhat sequential scenario in mind, if you are a software engineer interested in IAs from a general, informational perspective, you can just read sequentially through the whole book. A software engineer planning to develop a system incorporating agent technology should pay particular attention to Sections 3 to 7.

Software managers, corporate information officers, and those interested in a high-level understanding of IAs could read Sections 1, 2, 3, 5, 8, and 9. Those with only an academic interest in the topic (such as researchers in specific areas of computer science, or those interested in the social and operational aspects of computing technologies) should read Sections 1 to 5, 8, and 9.

May the Root Agent Be with You.
Enjoy!

Section

2

From Artificial Intelligence Come Intelligent Agents

In Section 1 we defined an intelligent agent as a software entity that can monitor its environment and act autonomously on behalf of a user or creator. To do this, an agent must perceive relevant aspects of its environment, plan and carry out proper actions, and communicate its knowledge to other agents and users. Of course, it is up to you, the agent developer, to incorporate these capabilities into your agents, using technologies that we believe currently exist in the realm known as artificial intelligence (AI).

2.1 Introduction

In this section we discuss important relationships between AI and IAs. After reading this section you will understand how 40 years of AI research, in a broad sense, can be incorporated into IA designs. We focus on ideas about brain and mind organization and function, and how these ideas give us insight into building IAs to work together. The material in this section is an attempt to give you some foundation in the progression of concepts of AI, to aid in understanding the sections that follow. It is not exhaustive, but is intended to highlight some of the major thoughts and thinkers whose work has led to the present time in which we (and you) can consider building software that, by traditional AI, exhibits some facets of intelligence.

Before we begin our short tour of AI, we want to acknowledge a particularly good recent work in the AI field—a work that positions AI efforts directly in support of IA construction. As a much more comprehensive treatment of AI in general, and of specific AI methods and algorithms used in making IAs, we recommend Russell and Norvig, *Artificial Intelligence: A Modern Approach* (AIAMA). In fact, "the unifying theme of that book is the concept of an intelligent agent" (Russell and Norvig, 1994). As we mentioned in Section 1, they posit that the quest of AI research **IS** the quest to create intelligent software entities.

Russell and Norvig have done a commendable job in presenting a wide range of AI concepts and techniques that you can use in designing your agents. They have also presented an example agent/agent system in Common-LISP, which threads throughout their book; capabilities are added to agents related to the topic being discussed in each section. So we heartily recommend their work for a one-stop, in-depth treatment of some of the AI-related material we present here, as well as of topics that are beyond the scope of our book.

Although we had similar goals as Russell and Norvig, our approach concentrates more on providing overviews and summaries of many different aspects of developing IAs, of which the use of formal AI techniques to provide the intelligence is but one. We also bring together and review in one place many non-AI considerations and technologies affecting the practical implementation of intelligent distributed systems. So without further ado, let's jump into the world of AI.

2.2 40 Years of Classical AI

In their introduction to the 1982 four-volume classic survey of AI, editors Barr and Feigenbaum proclaim that:

> Whether or not they [these AI technologies] lead to a better understanding of the mind, there is every evidence that these developments will lead to a new intelligent technology that may have dramatic effects on our society.

and:

> There is every indication the useful AI programs will play an important part in the evolving role of computers in our lives—a role that has changed, in our lifetimes, from remote to commonplace and that, if current expectations about computing cost and power are correct, is likely to evolve further from useful to essential. (Barr and Feigenbaum, 1981)

These contentions are being borne out. Indeed, given the information explosion, intelligent technologies, as encapsulated within simple IAs, are already here—and not a moment too soon. On the other hand, although it was once assumed that computer scientists doing AI work would be the creators of an artificial "mind," we shall see that it appears that neuroscientists discovering the workings of the natural human brain and mind are contributing much to the development of intelligent computer systems.

The list of contributors to the field of AI as a specialty of the computer sciences is very large, especially looking back over the last 40 years. Although the mathematical foundations of computation and logic herald from the first few decades of the twentieth century, the acknowledged birth of AI proper could be positioned in the 1950s. Among those important persons and ideas, Alan Turing, sometimes referred to as the father of AI, positioned computation as symbol processing. Influenced by Bertrand Russell's ideas on the mechanization of mathematical reasoning, Turing envisioned a machine that could, by following a finite set of logical rules or steps, transition from one state to another. This was the essence of symbol manipulation.

Turing was initially convinced that all of mathematics' axioms (composed of symbols) could be mechanized. He thought a machine could be built embodying all mathematical propositions and axioms. Indeed, many AI researchers maintain that given sufficiently complex symbols to process or manipulate, a computational device could be deemed intelligent. As computers became more adept at symbol manipulation, many workers in the field thought that, given almost infinite resources (memory, speed, "knowledge"), a computer could be thought of as intelligent.

In fact, Turing proposed a test (called, appropriately enough, the Turing test) that could determine whether or not a symbol-manipulating machine could be deemed intelligent. Turing's notion was that a machine is intelligent if it can fool you into thinking it is a human within some specified context. But this notion appeared to many to be too narrow a definition of intelligence; a key attribute of intelligence is understanding. Could a symbol-manipulating machine understand the symbols it was manipulating? Over the years, the debate over the validity of the Turing test being the unequivocal, necessary and sufficient determinant of intelligence still rages.*

As an alternative, consider the applicability, in this context, of a different, perhaps orthogonal, view of an intelligence test. Let's call it the Knapik test. It is sort of a reverse Turing test. Instead of a human deciding whether he or she is interacting with a computer or a human, how about a computer trying to determine whether it is interacting with a human or another computer. If it can tell the difference, then it is intelligent. This test assumes that all (or most) humans could, within a reasonably short time, pose a question or engage in a discourse such that all (or most) humans would "instinctively" know the proper conversational context and play their part appropriately. The test might involve a bit of trivia, a commonsense notion, or maybe a joke or a double entendre. It is this kind of understanding and reasoning, which we humans do on a day-to-day basis, that many AI folk are trying to get a computer to emulate.

During the midpoint of this century, other important contributions to theories of intelligence, the brain, and complex systems have influenced the direc-

---

*Indeed, even as Turing himself worked further on his theories, he realized that, in accordance with Godel's incompleteness theorem, a universal Turing machine that recognized undecidable propositions would be a contradiction. This followed the idea that a machine could not "get outside of itself" to analyze its analysis, resulting in an infinite loop, as it were. But this realization did not deter Turing from believing that the human mind functions in much the same manner. He reasoned that it, too, would be subject to these infinite regresses. However, Turing did not believe it followed that if humans are able to put a stop to the loop, they had possession of a "consciousness" that is not mechanizable.

tion of AI. For example, Wiener (1948) in his work on cybernetics and complex systems showed that, with sufficiently complex feedback mechanisms, a machine's control and feedforward-feedback predictive algorithms could be thought of as being akin to the brain's processes. The human body is composed of many feedback loops operating to control many subsystem processes such as hormonal concentrations, energy production and consumption, motor coordination and balance, and brain chemistry. If an IA system had such self-monitoring and compensating mechanisms, would the system take another step toward true intelligence?

In the 1940s Claude Shannon posited an information theory that defined information as the similarities and differences between a set of states. The structures that encode these states can be the electrical or magnetic thresholds in a computer's random-access memory (RAM) or on hard disk platters, or the long-term potentials encoded in the brain's neurons. Each of the states can be labeled and, as such, can be thought of as a symbol or model for the real or abstract thing. IAs have a state composed of all their relevant attribute values at a certain point in time. Some of these attributes reflect the agent's internal structure, whereas other attributes reflect what the agent knows about its environment and the domain for which the agent was devised. In hierarchical agent systems, higher-level or container agents that have responsibility for lower-level agents manifest a composite state encompassing ever more subsuming information.

John von Neumann applied mathematical logic to produce computational devices which, in turn, produced or proved mathematical theorems. Indeed, many of the early computers were used as dedicated theorem provers based on well-known algorithms. The important point here is that during the early days of computers, the conceptualization and the development of sequential, algorithmically based digital computers were thought to be following in the footsteps of, so to speak, the human brain's own architecture and operation. All these contributions, and many more, relied on the fundamental notion that intelligence, as embodied in the human brain or in computational machinery, could be modeled with an algorithm or systems of cooperating algorithms that manipulate symbols. Thus was born the information age—the age of computing.

The notion that understanding is a necessary precursor to intelligence is illustrated by the following. A major counterpoint to symbol manipulation as intelligence is called the Chinese room mind game, posed by John Searle. Searle (1980) rebutted the efficacy of the Turing test in determining intelligence by suggesting that a real human could be in a room with a book where, for every Chinese word, phrase, or sentence input or delivered into the room in Chinese, the human could look up the appropriate response in the book and output it appropriately. From the external observer's point of view it appears as though the human inside the room understands Chinese. The point is that the person in the room does not need to understand the language, in the usual sense of the way we understand language and meaning, to "process" the language. Now just replace the human in the room with a computer that does the same lookup task. Although responding appropriately, the computer cannot be said to possess any understanding either.

Various points and counterpoints regarding the Chinese room have been made over the years. As AI concepts, languages, and practical programs were being developed, certain philosophical arguments came to the fore. Still, the central issue is whether a symbol-processing machine that exhibited intelligent behavior could be said to understand the symbols in the manner a human subjectively understands symbols. AI practitioners try to correctly represent facts about the world through a language syntax (configurations of symbols) and semantics (symbols that represent the facts; meaning of a statement). Further, we also try to correctly reason about those facts (using logic or other mechanisms to generate inferences, that is, deduce new facts or whether the facts are true or false).

Many AI techniques are available that can help one agent "interpret" what another (human or computer-based) agent has "said." Reasoning with uncertainty is a big part of properly interpreting knowledge. Techniques include:

- *Syntactic analysis:* Using the structure of a sentence together with, for example, word order and reference preferences to resolve ambiguity
- *Lexical analysis:* Resolving ambiguity by looking at what sense of a word is most likely in the given context
- *Semantic analysis:* Using context and a priori knowledge stored in the receiver's knowledge base
- *Appropriate application of knowledge bases:* Colloquialisms, figures of speech, metaphors, similes, metonyms, homonyms, synonyms, and other idiosyncracies of the language

But is interpretation all there is to understanding? One aspect of understanding is surly to interpret correctly some bit of knowledge. But aren't there other aspects of human understanding that rely on grasping the significance or importance of knowledge; that this "grasping" is somehow linked to emotion and feeling as part of the context? What would it mean to represent these aspects of understanding in a computer? Suffice it to say at this point that the matter of understanding is central to our notion of what it means to be a human. We seem to have a conscious self that, whether via an apperception of qualia resulting from sensory input, or from a sense of "knowing that we know" (for example, knowing is different from saying; see the Turing test) regarding language and concepts, leads us to conclude that our understanding is of a kind that has not been achieved or even approached in the computer.

What does all this have to do with IAs? Even though the word "intelligence" is part of the IA phrase, the type of intelligence programmable into a single agent (however clever) is not to be confused with most aspects of human intelligence. This is especially so when considering that an agent is (at least as initially envisioned) an encapsulation of rather focused functionality. This functionality, however, can be greatly enhanced by the application of specific techniques that have come to us from research both in classical AI and in newer approaches to representation and computation (such as neural networks and fuzzy logic). In addition, several autonomous agents, each with embedded AI technologies, can collaborate to solve more complex problems. In a collaborative agent system, each agent contributes its own problem-solving, goal-seeking behaviors to a solution of a complex problem. Perhaps the word "intelligence," in an "emergent" sense that we will discuss further, can be applied to such a system.

The *Handbook of Artificial Intelligence* (Baar and Feigenbaum, 1981) delineates several areas in which classical AI has contributed to understanding how we may build "computers that think." These areas, still valid today (some of which will be reviewed in more detail later) include the following:

- *Problem solving:* Searching large problem spaces and problem reduction; for example, chess.
- *Logical reasoning:* Deduction, theorem provers, large database analysis, and maintenance.
- *Language understanding:* Translation, reading text, haptic input, word understanding, and expectations based on context.
- *Automatic programming:* Systems that can write programs to achieve some result based on descriptions of their purpose.
- *Learning:* Relatively recent progress using neural networks and genetic algorithms that emulate nature's feedback and feedforward control loops. These loops fine-tune system responses to produce expected or correct output more accurately and precisely.
- *Knowledge-based and expert systems:* One of the more fruitful applications of AI technologies to date. Knowledge engineering converts a human's expertise within a domain of knowledge to forms usable by some of the other, aforementioned computational capabilities. For example, the constraints imposed on assembling a real-world object together from parts can be stored as part of the assembly process and fired when the parts are configured or put together within the expert system. These systems also present feedback to the user as to whether or not that configuration is legal and why.
- *Robotics and vision:* From simple microcontrollers, seemingly omnipresent in our homes and vehicles, to sophisticated manufacturing robots on assembly lines.
- *Languages, operating systems, and tools:* Includes general AI languages, such as LISP and Prolog, to agent-specific languages such as Telescript. Object-orientated languages and distributed-object environments (such as

Distributed Smalltalk and CORBA) have become prominent factors in developing multiple cooperating distributed objects (agents by any other name).

One of the items on the list—knowledge-based and expert systems—plays a large enough role in practical computer applications today that we are going to take a closer look in Section 3. Indeed, over and above the use of domain-specific expert systems to which your agent may have access, we contend that agents in the real world need some semblance of what we humans take for granted, namely, common sense. Section 3 uncovers an amazing endeavor to do just that.

2.3 Hierarchy: Bridging the Gap between Natural and Artificial Computational Worlds

Some of the most intriguing work on intelligence that has been done over the last 15 years or so, has originated not from the computer sciences per se, but from the cognitive sciences, especially the neurosciences—the study of the architecture and functions of the brain—as well as the psychology and mechanisms of the mind. Neuroscience has transformed our understanding of the brain: from the naive view in the early days (1950s and 1960s) of the brain as a serial computer to the current thinking of the brain/mind as a multitiered hierarchy, in which each layer's functions are explained with language and concepts most appropriate to that layer. Hierarchy as an architectural pattern of complex systems figures predominantly in several popular theories about not only the brain and mind, but how nature and the universe are put together.

Science in general has separated the study of the natural world into subdisciplines, which investigate relatively well-defined "layers" of the vast hierarchy of our known universe. Mechanisms and entities at a lower level give rise to progressively more complex, composite entities with (in many instances) more elaborate operational mechanisms at the higher levels. We know that this natural hierarchy has given rise to intelligent systems (such as us).

Because we want to make IA systems ourselves, let's take a quick look at some of the more obvious layers making up naturally intelligent systems. Note that although each layer discussed in the following seems to fall within the purview of the rigid definitions of an individual physical science (such as biochemistry or physics), each layer has aspects whose complete description relies upon several of the subsciences. Indeed, some of the most amazing discoveries in the last decade or so have come from interdisciplinary research of the type going on at, for example, the Santa Fe Institute or the MIT Media Lab.

Possibly the "lowest" or most fundamental layer is captured within various speculative theories of the quantum mechanical aspect of consciousness. A number of books and papers have appeared over the last 15 years that explain consciousness with reference to aspects of quantum theory. Some theories posit that our consciousness (and possibly all of physical reality as well, since all matter is describable as interacting fields) is a holographic projection from some realm of reality governed and described by waves. David Bohm, the famous theoretical physicist, suggests the existence of an implicate and superimplicate order that, via hidden variables, affects our everyday world. Pilot waves and projections from the implicate order to the explicate order of sense-perceptible phenomena are inscrutable to our everyday consciousness, at least directly.

Work by Nobel Laureate Roger Penrose posits that we will never see machine-based consciousness. The reasons Penrose and others give are fairly complex. Suffice it to say here that our ability to understand ourselves fully, and thus create intelligences with consciousness like us, is limited by seemingly insurmountable walls. First, the implications of Godel's incompleteness theorem impede our ability to know ourselves completely. Second is the notion that our minds are nonalgorithmic in nature. Third is the notion that the very physical basis of consciousness lies in the interaction of microtubules (structures within each neuron) with the ineffable quantum realm. Quantum wave function collapse within these microtubules *is* the substrate of attention and thus consciousness.

The next layer of being, brain and mind, involves individual atoms and ions; classical physics and chemistry rule here. The specific densities and concentrations of atoms and ions in the different areas and organs of the body such as blood pH and the relative amount of potassium to calcium ions, which produce ionic pumps, all have great bearing on how we function. Then, progressing "upward" to intracellular aspects, electrochemistry and molecular chemistry take over. The interactions of proteins, enzymes, and hormones determine much of what we consider fundamental life processes and affect everything from our immune system to our moods.

As we progress to cellular and intercellular mechanisms, biochemistry and neurophysiological mechanisms explain what is happening. As cells die and are replaced, RNA and DNA, large proteins themselves, provide the information that encodes and determines the patterns of the physical stuff that we are. This encoded information is decoded continuously as replacement proteins and cells are produced to keep us functioning.

Further up the hierarchy, similar cells form organs. The human brain consists of cells called neurons, of which there are various subtypes according to specific functions. Neurons grouped in specific locations (such as vision and motor centers) or that appear configured similarly in the brain (such as the cerebral cortex) account for specific, discrete functions of the brain.

Finally, at the highest levels of organization, areas of the brain communicate with one another and, by some as yet unknown mechanism (if mechanism it is), give rise to our personality, or psychology and our consciousness. Communications mechanisms range from definite physical paths for the electrical, chemical, and electrochemical signals to a more speculative neuronal and neuronal-group entrainment to the various types of nonlocalized brain waves, such as the so-called alpha and beta waves.

Curiously the fundamental quantum-level layer seems to play a part at the top of the hierarchy as well. The seemingly holographic nature of some aspects of thought and memory has been likened to the everywhere and everywhen nature of quantum fields. Indeed, the debate on the connection between wavefunction collapse and the observer has attracted scientific luminaries like Stephen Hawkings and Roger Penrose. In addition, the informational encodings within DNA play such a great role in determining who and what we become that some, for example, physicist John Wheeler, speculate that information should be considered as the fundamental layer of reality—a layer that is actually orthogonal to all the other, merely physically, describable layers. Information is operational, not static; it is the coding, transfer, and decoding of information that determine the operational and behavioral characteristics of things. In other words, each layer discussed in terms of a specific discipline of science encodes information important to that layer, but also is transformed and communicated up and down the hierarchy to other layers.

2.3.1 Consciousness at the top of the hierarchy

Our mind, or consciousness, is one of the systems in the universe that, so far, defies explanation, even rigorous, well-defined modalities of exploration. There have, as of late, been a few conferences devoted to the study of consciousness within the scientific community. Certain universities and groups are spawning seminars, classes, and centers to foster the scientific approach to the study of consciousness. And we have been awash with books on consciousness over the last several years. But outside the realm of religious studies, especially those with an emphasis on an integration of Eastern and Western approaches to existence and consciousness, we are not aware of any courses of formal study leading to a Ph.D. in consciousness, the collective specialties known as the cognitive neurosciences notwithstanding.

It seems that each subdiscipline of science has attempted to describe certain aspects of the mind using the concepts applicable to its level of the hierarchy. What is ultimately needed, however, is a synergistic view of consciousness that subsumes and integrates all of the incomplete, fractured theories that have been put forth so far. When that theory of consciousness appears, we might then be in a position to create an artificial consciousness (à la Data of "Star Trek: The Next Generation"), possibly with emotions as well.

Part of the problem stems from the notion that we can truly understand consciousness only by identity. In other words, we know that we are conscious entities because of that seeming of an "I" within us. It is the I that understands, in the deepest sense raised earlier in this section. But we are hesitant to ascribe consciousness to an entity that we intuitively feel is a machine, or an artificial entity that cannot have that I seeming. But it is not necessarily the case that an entity, natural or artificial, must be conscious to be deemed intelligent. Sufficiently capable agent systems could be called intelligent (even without the

I-based "understanding" discussed earlier), especially if imbued with mechanisms and technologies described in this book.

2.3.2 Hierarchy and IA systems

Now what does all this talk of hierarchies and the various layers of description have to do with IAs? We want to expose the great similarities between the organization of the natural world—intelligent entities like ourselves—and the organization of computers. Many of those in the so-called hard AI camp profess that intelligence—indeed consciousness itself—does not depend on the type of instrumentality used, that is, intelligent behavior and consciousness are, in their opinion, independent of the physical substrate. "Only" organization, structure, and, of course, information content are important. Some hard AIers profess that, given a sophisticated enough structure, intelligence and even consciousness will arise from silicon-based mechanisms. (But the key question will remain: how will we know such a mechanism is conscious?)

Further, the manner in which computer science conceptualizes and builds computers has followed, in many respects, what is conceptualized in other areas of endeavor—especially the neurosciences and recently the more expansive cognitive sciences. Just as we have, in some dualistic philosophical camps, posited a hardware substrate (the brain) and software (the mind) that constitute our being, so have computers followed suit with physical hardware (such as silicon RAM and processors) and the otherwise ethereal (some might say platonic) software that really "exists" and is operational only when embodied and running in or on the silicon.

These two layers—hardware and software—are only the most gross description of modern computers. There are layers, or functional "modules," within a computer's hardware—the input-output (I/O) subsystem, RAM, read-only memory (ROM), processors, and so on. There are layers of functional "modules" that compose a computer's software, from the register-transfer logic and ROM-loaded instruction set to the various operating system (OS) layers, to the application "on top of" the OS. The application is in most cases, broadly speaking, an agent.*

---

*As computer science makes progress, we will see increasingly flexible hardware circuits used in information systems. First, application-specific integrated circuits (ASICs) and digital signal processors (DSPs) are being used to perform functions typically done in software in the past. Field-programmable gate arrays (FPGAs) are the next step toward making hardware reconfigurable on the fly to meet changing field requirements. As these trends continue, the normally distinct lines between functions embedded in software versus hardware become increasingly blurred. (The reverse of this trend also seems evident as main processors become ever more powerful. For example, Intel's Pentium MMX chip now has on-board graphics processing—but the point made is still valid.) Interestingly, this very blurring of where and how things get computed is the same kind of problem the cognitive neurosciences are wrestling with in developing a satisfying explanation of where and how the mind and consciousness arise and reside. In other words, is consciousness purely physical; or is there a nonphysical mind (à la software) that resides in or on the physical brain? And in what sense is the term "nonphysical" used; in other words, the electrons representing software's "bits" are physical too—aren't they?

The point we are making is that one of the most evident characteristics of naturally and, now, artificially intelligent systems is that they have a hierarchical organization of increasing complexity and sophistication. Each layer within the hierarchy contains "agencies" that communicate information between themselves on the same layer (intralayer) or transmit and receive information across layers (interlayer) to achieve their local goals. Interlayer communications, both upward and downward in the hierarchy, facilitate and are the mechanism enabling information compression (that is, agents "up" the hierarchy receive generalized information from lower-level agents dealing with more specific information) and delegation (that is, higher-level agents transmit instructions to lower-level agents). In Section 5, which discusses architecture, including the organization of agents within a system, we'll see some systems constructed from multiple layers of agents. Even though those systems include simple hierarchies, much more elaborate hierarchies with many layers are probable once agents (and the practice of modeling and implementing problem spaces and domains with objects in general) come to be used more often.

2.4 The Connectionist Revolution: Human Brain as Inspiration

Many of the concepts discussed earlier were initially developed assuming implementation on a serial computer using discrete algorithms. This was the then prevalent view of how the brain did its processing. But with the advent of new connectionist theories, based more closely on the discovered parallelism of the human brain, new computer architectures, referred to broadly as parallel distributed processing (PDP), started to evolve.

An important step in the connectionist revolution was the notion that learning in the brain is correlated with physical changes in the brain. For example, altering of synaptic connection strengths between neurons is believed to play a part in the brain's learning mechanism. In 1949 D. O. Hebb proposed rules for determining the synaptic connection strengths, or weights, in a neural network such as our brain. Hebb's analysis of the mechanism and rule(s) are referred to as hebbian learning. His simplest rule was: When unit A and unit B are excited simultaneously, increase the strength of the connection between them.

To account for both positive and negative adjustments, a rule that states: "Adjust the strength of the connection between units A and B in proportion to the product of their simultaneous activation" was a natural extension (Rumelhart, McClelland, et al., 1988).

This rule along with more sophisticated versions, like the delta rule, are used to train artificial neural networks (ANNs), as you will see in Section 3. Hebb's other ideas, such as "...the concept of cell assemblies—a concrete example of a limited form of distributed processing...and reverberation of activation within neural networks...capture some of the flavor of parallel distributed processing mechanisms" (Rumelhart, McClelland et al., 1988).

Schwartz (1988) relates how the brain differs from the serial-processing paradigm so far prevalent in modern computers:

1. The brain operates not as a serial computer of conventional type but in enormously parallel fashion. The parallel functioning of hundreds of thousands or millions of neurons in the brain's subtle information-extraction processes attains speed. Coherent percepts are formed in times that exceed the elementary reaction times of single neurons by little more than a factor of ten. Especially for basic perceptual processes like sight, this observation rules out iterative forms of information processing that would have to scan incoming data serially or pass it through many intermediate processing stages. Since extensive serial symbolic search operations of this type do not seem to characterize the functioning of the senses, the assumption (typical for much of the AI-inspired cognitive science speculation of the 1960–1980 period) that serial search underlies various higher cognitive functions becomes suspect.

2. Within the brain, knowledge is stored not in any form resembling a conventional computer program but structurally, as distributed patterns of excitatory and inhibitory synaptic strengths whose relative sizes determine the flow of neural responses that constitute perception and thought.

In contrast with classical computer architectures, the PDP class of computational approaches focuses on the study of parallelism in computational systems. This connectionist approach posits that intelligent behavior could emerge from the activity of many highly interconnected (possibly very simple) parallel computational units. In addition, connectionism focused on the mechanisms needed for adaptive learning. Initially this approach foresaw the *current* insights into how brains work and argues convincingly against the strictly serial, algorithmic approach of the classicists.

Rosenblatt dubbed the aforementioned highly interconnected network of units perceptrons. He defined the principles behind perceptrons in what some consider a landmark book, *Principles of Neurodynamics* (Rosenblatt, 1962). Operationally, perceptrons made decisions as to whether an event, piece of knowledge, or item of interest fit a pattern by examining a set of inputs into simple representations of neurons—the guts of the neuron performed a basic calculation, like a summation of the inputs multiplied by the weights attached to the inputs. All the inputs taken together represented the event of interest. The neurons were connected in various ways, but a network produced one output.

A perceptron compared the output of this "network" with a desired result. The result of the comparison was fed back to the inputs in the form of altered weights or strengths on the inputs. This process iterated until the perceptron's output converged on the correct or desired result. The network now represents the event, or item of knowledge.* This mechanism is similar to what neuroscientists have associated with the brain's learning processes.†

One of the things we emphasize in this book is that to make an agent intelligent, you have to use certain techniques. We profess that the use of ANNs will increase substantially and that, as an agent developer, you should seriously consider the use of ANNs as discrete entities within IAs, or the use of the connectionist and PDP paradigm that ANNs exemplify. More discussion on ANNs in IAs is presented in Section 3.

2.5 Agents of the Mind

We discussed both the natural brain's structure and the ANN's structure (and mode of representation and computation) as composed of groups of highly connected, simple processing units. Now consider the hierarchies that we talked about earlier. In the brain's case this may include hierarchies of groups of subnets. It is just this set, or society, of many different, localized neural nets (or, rather, their behavioral, mental correlates) that makes up Minsky's model of what happens in the human brain and mind. In Minsky's book (1986) about the subject, appropriately entitled *The Society of Mind,* each of these smaller networks acts, or has correlated mental behavior that acts, as an agent within a defined context and performs discrete functions related to that context. In addition, each of these network subsystems can have its own, internal representational scheme.

---

*Note that in contrast to discrete approaches of representing knowledge born from classical AI, the ANN represents knowledge *as* the interconnections between parts (the neurons). For example, there is no specific "place" to point to in the ANN that has the word "blue," whereas in a typical knowledge base created with some relational database management systems (RDBMS) or object-oriented database management systems (OODBMs), I can find "blue" represented as a discrete element (for example, an object or an attribute of an object).

†Rosenblatt professed that it was "...feasible to prove a theorem which stated that a perceptron would learn to do anything that it was possible to program it to do" (Minsky and Papert, 1988). Various criticisms of this and other connectionist claims appeared. Minsky and Papert pointed out that the connectionist camp was unable to get simple, single-layer networks of perceptrons to learn to do certain things, while other things were possible. Different problems made a difference.

Many believed that Minsky and Papert's *Perceptrons* was an attempt to discredit the connectionist approach to computing. But as Donald Norman states: "The critique of perceptrons by Minsky and Papert (1969) [in the original 1969 edition] was widely misinterpreted as destroying their credibility, whereas the work simply (but elegantly) showed limitations on the power of the most limited class of perceptron-line mechanisms, and said nothing about more powerful, multiple layer models" (Rumelhart, McClelland, et al., 1988). Indeed, in the updated, expanded edition of *Perceptrons,* Minsky and Papert maintain that their critique of connectionism was a rigorous treatment of perceptrons, and their goal was in fact "to develop analytic tools to give us a better idea about what made the difference" (Minsky and Papert, 1988). Minsky and Papert point out that specifically interconnected arrangements of networks of perceptrons are possible models for parts of the brain and that an aim of future research will be to "develop networks that can learn to embody wide ranges of different, context-dependent types of matching functions" (Minsky and Papert, 1988).

Of course for the brain and mind to work as a society, the networks must communicate in some fashion. It is here that, in the opinion of Minsky and others, symbolic computation (that is, representation and manipulation) is used. Symbols are communicated between networks via limited pathways. The symbol encoding occurs in the sending network and the symbol decoding and recognition occur in the receiving network or networks. Evidently, there are "standards" in the brain and mind that are used to ensure proper functioning. Neuroscience is about discovering what these standards are.

Using representational and communications standards, a receiver of knowledge need not know intricate details of what is occurring in the generator of that knowledge. Only an agreed upon (or learned) protocol for sharing a sort of shorthand (the encoded symbols) is necessary for effective communications and collaboration. How does this relate to agents? Each network in the brain/mind and the context or domain it serves are analogous to an agency or, anthropomorphically speaking, an agent assigned to perform the specialized function that it learned to do.

Minsky (1986) posits that the mind is "a society of ever-smaller agents that are themselves mindless." Although Minsky is proposing a theory of the human mind, it is interesting to note the similarities to current theories of network intelligence. That is, networks of simple software entities often give rise to surprisingly robust, complex, and synergistic behaviors.*

The simplest "agent" may be the function or functions that operate within the neuron, determining whether the neuron fires or not. As we mentioned earlier, even lower-level activities, such as those at the atomic or quantum layers, could be thought of as agents. After all, information is processed and communicated at these layers, just as it is in layers that are more semantically recognizable to us, such as in the language- or vision-processing systems.

Minsky starts by breaking down what seems to us as a monolithic I into individual functions and behaviors. He then assigns a type of agent to take care of that function. For example, there are agents for motor skills, agents that keep your balance, agents that interpret language, agents that interpret visual input, and agents that recall a series of memories when you experience something similar.

Groups of agents that perform similar functions could be called agencies. In a familiar context, a real-estate agency has multiple real-estate agents, each dedicated to a specialty, such as commercial or residential transactions. In a like manner, the mind has multiple agencies, each with dominion over a specialized function of the human. For example, one could suppose there is a motor-control agency, or many of them at different levels of the hierarchy corresponding to the finer to courser levels of motor control. The motor-control

---

*This is so especially in those systems where chaotic, far-from-equilibrium conditions prevail. An order, often unpredictable from an analysis of the components, emerges.

agency could also be viewed as still another classification structure (orthogonal to the fine to course hierarchy), corresponding to the different parts of the body.

It can be argued, at least metaphorically, therefore, that the mind is a cacophony of multitudinous agents, all clamoring for attention. So what keeps us from experiencing all these individual agents and the resulting chaos? Because the lowest-level agents take care of their relatively simple, well-contained environment (or layer) and perform relatively simple tasks. Likewise, higher-level agents sort through multiple lower-level agents' contributions to a particular function, generate some "value added" themselves, maybe just a transform or filter, and direct the result to yet higher-level agents. In this way, many hierarchies and other types of networks of agents are composed within the brain and mind. And the answer to the question is that the topmost agents are the ones that we respond to consciously; they get our attention. Indeed, it may even be the case that the I *is* the topmost agent at any one slice in time.

A related issue is whether the very idea of a "self" that does the understanding in the brain and mind is valid. Some have suggested that there is no homunculus (a discrete doer or entity) in the brain. This idea of a central, real self as an entity unto itself gained credence from the apparent localization of the processing of perceptions and thoughts. There seems to be an I that does the perceiving and organizes behavior. In other words, it may be that the I *is* the topmost agent. This gives rise to the illusion that there is a central focus, or cartesian theater, as Searle (1994) has called it, upon which all our thoughts play their parts.

Another, opposing view is that, as multiple parallel physiological processes (agents), which are correlated with thoughts and behaviors, arise and compete for attention in the brain, the one that gains immediacy *is* the mind at that moment. ["You are your thoughts" (Krishnamurti (1982).]. We *are* what we are thinking at the moment—from second to second the topmost agent changes. The manner in which immediacy or attention is gained is unknown.

Suppose we *are* what we are thinking, moment to moment. Further suppose thinking can be very complex and multifaceted. A question immediately comes to mind: how does one topmost agent (of which we are or are not aware) deliver the richness of experience? Gestalt theory conceives of events and entities as unified wholes that cannot be derived or composed by a simple summation of the (perceived or conceptualized) parts. The richness of a thought, the complexity of an event is an emergent, synergistic phenomenon. Several experiments, at the Santa Fe Institute and elsewhere, have shown gestaltlike, emergent behaviors. To reiterate a point, in most cases, emergent behavior is unpredictable from knowledge of the parts (graph theories notwithstanding). This makes it very difficult to "reverse-engineer" the behavior such that interacting parts produce it.

Human intelligence, mind, and consciousness are examples, in many respects, of *patterns* of behavior and phenomena that emerge from billions of interconnected parts. Will we ever create artificial gestalts? Will millions or billions of relatively simple agents, all interacting on several layers of a vast, network-based hierarchies, produce a cyberconsciousness? Stay tuned (or connected!).

2.6 Agents of the Computer—AI Begets IAs

As we imagine the analogy of mind as a society of agents, we can imagine a cyberspace (local on one machine to global as in the Internet/World Wide Web) composed of many software agents. There might be repositories of agents categorized first by domain and thence, within each domain, by function and further by sophistication—emulating such an organization in the mind.

As we progress in developing distributed computing architectures with agents and other types of applications and functions, we are in fact approaching the manner in which some believe the brain and mind are organized and carry out certain of their functions.

In Section 5 we discuss agent architectures. There you'll see parallels between the proposed hierarchies and federations of neuronal groups composing the physical brain (which give rise to a concomitant or parallel hierarchy of agents that compose the mind) and the possible organizations of computer-based agents.

These mind processes we discussed are akin to the relatively independent "agents" that may be traversing the cybersphere soon. Does that imply some sort of "self" emerging from their synergy? "Sometimes a system with many simple components will exhibit a behavior of the whole that seems more organized than the behavior of the individual parts." (Hillis, 1988). If we are to take our cue from the brain's organization, then how could we incorporate this sort of connectionism into agent systems? "It would be very convenient if intelligence were an emergent behavior of randomly connected neurons.... It might then be possible to build a thinking machine by simply hooking together a sufficiently large network of artificial neurons" (Hillis, 1988). So could we interconnect large numbers of relatively simple agents and have some sort of intelligence emerge?

But what would those simple agents consist of, and what would they do? First of all you could "build a model of the emergent substrate of intelligence...[that] would not need to mimic in detail the mechanisms of the biological system, but it would need to exhibit those emergent properties that are necessary to support the operations of thought" (Hillis, 1988). Now all we need do is find out what those operations of thought are and model them within a computer. That is just what the cognitive neurosciences and AI folk have been doing over the last 20 or 30 years. But even as all this knowledge of cognition, and its emulation in artificial systems, is coming to fruition as our machines seemingly become more and more "intelligent," there still *seems* to be something missing. And that something points us back to earlier discussion on *understanding*. "Although the questions of capacity and scope are necessary in determining the magnitude of the task of constructing an emergent intelligence, the key question is one of understanding" (Hillis, 1988).

Yet even as we don't understand understanding yet, at least enough to emulate it artificially to the extent that we can say convincingly "Ah, my computer *understands* that I am lonely, not only because it behaves towards me in a manner programmed by someone, but I *feel that it empathizes with me because it is capable of loneliness too.*" Maybe this is going a bit too far, but it seems that the key to believing something others than myself can understand, is that I believe that thing *has* those (perhaps nonovertly behavioral) same *feelings* too. Can understanding be correlated with a feeling of certainty or identity (in the semantic sense) with what we understand? As we progress in developing machines based on biological models, perhaps we'll gain an understanding of what it means to understand and to feel. After all, Data kept professing that the lack of emotion kept him on the other side of the chasm separating him from true humanity. It is possible that, as we learn to emulate, on nonbiological substrates, the hormonal and other biochemical interactions that occur when we feel, we'll be a bit closer to the machine that understands in the same way that humans understand. Even further, as we develop biologically based intelligences, complete with the complex biochemistry involved, we'll have provided at least the substrate of biologically based intelligence that Hillis speaks of.

In fact, certain technologies presented in Section 3 that can be incorporated into agents are based on biological models:

- *Artificial neural networks:* Based on the brain's connectionist architecture and learning methods
- *Genetic algorithms:* Based on nature's evolutionary processes
- *Fuzzy systems:* Based on the human's ability to reason and communicate via linguistic modalities that allow for uncertainty, incompleteness, and ambiguity

These AI technologies all show great promise for making computer systems that seem more intelligent and better able to deal with humans.

In addition to the necessity (or desirability) of employing one or more of a multitude of existing AI techniques and so-called soft-computing technologies for the creation of IAs, we propose that the manner in which the agents themselves are interrelated and organized (in a multiagent system) is very important. The primacy of architecture is evident when we consider the different perspectives on how our brain may be organized.

The apparent hierarchy of a "society of agents" all clamoring for attention in the brain, all relaying their information upward and downward throughout the layers of the hierarchy, is one way in which computer-based agents could be organized. We'll see in Section 5 how certain agent architectures involve rudimentary hierarchies by their use of coordinating or controlling agents that are responsible for simpler agents. Again, perhaps our perception (or introception) of a singular I *is* a single higher-level "agent," at the topmost level of the hierarchy, together with other ancillary agents "being" us for some slice of time.

An alternative view involves the notion that our brain's functions are organized in terms of a rather flat, nonhierarchical architecture. This view is sometimes referred to as "modular theory." This theory posits that there is no place where everything comes together—Descartes's homunculus. In a conversation Richard Restak had with renowned neuroscientist Vernon Mountcastle on the subject of modularity, Mountcastle said: "The brain is a complex of widely and reciprocally interconnected systems. The dynamic interplay of neural activity within and between these systems is the very essence of brain function." This supports the contention that the brain is "arranged according to a distributed system composed of large numbers of modular elements linked together" (Restak, 1994). This has the further implication that information does not flow between modules through fixed pathways.

Most importantly, this distributed peer-to-peer architecture we mentioned in Section 1 in the context of distributed computing paradigms has implications for creating agent architectures. The brain is known for its robustness, partly because it has redundancy built into its connections. This is similar to distributed computer systems. In fact, one of the tenants of the design of mission-critical systems where life or limb is at risk is that there be no *single* point of failure. Another tenant is that such systems incorporate redundancy for the critical functions. Modular design is one factor that enables such tenants to become operational within a system. When you create agent systems patterned after this nonhierarchical conception of the brain, you may find your system capable of more graceful degradation and less prone to catastrophe by not having a single point of failure (that is, no "controlling" agent). Supporting this architectureal approach is the software engineering equivalent dictum in object-oriented systems: eschew controllers.

In addition, with respect to the brain, modularity theory goes a step further. It posits that this modular reorganization involves "…changes in the effectiveness of existing but formerly unused connections" (Restak, 1994). This means that other parts of the system can pick up the functions of lost or degraded parts not by brute force redundancy, but by reconfiguring themselves to do the new job, over time.

One last aspect of brain research that we'd like to mention as relevant to IA creation is the notion that much of what the brain does in its processing is, mathematically speaking, a form of compression. University of Wales researcher J. Wolff theorizes: "There is solid evidence that first language learning by children may, to a large extent, be understood as information compression" (Wolff, 1996).

One of the reasons you want IAs making use of compression techniques is that a large amount of the information out there is redundant, and by first eliminating redundancy the agent improves its efficiency over a wide range of tasks. Many agent tasks involve getting and processing lots of data and information. If an intelligent search agent (IA), for example, could prune the redundancy from all the search results returned to a user, the user would save time. Also, considering the same search example, eliminating redundancy saves bandwidth (which equals dollars for most of us) when the agent only has to return a fraction of what otherwise would be mountains of, typically, redundant information. These principles follow those the brain uses in its own attempts to deal with information overload. Wolff states that the brain invokes algorithms based on "...cognitive economy and principles of information compression which have been recognised by neurophysiologists for some time, especially the importance of inhibition in nervous tissue as a means of extracting redundancy from information" (Wolff, 1996).

Agents making use of compression techniques open a wide realm of possibilities for incorporating intelligence into their behavior. Indeed, Wolff's SP theory conjectures that "all kinds of computing and formal reasoning may usefully be understood as information compression by pattern matching, unification and search" (Wolff, 1996). How can an agent make use of this theory? By incorporating the techniques as Wolff elaborates them: "In this connection, unification means a simple merging of matching patterns, a meaning which is related to but simpler than the meaning of that term in logic. Information may be compressed by searching for redundant information and removing it wherever it is found. This means searching for patterns which match each other and merging or 'unifying' repeated instances of any pattern to make one. Since there is normally an astronomically large number of alternative ways in which patterns may be matched and unified, it is necessary to use some kind of metrics-guided search ('hill climbing,' 'beam search,' etc.) or otherwise to restrict the search space in some way" (Wolff, 1996).

Agents that transform (via compression, in some cases) raw data into information, thence into knowledge, and finally (some day) into wisdom, can make use of compression, just as the brain does. Wisdom, knowledge of that which is true or right, coupled with good judgment, while not usually reducible to an algorithm, can be approximated, at least relatively, by a sufficiently sophisticated agent in a narrowly enough defined context.

Certainly, compressing raw data by removing redundancy and eliminating irrelevancies creates more useful data or information. By creating information in context, relating that new information to other relevant information (perhaps by automatically creating links from significant terms within a returned web page and other web pages with expansions on those terms), and communicating that information to a human user, an agent can thereby impart knowledge about itself or a domain. Further (stretching), an agent that is able to "judge" what may be either true, relevant, or essential to its master, and acts accordingly, has, by definition, some degree of wisdom, especially if said agent learns, or otherwise acquires, this ability, through interaction with its world.

2.7 AI and Agents—the Specifics

What specific concepts within AI proper contribute to your ability to get intelligence into your agents? Several of the AI techniques that Russell and Norvig discuss at length in their book as useful for making IAs are summarized in this section. For much more detailed information on any of these topics and their relevance to agent design and implementation, see Russell and Norvig's AIAMA.

2.7.1 Solving problems with search methods

Programming an agent to solve a problem using search techniques involves setting goals, and if the agent is unsure of what action to take to reach the goal, it can start by using any of a number of algorithmic search techniques (breadth first, depth-first, bidirectional, uniform cost, and so on) to take steps toward the goal. Many times searches of potential state space (search space) can be pruned to eliminate subspaces that cannot affect the outcome. This is done in many game-playing applications, such as chess.

As an agent gains more access to its environment and the search space, it can use more sophisticated AI techniques such as heuristics and iterative improvement algorithms to optimize search costs. A heuristic is a method of estimating and providing a priori knowledge of a search method's chances of success before embarking on that route—sort of like an intuitive guess for humans. Iterative improvement algorithms include the hill-climbing algorithm and simulated annealing, both of which attempt to ensure that nonoptimal solutions are not just locally applicable. Hill climbing is a way to ensure that when you reach a minimum or maximum of a function that is optimal or global, and not just local, the algorithm keeps climbing the hills, comparing the results until a global optimum is found. Simulated annealing is a way to iteratively perturb and then relax a system described by equations such that the relaxation is a systemwide optimal solution. This technique is related to the way natural, physical systems react and cope with stress. Glass, for example, is "annealed" and made stronger by eliminating local weaknesses by repeated heating (perturbation) and cooling (relaxation).

2.7.2 Knowledge-based agents and agents that can reason

Many agent-based systems involve existing or dynamically created or acquired knowledge. Knowledge about something, for example, a domain like the stock market, and the use of that knowledge involve ways in which agents can represent their world [referred to as a knowledge base (KB)] and various types of logic (such as, Boolean, propositional, first-order predicate calculus, modal, temporal, fuzzy) that can be used to perform inferencing on the KB. Inferencing or reasoning over a KB generates beliefs. Reasoning capability is important for the agent trying to ascertain the truth value of its own beliefs or assertions, or those made by other agents. Inferencing is also used to determine a time sequence of actions needed to achieve a goal.

Temporal logic is especially important in those domains where facts are likely to change their truth values over time, such as in scheduling and planning activities. Modal logic can be used to differentiate between the truth values of an assertion over different contexts. Something true for an agent in its world, defined by its KB, may not be directly transferred to another domain or context and remain consistent. Some sort of transform or qualification may have to be considered. This is important when dealing with distributed agent systems, where an agent may find itself in an environment different from that for which it was intended.

A KB is a collection of concepts about a domain or domains. Knowledge engineering is the process by which those concepts are elicited, codified, and related to each other. The way in which knowledge is organized and the vocabulary used to describe the domain's concepts, together represent a theory or model of existence about that domain. This theory of existence is called an ontology. We will see an example of a very large, general-purpose ontology later when we look at Cyc. KBs and ontologies, whether local and domain-specific, or large, global, and ranging over many domains, are some of the most important resources developers can exploit in their quest to build truly intelligent agents.

"In multi-agent domains, it becomes important for an agent to reason about the mental processes of other agents" (Russell and Norvig, 1994). For example, if my agent can infer or believe that your agent can help in setting up a vacation, it is more apt to request that help.

Reasoning capabilities can be built into agents via a plethora of means: logic programming languages (such as Prolog), production systems (OPS-5, CLIPS), frame systems (such as OWL), semantic networks (conceptual graphs), and others. Semantic networks and frames are similar in that things are represented as nodes, and relations are the links between the nodes.

2.7.3 Goal-driven agents and agents that can plan

"Planning agents use look-ahead to come up with actions that will contribute to goal achievement. They differ from [simpler] problem-solving agents in their use of more flexible representation of states, actions, goals and plans" (Russell and Norvig, 1994). Planning is done in many cases using situational calculus, which involves representing the current and desired situations to an agent, as well as a sequence of steps (the plan) to get from an initial situation to the desired one.

Many times it is not practical to initially codify (and then search) all possible paths (or plans) from one state or situation to another. In this case an overall plan can be decomposed and mapped to subgoals. An agent can then attempt to achieve the subgoals and then extend the plan as appropriate until the final, desired situation or goal is reached. As subgoals and the plans that reach them are analyzed, it is useful to represent and account for resources such as time, money, and raw materials. Constraint satisfaction routines can be used to determine if a plan's actions overcommit resources.

As plans unfold, your agents should determine whether actions taken or to be taken conform to some optimal criteria. As plans are produced, certain actions are more optimal than others when that stage of the plan is ready to be carried out. If you have bifurcations (many paths) in the plan based on extant conditions, your agent is engaging in conditional planning or replanning. New subplans can be switched dynamically and linked into the overall plan in situ.

2.7.4 Agents that can reason under uncertainty

As Russell and Norvig point out, agents' actions based on first-order logic alone "…almost never have access to the whole truth about their environment" (Russell and Norvig, 1994). Therefore agents cannot always know for certain what is the correct, rational action to take in the real world. There are too many uncertainties: environmental factors such as location, where to go next in case of a mobile agent, resource uncertainties, unclear or wrong objectives and goals, faulty communications links, and so forth. A rational, intelligent agent must be able to deal gracefully with these uncertainties.

There are many methods that help in this respect: probability theory and utility theory, together referred to as decision theory, provide guidance in building agents able to deal with a sometimes irrational world. Probability theory helps in that you can specify a degree of *belief* in a relevant fact.* "Utility theory says that every state has a degree of usefulness, or utility, to an agent, and that the agent will prefer states with the higher utility" (Russell and Norvig, 1994). Utility functions represent the relative preferences of the agent for the different possible states the agent is in, or trying to reach, as part of its plan.

When techniques from probability (such as Bayes' rule and bayesian or belief networks) and utility theories are combined, we have an overall decision theory whose primary concept is the principle of maximum expected utility (MEU). "The fundamental idea of decision theory is that an agent is rational if and only if it chooses the action that yields the highest expected utility, averaged over all possible outcomes of the action" (Russell and Norvig, 1994). In other words, decision theory is used to define what the agent should do to "be the best that it can be."

To use techniques from these theories, you must be able to represent probabilistic beliefs your agents have about their world. In addition, you must provide methods that update the probabilities, determine outcome probabilities for various actions the agent can take, and also select that one action which conforms to the MEU. A utility function is used to "assign a single number to express the desirability of a state" (Russell and Norvig, 1994). For example, a utility function can map an agent's preference to buy your plane tickets for $125.00.

But mappings from simple state variables rarely capture the complexity of an agent's decision-making parameters. In the plane ticket example, while the specific amount of money to be spent expresses a preference for money, there

---

*Similarly, fuzzy logic, which we discuss in Section 3, allows you to specify a degree of *truth* about a fact.

may be several other factors, attributes, or states in the agent's world that are of equal or greater importance. For example, maybe the time you spend traveling is at least as important, or maybe more so, than the money spent. Since dealing with multiple attributes, and thus multiple utility functions, is usually the case (not always—for example, a gambling agent might care *only* about maximizing it winnings), you need to consider how to deal with possible conflicts. Multiattribute decision theory helps out here. Sometimes simple addition of the individual utility functions serves to arrive at an overall, maximized value, indicating the preferential state to be sought. Other times, one or more of the selected attributes is more important than others, relative to the overall goal, and you should program your agent to assign greater relative weights to those attributes accordingly. This is also one of fuzzy logic's strengths—the ability to ascribe weights (in the form of a percentage or degree) to the truth or falsity of an attribute or proposition.

The importance of decision theory to IA design is evident. The MEU principle helps us to understand what it means to use AI to make IAs. "If an agent maximizes a utility function that correctly reflects the performance measure by which its behavior is being judged, then it will achieve the highest possible performance score, if we average over the possible environments in which the agent could be placed" (Russell and Norvig, 1994). That is what you are trying to achieve—the highest possible performance for your end users.

Of course, agents perceiving and acting in the real world are dealing with a dynamic world. Therefore uncertainty is also dynamic. Agents need a way to look into the future and update their plans to reflect a possible future MEU. Dynamic belief networks and dynamic decision networks are techniques that handle attribute updates and solve sequential decision problems.

2.7.5 Agents that learn

Unless an agent's behavior is fairly trivial, an agent's designer can only go so far in building or programming "correct" behavior into an agent at the start. Even with the techniques mentioned, which help ameliorate uncertainty, an agent's mapping to its environment is almost always incomplete or inadequate, except in the most trivial situations. If an agent is to weather the vicissitudes of the real world, and to fulfill one of the main definitions of true intelligence—autonomy—it shall have to come to its environment as a child, that is, ready and able to learn.

Many of the AI techniques you've already read about involve programming in whatever actions you want your agent to perform, based on certain conditions. If you want your agent to learn by itself what actions to perform, you need to give it the capability to monitor its performance in achieving its goals. This is called feedback. You also need to provide a method by which the agent, upon comparing its performance against desired performance, can change its own internal workings such that the next iteration of perceiving and action comes closer to the goal than before.

Of course, all of this implies that you can represent, with some mapping or function, what the agent is doing in relation to a goal state. "The key point is that all learning can be seen as learning the representation of a function. We can choose which component of [an agent's performance] to improve and how it is to be represented" (Russell and Norvig, 1994). Functions can be logical sentences or assertions, belief networks, or neural networks (discussed in more detail in Section 3), among others. When such a function is learned by comparing inputs and desired outputs, the learning is called inductive learning.

Another type of learning, common to humans, is to provide positive feedback to reinforce positive behaviors—behaviors that reach a goal state successfully. But how do you get a software agent to recognize whether some feedback percept is pain or pleasure? One way is to make use of the MEU to select actions. When the correct action is taken, a reward is given in that the agent is designed to recognize that action as better than other possible actions it could have taken. The reward can be increases in a value of another utility function the agent is programmed to maximize.

Another type of learning mechanism is embodied within the AI subdiscipline of evolutionary programming and genetic algorithms. A system of agents or functions and methods within an agent learns what is an MEU via a fitness function that works like natural selection in nature's evolutionary schema. "Genetic algorithms achieve reinforcement learning by using the reinforcement to increase the proportion of successful functions in a population of programs" (Russell and Norvig, 1994). We discuss evolutionary programming in more detail in Section 3.

In humans, learning occurs based on a substrate of previous knowledge. In addition, much of what we learn is learned via analogy or recognition of patterns which connect similar scenarios to each other and to what we already know. This is referred to as cumulative learning. Explanation-based learning uses single examples to derive a general rule by explaining the example and then generalizing the explanation. This is known as deduction. Knowledge-based inductive learning uses prior knowledge, coupled with examples, to explain and thus learn from the examples. This enforces consistency on any resulting hypothesis the agent might try to incorporate into its knowledge base.

2.7.6 Agents and communications

To be useful, any agent, whether intelligent or not, cannot be an island unto itself. Agents, by definition, act as a proxy for someone or something. Acting, to be useful, means acting eventually on something other than its own state, that is, its environment. An IA's environment could consist of other IAs, other types of software, and hardware, including control devices such as valves and toasters. Environmental entities can be local to the agent (such as the same machine or platform on which the IA resides) or remote (on other machines connected to the IA via some type of network).

Acting further implies, for an intelligent, rational agent, acting according to the IA's beliefs and goals, garnered through designed-in methods and sensed data. Both sensing and acting are forms of communications. "In general, communication is the intentional exchange of information brought about by the production and perception of signs drawn from a shared system of conventional signs" (Russell and Norvig, 1994). A shared, structured system of communications, whether between animals, between machines, or between machine and animals, is a language.

Russell and Norvig point out that an IA's communication with its environment can take many forms:

- Inform other agents about itself and its knowledge of its environment
- Query other agents about their state
- Answer queries
- Request or command other agents to do something
- Give a promise or offer to do deals; engage or enter into contracts
- Acknowledge communications with other agents

One of the more difficult aspects of IA design involves giving the agent the ability to determine what to communicate, and when. A corollary is the difficulty of designing agents that can understand each other's communications when they take place. Understanding, in a rudimentary sense, makes use of some language or protocol: a formal specification of the syntax and semantics of a message. Many IA researchers are basing their work on natural-language processing for interagent communications, rather than formal computer languages like LISP or Pascal.

If communicating agents share the same internal representation of knowledge, a simple, direct-access message interface of the form: TELL (AgentX, SomeKnowledge) or ASK (AgentX, SomeQuestion) can be used. In most complex agent environments, agents need to communicate with other agents not having the same internal representation of knowledge. Unless developers carefully encapsulate agents' knowledge, you may have to use more complex external languages to communicate with other agents. Unfortunately this may require parsing messages, performing syntactic, lexical, and semantic analysis, and performing disambiguation—a technique used to diagnose or interpret a message in relation to a particular world model.

All these natural-language processing techniques and many more complex issues are involved in developing agent-communication languages. In Section 5 we talk more about the issues relevant to agent communications and interoperability. In particular we look at a particularly promising agent language dubbed, appropriately enough, Agent Communications Language (ACL), based on evolving standards such as KQML and KIF.

So far we have discussed agents acting within their environment via communicating signals, or information. IAs in full dress are otherwise known as robots. These IAs have a physical embodiment equipped with sensors and mechanical effectors or actuators that affect the physical world. Of course, since the embodiment of any software entity in a physical instrumentality can be said to affect the physical environment by pushing electrons around, these software-based bots are appropriately called softbots. When we speak of robots, think C3PO, Gort, or Robbie!

Effective IA-based robots are equipped with a representation of their physical environment as well as their own software environment and physical embodiment. Analysis of that physical environment takes place via modeling relevant artifacts within a configuration space. Navigation through the configuration space produces equivalent motion in real space.

Of course, if an agent is to act "correctly" within its environment (cyberspace or real world), it needs to sense or perceive relevant aspects of that environment and, further, to understand what those percepts mean. The operative questions is "…if sensory stimuli are produced in such a way by the world, then what must the world have been like to produce this particular stimulus" (Russell and Norvig, 1994). Depending on the stimuli, IAs may have to perform optical processing, speech processing, and recognition and even be able to process percepts from sense modalities such as touch, taste, and smell.

Now that you have a grasp of some of the important AI-based techniques that can be used to help create IAs, Section 3, among other things, presents some of them in more detail. In addition, other sections, especially Section 6, bring up some of these techniques again.

Section

3

Converging Technologies that Facilitate and Enable Agents

This section is an overview of several computing technologies that, we believe, can facilitate the development and use of IAs, independent of the infrastructure (such as CORBA, DCE) or specific language (such as Smalltalk, Java) you may decide to use. What we have chosen to describe here are several technologies as well as specific instances of technologies that have proven their practical use in commercial applications and for which sufficient, mature development tools are available.

3.1 Introduction

As you begin the analysis and design phases of your agent project, there may come a point where you simply don't know where to start—especially when you consider how to imbue your agent with intelligence or what software development paradigm is best suited to create relatively independent entities existing throughout a distributed system. When you finish reading this book, we hope you have a grasp of most, if not all, of the concepts and techniques involved in building useful IAs. Although no single book can both present all the tools and techniques and at the same time delve deeply into each of them, we believe we have achieved a reasonable compromise with this work. Part of Section 2 presented, at a high level, several AI concepts and techniques that can be integrated into an agent. This section discusses in more depth those technologies [both AI-related and object-oriented (OO) software construction methods] that, in our opinion, should be considered seriously in any attempt to construct IAs. While technologies and concepts in Sections 4 and 5 are concerned with make versus buy decisions regarding an agent-support infrastructure, or how to structure several agents in a collaborative architecture, this section, as well as most of Sections 6 and 7, deal with the agent itself. These technologies are:

- Knowledge-based and expert systems
  - Cyc
- OO software development
- "Soft" computing technologies
  - Fuzzy systems
  - Neural networks
  - Evolutionary computing, genetic algorithms

In addition to the specific technologies we cover in this section, it is worth mentioning here that less is sometimes more with respect to the creation of intelligence. Instead of immediately implementing a technically advanced, or perhaps unfamiliar, technique such as neural networks or genetic algorithms, perhaps your application can benefit from some simple but clever constraint-based or goal-directed programming. For example, intelligent behavior can emerge by implementing fairly simple rules or algorithms within software entities. As these simple automatons individually seek local goals, and collectively seek group goals, novel solutions to problems emerge.

One popular example of emergent collective intelligence involves programming individual, differently shaped three-dimensional blocks to obey simple physical laws related to gravity, tumbling constraints, acceleration, and the ability to form connections with other blocks. An important component of the system of blocks was a collective goal. The goal was to move from one place to another while observing the blocks' individual goals such as minimizing their own expenditure of energy. As the blocks engaged each other in various combinations, certain very novel, even surprising solutions emerged.

Whether or not this constitutes real intelligence is an open question. Some would say this is an inevitable result of simply trying many combinations of solutions; some combination will arise as clearly superior to most others. But the point here is not a definitional dilemma vis-à-vis intelligence. Rather it supports the notion that you don't necessarily need a heavy hammer to drive a nail. Indeed, certain architectures, or ways of combining societies of collaborating agents, may lead to intelligence emerging synergistically from the whole (the whole being greater than the "sum" of the parts).

The first part of this section presents a quick overview of one important commercially successful application of AI—expert systems. We then discuss one particular approach to imbuing software with the ability to become expert in a particular domain, as well as having commonsense reasoning abilities. We then present an overview of object orientation, and why it is important for developers of IAs, especially within distributed systems. Next we present certain AI technologies sometimes referred to as "soft" computing, which, when integrated into your agent system, can help provide the "I" in IAs. These technologies include fuzzy systems, neural networks, and genetic algorithms. Together, the technologies presented in this section, along with the AI techniques in Section 2, can provide agents with intelligence having the following qualities and attributes:

- The ability of agents to reason about the world or domain in which an agent finds itself, via the use of (inherently within the agent's code or externally) expert systems
- The ability of agents to reason under uncertainty and with imprecise or incomplete data via the use of fuzzy systems techniques
- The ability of agents to discern patterns, learn, and generalize via the use of neural networks
- The ability to evolve software agents to best fit or deal with the situation at hand via the use of genetic algorithms
- The ability to structure agents such that their world view, communications, and internal mechanisms are based on a strong notion of modularity, avoidance of single points of failure, relative autonomy, and various types of reuse via OO methodologies of design, development, and execution

Your decision to further investigate and use any of these technologies depends on your specific application. What we want to do here, is expose these technologies to you and make the case that if agents are to be considered intelligent, their intelligence has to come from somewhere.

3.2 Expert Systems and Knowledge Bases

There are many subdisciplines of AI that have proven fruitful in the move from the research lab to commercial use. None is more ubiquitous than expert systems. An expert system and the knowledge base it encapsulates may be the easiest way to give your agents some intelligence. This section tells how agents can make use of expert systems that are already up and running in most domains in which agents are likely to be used.

What we do not do here is tell you lots of details about how expert systems are created, or the different ways of representing knowledge objects (such as frames, rules, constraints, and other relationships). For details on expert system construction, refer to the extensive literature on that subject.

3.2.1 "Classical" expert systems

The last two decades have seen an increased reliance on the computer as expert. Expert systems have been used for:

- Interpreting data and identifying significant features
- Predicting consequences or trends

- Diagnosing and aiding in troubleshooting problems
- Product configuration
- Planning and scheduling
- Process and production monitoring and alarm management
- Developing tutoring and help systems
- Searching text for meaning, performing rough translations

In sharply delimited domains, the functions listed have been performed by computer-based expert systems over the years with remarkable success. Expert systems codify knowledge about a domain. The codified knowledge is sometimes referred to as an ontology (which means the systematic study of being, that is, the study is in the form of names of things and their interrelationships—an expert system). Some expert systems codify a domain's ontology via rules, constraints (and other relationships), frames (encapsulation of an object's attributes and operational characteristics), and other methods.

Possibly the first rule-based expert system was called DENDRAL. Produced by Joshua Lederburg in 1965, DENDRAL determined molecular structures from spectroscopic data. Rules were made by encoding knowledge from a chemist in if-then constructs using the LISP (list programming) language. This expert system performed very well. In fact, "DENDRAL outperformed some of the finest human experts in the field" (Wolfram, Dear, and Galbraith, 1987). By 1972 three other notable expert systems had been developed and were quite successful:

- CADAUDEUS: Diagnosed internal diseases
- MACSYMA: Solved calculus problems
- MYCIN: Diagnosed and suggested treatment for blood diseases
- DEC's XCON: A system configuration tool

These early successes contributed a great deal to the hype in late 1970s and early 1980s about the imminent coming of machines that would be smarter than humans. These knowledge-based systems were (and still are) composed of discrete pieces of knowledge about some well-defined domain, such as diagnostic medicine in MYCIN's case. MYCIN is also historically important in that it led to the development of the first expert system shell. An expert system shell is "a program containing all the logical structures and thinking strategies, but without the knowledge base of a specific domain" (Wolfram, Dear, and Galbraith, 1987).

An expert system is typically composed as follows. Using an expert system shell, domain knowledge is partially represented with attributes describing or defining the different types of (relevant) things residing in the domain. Expertise is encapsulated using "production" rules—if-then constructs that can be executed in some order to arrive at some conclusion. In addition, rules are coded to reflect possible legal (or illegal) relationships between the things, as well as possible externally imposed constraints. An inference engine is used as a way of executing the rules to resolve a problem statement. The system usually has a way to create new rules and things, and of posing questions or scenarios to the system for resolution.

For example, a system configuration tool like DEC's XCON is given a possible combination of products that may or may not fit or work together properly. The configurator then inferences or executes its relevant rule set and tells the user what is wrong with the configuration, and even fixes it if possible. Ideally such systems are extensible such that new sets of products can be added or deleted, including their own rules and constraints, and fit right into an existing configuration engine.

The advantages of using existing expert systems to an agent developer are many. You have ready access to knowledge and "intelligence" about a wide variety of subjects within many domains. If you can gain access to a knowledge base in your domain, and can formulate appropriate interfaces, your agents can leverage that extant knowledge in performing their tasks. The question is, how does your agent-based system gain access to this expertise?

Unfortunately many expert systems are one of a kind, based on proprietary "engines" or development environments, and are thus difficult to integrate with other software. In addition, the raison d'être of expert systems—the encapsulation and timely transportation and dissemination of expertise or knowledge—relies on an inexact and problematic exercise of knowledge acquisition. In addition, even the most sophisticated expert system is really an "idiot savant." They are constrained to an area of specialty and therefore brittle and unable to share information.

Most existing expert systems would be too large to encapsulate within a lightweight mobile agent. Therefore some way of communicating with the expert system must be devised, if not already in place. However, if the amount of expertise needed is very focused and expressible with a relatively minor amount of code, you could program a small set of rules or constraints within one or more agents, using the language in which you developed the agent.

For more complex or larger rule bases you can use an expert system package to generate the rules. Of course, the agent has to be able to execute the rule base when called on. One way to do this is with packages that allow you to develop the rule base and embed the rule base into the agent, while the inference engine runs separately when a rule is fired. One example of this approach is Production Systems Technologies' CLIPS/R2 product. The rules are embeddable within C or C++ programs, which, in turn, can be used by agents built using other languages such as Smalltalk via in-line calls.

Another way to do this is to embed the rule *and* the inference engine itself in the agent. Agent OCX from Haley Enterprise* is an OLE-based custom control implemented in C++ that encapsulates Haley's Eclipse inference engine. Using

---

*http://www.amzi.com.

their suite of development tools, you can create the rule base and embed both the rules and the mechanism for running them right in your agent.

Or if the expert system was or is being built with the same development environment as the agents, then you would have access to the expert system's application programming interface (API) directly. Your agents could make use of that API to pick and choose relevant functionality from the export system. In many cases you may find an expert system that provides a limited API that your agents could use; a few in-line C calls might do the trick. Still another case is when the infrastructure components you choose provide agent-to-object adapters for various services. For example, in IBM's common object model (COM) architecture such facilities are being developed to enable the integration of agents with databases, communications services, and various applications, including, presumably, expert systems.

Another example of embedding (or in this case wedding) expert system technology into IAs comes from WEBLS from Amzi.* Their environment lets you embed rules in web sites. A HyperText Markup Language (HTML) document collects information from a user. Then, via common gateway interface (CGI) the IA, in the form of the rules, can manipulate the data and give feedback to the user, or direct the user to take other actions.

Another example of an environment that integrates expert system technology and agents is an IA developed by IBM/Lotus Notes. This agent automates the dissemination of information, in the form of Notes databases, by combining rule-based reasoning and scripting. Normal Notes agents are built as part of a Notes database and are launched at a user's request, at certain configured intervals, or via triggers. They work in the background, essentially alone, performing routine tasks like sending E_mail, automatically filing or archiving documents, bringing in data from other applications, manipulating the values of fields in individual Notes database documents, and searching for particular topics in a database. The ES enabled agents, on the other hand, are being developed to collaborate; using a common rule-representation schema and a common inferencing engine, the agents from different databases, representing different users or domains (such as in a work-flow-like application) can work together to provide total solutions.

An expert system's capabilities thus distributed throughout your agents, or accessible to your agents, can make them more autonomous when dealing with their specific tasks. This approach might be preferable to the chore of interfacing to a foreign (in terms of programming language/API, semantics, fitness, and so on), though probably more robust, existing expert system.

The ideal approach to accessing expertise would be a distributed agent-execution environment that could provide uniform access to expert systems wherever they reside. Expertise on remote nodes could be accessed by moving the agent to that node where it can continue execution. This allows agents to chain together results from many expert systems, but the execution state of the agent

---

*http://www.amzi.com.

must move along with the agent itself. As of this writing, the only commercial agent system capable of this is Telescript.

What if an agent could make use of a great storehouse of knowledge when needed; indeed, what if an agent could even make use of resources that include the ability to use logic, reasoning, commonsense information about the universe...from places other than its own locale—the essence of a distributed mind arising out of independent entities "collaborating" to fulfill a goal of a part of the mind?

As Lenat (1995) states it: "People share knowledge so easily that we seldom even think about it. Unfortunately, that makes it all the more difficult to build programs [agents] that do the same. Many of the prerequisite skills and assumptions have become implicit through cultural and biological evolution and through early childhood experiences."

In addition to being able to share information, expert systems and the agents that use them should be capable of varying levels of commonsense reasoning. Instead of requiring that users speak computerese, it is desirable that computers be able to take natural-language expressions and translate them into actions commensurate with the user's desires. Commonsense reasoning skills, as a prerequisite to domain-specific functionality, embedded, or accessible by a given software application, would go a long way toward making computers more accessible and acceptable.

Why should we expect it to be any different with AI and its carrier—IAs? Lenat (1995) continues: "Before machines can share knowledge as flexibly as people do, the prerequisites need to be recapitulated somehow in explicit, computable form." The next subsection brings you up to date on a project that has been ongoing for almost a decade. Its aim is to encode and provide access to an enormous corpus of generalized and specialized knowledge about *every* thing.

3.2.2 Agents with common sense: Lenat's universal expert system*

An ambitious project is just now coming to commercial fruition. It is the construction of a very large corpus of universal knowledge called Cyc. Cyc is intended to include "such common sense notions of time, space, causality, and events; human capabilities, limitations, goals, decision-making strategies, and emotions; enough familiarity with art, literature, history and current affairs..." (Guha and Lenat, 1994).

The Cyc system comprises a very large, multicontextual knowledge base, an inference engine, a set of interface tools, and a number of special-purpose application modules, running on a variety of platforms. The knowledge base is built upon a core of approximately 400,000 hand-entered assertions (or "rules") designed to capture a large portion of what we normally consider consensus knowledge about the world. (For example, Cyc knows that trees are usually outdoors, that once people die, they stay dead.) The Cyc system is composed of:

---

*We thank Doug Lenat and Cycap for providing much of the information in this section.

- The Cyc knowledge base
- The CycL representation language
- The Cyc inference engine
- Cyc interface tools
- Cyc application modules

3.2.2.1 Cyc knowledge base. The Cyc knowledge base is a formalized representation of a vast quantity of fundamental human knowledge: facts, rules of thumb, and heuristics for reasoning about the objects and events of everyday life. The medium of representation is the formal language CycL, described in this section. The knowledge base consists of terms, which constitute the vocabulary of CycL, and assertions, which relate those terms. These assertions include both simple ground assertions and rules. Cyc is not a frame-based system: the Cyc developers think of the knowledge base instead as a sea of assertions, with each assertion being no more "about" one of the terms involved than another.

The Cyc knowledge base is divided into many (currently hundreds of) "microtheories," each of which is essentially a bundle of assertions that share a common set of assumptions. Some microtheories are focused on a particular domain of knowledge, a particular level of detail, a particular interval in time, and so on. The microtheory mechanism allows Cyc to independently maintain assertions which are prima facie contradictory and enhances the performance of the Cyc system by focusing the inferencing process.

At the time of this writing, the Cyc knowledge base contains tens of thousands of terms and several dozen hand-entered assertions about or involving each term. New assertions are continually added to the knowledge base by human knowledge enterers. The aforementioned numbers do not include (1) nonatomic terms (such as `#$LiquidFormOf #$Nitrogen`) or (2) the vast number of assertions added to the knowledge base by Cyc itself as a product of the inferencing process.

3.2.2.2 CycL—the Cyc representation language. CycL, the Cyc representation language, is a large and extraordinarily flexible knowledge representation language. It is essentially an augmentation of first-order predicate calculus (FOPC), with extensions to handle equality, default reasoning, skolemization (the removal of all existentially quantified variables), and some second-order features. (For example, quantification over predicates is allowed in some circumstances, and complete assertions can appear as intentional components of other assertions.) CycL uses a form of circumscription, includes the unique names assumption, and can make use of the closed-world assumption where appropriate.

3.2.2.3 Inferencing in Cyc. The Cyc inference engine performs general logical deduction (including modus ponens, modus tolens, and universal and existential quantification), with AI's well-known inference mechanisms (inheritance, automatic classification, and so on) as special cases. Cyc performs best-first search over proof space using a set of proprietary heuristics, and uses microtheories to optimize inferencing by restricting search domains.

Because the Cyc knowledge base contains hundreds of thousands of assertions (also called "rules"), many approaches commonly taken by other inference engines (such as frame-based expert system shells, rete match,* Prolog) just don't scale up to knowledge bases of this size.

Cyc also includes several special-purpose inferencing modules for handling a few specific classes of inference. One such module handles reasoning concerning collection membership and disjointness. Others handle equality reasoning, temporal reasoning, and mathematical reasoning.

3.2.2.4 Cyc interface tools. The Cyc system also includes a variety of interface tools that allow you to browse, edit, and extend the Cyc knowledge base, to pose queries to the inference engine, and to interact with the natural-language and database integration modules.

The most commonly used tool, the HTML browser, allows you to view the knowledge base as the WWW is viewed (in a hypertexty way). HTML pages describing Cyc terms are generated dynamically (on the fly) by the Cyc system. Each page describes a Cyc term by showing all the assertions in which it is involved, organized according to a standard schema. Every occurrence of a Cyc term is an HTML link to a dynamically generated HTML page describing that term, so that it is easy to surf around the knowledge base following a network of relationships. The HTML browser also includes facilities for searching and editing the knowledge base and for posing queries to the inference engine.

Other HTML interface tools include:

- A "knowledge-base file" tool, which allows you to batch-process text files describing new material to be added to the knowledge base.

- A hierarchy browser, which displays any desired subtree of the Cyc subset tree in outline format.

- A lexicon editor, which provides a user-friendly way to edit and extend the Cyc lexicon.

- An English-to-CycL parser, which lets you experiment with Cyc's natural-language facilities by parsing arbitrary English strings.

- A database tool interface, which provides an interface to Cyc's database integration module.

---

*A rete algorithm is used in forward-chaining logic languages (production systems) to apply inference rules (Condition if true causes Action) to a knowledge base. The rete algorithm and the rete network that represents actions, constraints, and tests, can be used within an IA system to maintain an agent's knowledge bases.

3.2.2.5 Applications of Cyc. The development of Cyc was a very long-term, high-risk gamble that has only recently begun to pay off. Begun as a research project in 1984 as part of the MCC initiative,* Cyc is now a working technology with potential applications in many real-world business problems. Cyc's vast knowledge base enables it to perform well on tasks that are beyond the capabilities of other software technologies.

The following is intended to give an indication of Cyc's potential by describing some tasks to which Cyc is currently being applied and by suggesting others to which Cyc might be applied in the future. Applications currently available or in development include:

- Natural-language processing
- Integration of heterogeneous databases
- Knowledge-enhanced retrieval of captioned information
- Distributed AI
- WWW information retrieval

Potential applications include:

- On-line brokering of goods and services
- "Smart" interfaces
- Intelligent character simulation for games
- Enhanced virtual reality
- Improved machine translation
- Improved speech recognition
- Sophisticated user modeling
- Semantic data mining

Although space limits a full description of all these applications, let's look at some of the efforts. One such effort involves modeling people or organizations. These models can then be used in a variety of scenarios such as job matching, personal assistance or services, and product marketing.

Another broad category of applications amenable to Cyc's talents is information retrieval, specifically semantic (as opposed to syntactic) information retrieval. This is a much more powerful and useful way to find knowledge. If your agent actually understands what it is looking for, it is apt to be much more accurate and complete.

Another interesting application would use Cyc-ic agents to provide consistency checking across various knowledge sources. Briefly, you can do this by

---

*MCC stands for the Microelectronics and Computer Technology Corporation. It is a consortium formed to counter the Japanese Fifth Generation Project, which was dedicated to producing intelligent computers.

forming some axiom that includes two different columns or tables. Then Cyc's inferencing capabilities can run through actual values in the column or row (cells) to find any inconsistencies. You could also use this method to eliminate redundancies. As you can see, this capability of fusing multiple sources of information points toward enterprise knowledge integration. Just replace the discrete information repositories or formats with schemas and metaschemas to enable translation between them using Cyc.

Since Cyc can deal with nontextual information-bearing objects, semantically based image (and other object types) storage is a natural application area. If stored images had an appropriate caption, Cyc could use its abilities to "stretch" the meanings of words to find many more possible relevant images from an initial query.

Of course, this implies some preexisting textual-based element (a caption) such that Cyc can deal with it. However, we posit that with sophisticated image processing, certain image elements could be matched with a suitable library of image elements, thereby enabling a preprocessing function to effectively generate a description of the image to some degree of accuracy—in effect, automatically generating a Cyc-usable caption.

Using agents to reorganize all of a system's information-bearing objects into a semantic file system is also a possible application. To some degree of granularity, an agent can (virtually) "dismember" all the files from a source and rearrange those information-bearing objects into a pile. Recovery of any bit of information is now possible by a content-full query.

So you see, you can immediately incorporate Cyc-ic-like capabilities into your agent-based systems. Indeed, as Lenat professes: "The prospect is there, finally, for Cyc to be the vector of IAs in the next few years" (Guha and Lenat, 1994).

Lenat suggests that until more experience is gained in the practical use of Cyc, only certain types of applications are amenable to being Cyc enabled (or, as we put it, "Cyc-ic"). The following are some things to consider if you think your agents should be Cyc-ic:

- Use Cyc in a pilot or prototype system that does not require a large commitment of resources.

- Don't let your agents rely totally on knowledge from Cyc; enable your agents to recover gracefully from a Cyc failure.

- If your agents are "real time," don't expect Cyc to keep up all the time.

- Optimize the use of Cyc by your agents, that is, isolate and characterize the types of queries your agents might initiate. Then add heuristic algorithms to speed any required inferences.

- Unless your agents need access to a broad spectrum of knowledge with rich commonsense representation and semantics, your agents might better profit from a more narrowly focused expert system or other AI-like applications.

- Because of the resource requirements of Cyc, in a multiple-node system assume Cyc will run on only one node.

One of the applications mentioned, distributed AI, is particularly applicable to our quest for IAs. As such, it is discussed in more detail next.

3.2.2.6 Distributed IAs using Cyc. Within the next few years, Cyc may be installed at sites where even thousands of users need to harness its power simultaneously. An important question is: what should the architecture of such an installation look like? While one Cyc server can easily support a dozen clients at the same time, handling operations from thousands of clients would bring a Cyc server to its knees. How do we address this scalability issue?

The obvious alternative is an architecture in which each user (or work group) runs its own local Cyc server, with its own copy of the Cyc knowledge base. But this scenario does little to alleviate the scalability problem. Now each user is saddled with the responsibility of hosting a knowledge base that already contains nearly half a million assertions, and could easily grow into the millions as more and more specialized knowledge is added. Moreover, each user must tolerate the computational and communicational overhead associated with keeping each of the thousands of copies of Cyc in sync with each other.

A great deal of the redundancy and inefficiency of such an architecture can be eliminated by adapting Cyc to work in a distributed fashion. In fact, this is how humans work: we specialize and collaborate. Although most humans share a core of common knowledge which allows them to communicate, no one human possesses all the knowledge of the human race. Instead, we have experts in law, medicine, construction, automotive repair, and software design. When a human lacks the knowledge to solve a problem, he or she can often find a solution by collaborating with an expert. When I go to the doctor with a persistent headache, the doctor's knowledge of medicine combines with my knowledge of my symptoms to produce a solution that neither of us could have reached alone.

In a distributed Cyc architecture, the network is populated with Cyc agents, each of which shares a common core of knowledge, but possesses also one or more additions to the core knowledge base that extend its expertise into new domains. Most importantly, the various Cyc agents are endowed with the ability to communicate with each other and to perform inferencing in a collaborative fashion. The interagent communication can be handled flexibly by using KQML (refer to Section 6) or some other knowledge-sharing protocol, or it can be implemented using a more efficient Cyc-specific protocol.

Cyc agents in a distributed architecture share knowledge describing how they can be reached (via which network address, port, and protocol) and what their areas of expertise are. During inferencing, when an agent tries to expand a formula whose content lies outside its area of expertise, it determines (by consulting its own knowledge base or a central knowledge-broker agent) whether another agent is available that might be able to help. If so, the agent sends a message to its remote counterpart asking it to help expand the formula in question. The answer may be a complete set of bindings, but more commonly it is simply a partial result, still containing unbound variables. The local agent incorporates the result into its own proof tree just as if it had done the work itself, and continues its task. In many cases, the local agent may need to consult the remote agent, or several remote agents, multiple times in the course of its inferencing process.

Cycorp has cooperated with the Computer Science Department of the University of Maryland in Baltimore County (UMBC) to develop a demo of such a distributed architecture. In the demo, three Cyc agents communicate with each other using KQML. While all three agents possess the same core knowledge base, each possesses additional knowledge about an additional domain in which it is considered expert: GeoAgent in geography, PolAgent in politics, and EcoAgent in economics. Working together, they can answer queries that no one of them could have answered alone.

For example, suppose a user asks GeoAgent for "elected heads of government of countries north of the equator." This might be represented as follows:

```
(#$LogAnd
  (#$headOfGovernmentOf $x $y)
  (#$hasAttributes $x #$Elected)
  (#$northOf $y #$Equator))
```

GeoAgent is able to find bindings for the third clause by using its own knowledge of the geography domain:

Britain is in Europe.

Europe is in the northern hemisphere.

The northern hemisphere is north of the equator.

If region A is part of region B, and region B is north of region C, then region A is north of region C.

Therefore, Britain is north of the equator.

But to find bindings for the first two clauses, GeoAgent must enlist outside help. It sends these as queries to PolAgent, which is able to find bindings for them using its knowledge of the politics domain:

Heads of government of democratic countries are elected.

Great Britain is a democratic country.

Tony Blair is the head of the government of Britain.

Therefore Tony Blair is the elected head of government of Great Britain.

When GeoAgent agent receives this answer back from PolAgent, it combines it with its own partial answer to produce the final result: Tony Blair.

An important research focus of the Cycorp/UMBC project has been optimizing the implementation of collaborative inferencing so that the costs of managing the necessary communications overhead are significantly outweighed by the increase in scope and efficiency which results from sharing the inferencing workload.

Cycorp has been sufficiently pleased with the results that they plan to convert the network in the Austin office to a distributed architecture in the near future.*

It should also be pointed out that while the foregoing description assumes that all the agents participating in the distributed architecture are Cyc agents, this is not a requirement. Cycorp has defined a very simple protocol for implementing cooperative inferencing, and any agent that has been taught to adhere to this protocol, and which possesses knowledge of interest to other agents, can meaningfully participate in such an architecture. For instance, the role normally played by a Cyc agent in a distributed architecture could equally well be filled by a gateway to:

- An expert system implemented in Prolog
- A standard query language (SQL) database
- A special-purpose inferencing tool (such as for spatial inferencing)
- A human expert

Cyc can be used by agents, for example, to retrieve something that matches a user's criteria that wouldn't be explicitly contained in the knowledge being searched. To show you the power of Cyc to augment the capabilities of your agent, consider the following example.

A user asks his agent to find a picture containing seated persons from some huge database. The Cyc-enabled agent finds a picture with the following description or caption: "There are some cars. They are on a street. There are some trees on the side of the street. They are shedding their leaves. Some of them are yellow taxi cabs. The New York City skyline is in the background. It is sunny." Cyc used its common sense about cars—they have a driver seat: on street, so in motion, so probably a *sitting* driver.

When you start making architectural choices and agent-design decisions, one of the most important aspects to consider is: on which of the two major agent paradigms do you intend to pattern your development? These paradigms exert a great deal of influence on the direction your efforts take and on your technical approaches and assumptions:

Paradigm 1. Competence emerges from a large number of relatively simple agents integrated by some cleverly engineered architecture. The choice of architecture is the make-or-break theoretical part of this; the detailed characteristics of the implementation of the architecture (and the algorithms that crawl around it) are the make-or-break pragmatic parts. The archetype of this paradigm is SOAR (Laird et al., 1984); its forerunners were the early "pure production systems."

---

*For more details on the Cycorp/UMBC demo of the distributed Cyc architecture, see "The Cycic Friends Network: Getting Cyc Agents to Reason Together," a paper describing the project that was presented at the 1995 CIKM conference. Also see the Cyc KQML Project (1996) describing that work.

Paradigm 2. Competence emerges from the aggregate system possessing a large amount of useful knowledge. For most real-world tasks this includes a dauntingly large fraction of what might be termed "general common sense." In this paradigm the architecture is relatively unimportant, and building the system as a set of "agents" is little more than a form of scaffolding, reducing the cognitive load on the human builders. The archetype of this paradigm is Cyc; its forerunners were the early expert systems (Guha and Lenat, 1994, p. 127).

With either paradigm, the need for agents to share their knowledge is important, especially for systems in which intelligence is to emerge from the collective behavior of the individual agents. But with Paradigm 2, individual agents and federations of agents can do complex and useful tasks without as much dependence on a clever or complex architecture. If agents can partake of a large enough body of knowledge (indeed, even wisdom), then they can become universally useful, and do far more than the narrowly outlined tasks for which they are now responsible. In large part, many of the architectures we present in Section 5 are of the Paradigm 1 form.

The issue for agent developers is, once Cyc and other knowledge bases exist, how do agents share in the knowledge? How do they communicate their expertise? Even as agents learn from their own experiences and are able to codify their new knowledge, how do they share it? There are basically two answers: either architectural and language standards are imposed, and the agents are engineered to comply with the standards, or the agents themselves are endowed (by their creators) with understanding.

Lenat's position is clear; the second alternative is preferable. Further, he believes that agents should share most of the meaning of most of the knowledge, most of the time; and in order for agents to cooperate optimally by sharing their understanding, they need access to the kind of facility that *is* Cyc.

3.2.2.7 Accessing Cyc. Cyc is being developed by Cycorp, Inc., based in Austin, Texas. Founded in January 1995 by AI pioneer Doug Lenat as a spin-off from MCC, Cycorp has inherited from MCC's Cyc project not only the Cyc technology but the majority of its staff. In June 1996 Cycorp released a subset of the Cyc ontology, called the upper-level ontology, to the public, together with tools to permit the specialization of the existing knowledge into specific domains. Cycorp sponsors a program that, for a fee, gives developers access to the complete Cyc ontology, together with help for specializing the knowledge into your specific domain of application.

Although the information that follows is correct as of this writing, as with all technologies that are in continuous development and enhancement, please obtain the most recent information on Cyc yourself. You can contact Cycorp directly, or cybervisit their WWW site.*

---

*http://www.cycorp.com.

Most of Cyc is the Cyc knowledge base. That in turn is expressed as a large set of assertions in the CycL language. CycL, in turn, can run on top of Common-LISP or C. In principle, therefore, virtually any current hardware or OS platform should be capable of acting as a Cyc server, and almost any platform can serve as a Cyc client.

As of this writing, the Cyc system is available in both Common-LISP and C versions, allowing a broad variety of platforms to host Cyc servers. The Common-LISP version of the Cyc system is compatible with, and is currently in regular use under, the following platforms:

- Symbolics LISP machines
- Lucid Common-LISP running on SPARC-based machines under SunOS 4.1
- Macintosh Common-LISP 3.0 running under Macintosh System 7

The C sources can be compiled and run on any system that provides an ANSI C compiler, at least 150 MB of virtual memory, and at least a 32-bit flat virtual address space. C versions of Cyc have been compiled and tested on the following platforms:

- SPARC SunOS 4.1
- Alpha AXP OSF/1
- Macintosh System 7

Cyc client and user interfaces include an HTML interface. Any platform for which a WWW browser is available can be used as a Cyc client to browse and edit the knowledge base of any network-accessible Cyc server. For example, a site might run a C version of the Cyc server on a UNIX box and access it using a mix of Windows- and Macintosh-based Netscape clients.

Client programs and Cyc servers communicate via a TCP-based network protocol. The protocol provides access to the full Cyc functional interface, which allows them to query and update the Cyc knowledge base.

3.3 Object Orientation—An Overview

In this subsection we take you on a short tour of the world of object orientation. First we review important OO concepts and terms independent of the potential use in agent system design and development. Then we explain how these OO principles can be used to produce agent systems, together with reasons why such an OO approach can prove very beneficial. Object orientation is very important to IA developers because agents are, after all, encapsulated packages of functionality and data—the very definition of objects. Moreover, most of the modern development environments are OO; in addition, the infrastructures that support distributed computing are designed with objects in mind, for example, CORBA.

There are a number of good books [see, for example, Booch (1994), Rumbaugh et al. (1991)] that explain various OO methodologies from a relatively formal point of view. What we offer here is a very brief introduction to the OO paradigm, as well as some of the insights we've gained from life in the object-technology trenches. Learning theories and notation is a good start, but it often takes ingenuity and compromise to make things work in the real world.

3.3.1 OO concepts and terminology

The following are some of the many terms that comprise the jargon of OO computer science. There are, of course, specializations of each of these terms which increase their breadth and depth of meaning. In addition, there are many ideas of what it means to be OO, and of the elements required for a system or language to be truly OO. Some, for example, maintain that only abstraction and encapsulation are needed for true object orientation. Others say that inheritance, polymorphism, dynamic binding, and persistence must also be present.

3.3.1.1 What is an object? In the most basic sense, an object is a person, place, or thing. Indeed, candidate objects within a system are often determined by finding the nouns in the requirements and identifying the possible relationships between them. In the OO paradigm, an object is generally defined with a class. A class is an encapsulation of the data and all of the routines or "methods" that operate on the object when it is instantiated from the class. An object has specific states, such as "open," "closed," "on," or "off." An object also has behavior such as "start computation," "close connection," or "send message." Finally, an object has an identity which makes it unique within a set or system of related objects. For example, "books" is a class of objects, and within that class a single book object may be uniquely identified by a number of attributes such as title, subject, author, and publisher.

The term "object" often has a dual meaning in OO circles, since it can mean either one specific instance of an object within a class of objects, or the class itself. In analyzing requirements, it is common practice to initially identify representatives, or types of, objects. These usually become classes as the analysis process proceeds.

3.3.1.2 Abstraction. Objects, as we know, model persons, places, or things in the real world. For example, an object could be a "customer." A customer, as a human being, has hair color, height, weight, hopes, dreams, loves, and many other attributes. In a customer's system, however, the developers are probably only interested in how much money a customer has, how much the customer has purchased in the past, whether or not the customer pays bills on time, and perhaps the customer's name, address, and telephone number so he or she can be billed.

In other words, OO system developers are usually not interested in a "real" object, but are instead concerned only about an "abstraction" of a customer that fits into the design. These abstractions of entities in the real world are often called "abstract data types." Here is an example of an abstract data type in pseudocode:

```
customer class
  attributes
    total amount purchased to date
    account balance
    total number of past-due billings
    string name
    string address
    string phone
  methods
    get name
    edit name
    get address
    edit address
    get phone
    edit phone
    total amount purchased to date
    account balance
    total number of past-due billings
    determine if this is a good customer
end
```

3.3.1.3 The states of the art. One way to look at the behavior of an object is through its states. An object changes states according to a set of events that may occur. For example, an object modeling a switch may have two possible states, OFF and ON. If the current state of the switch is ON, the state may be changed to OFF by an event such as "turn switch off." Conversely, the state can be changed to ON by the "turn switch on" event. This simple example can be illustrated by means of a state diagram, as shown in Fig. 3.1.

In this switching example the state of the object could have been dependent on the value of only one attribute, but the state of an object is often an aggregation of its attributes. A complex object may have many attributes, and several of them may be combined to determine a state. For example, as Rumbaugh

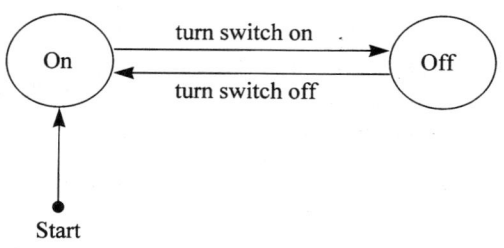

Figure 3.1 State diagram

et al. (1991) explain, a bank may be in a state of either solvency or insolvency. The state of the bank depends on whether the sum of its assets exceeds the sum of its liabilities. In the customer abstraction in the preceding section, whether a customer is in a "good" or a "bad" state at any given moment is determined by a combination of monetary factors. A good customer buys a lot of merchandise, has a lot of money, and pays bills on time.

The analysis of a model through states and events is invaluable in designing complex control systems, but may be of questionable usefulness for data-centric systems such as those focused on interactions with a database. Many systems lie somewhere in between these two extremes, and in these cases state analysis is a useful tool for parts of a given system and not for others.

3.3.1.4 Attributes. Structured design is procedure-driven, and OO design is focused on behavior and the data that directly support that behavior. The data on which an OO design focuses are represented in objects as "attributes." Attributes are implemented as variables or parameters. In this sense, a class is like a C "struct," or a COBOL or Pascal "record." Of course, classes in C++ and other OO languages also contain member routines which operate on the object and may change its state by manipulating its attributes. In a pure OO system, attributes can only be accessed through defined operations on the class.

Class attributes define aspects of the class itself, such as when the class was created and the class name. Class attributes can be almost infinitely complex, since they are usually classes themselves, as in the following example:

```
some class
  attributes
    another_class ac;
    yet_another_class yac;
  methods
    do_something_with_another_class();
    do_something_with_yet_another_class();
end
```

3.3.1.5 Methods in the madness. The work an object does is accomplished by routines or "methods" made visible to other objects through a public interface. In general, methods fit into an object like this:

```
a_demo class
  attributes
    data element A
    data element B
  methods
    function A
    function B
end
```

Methods generally fall into four general categories:

1. *Destructors,* which delete an object of a specific class
2. Routines, called *constructors,* which set the values of data elements in an object
3. Routines which retrieve values from the data elements within a class, generally called *selectors.*
4. *Modifiers,* which change data elements in the class

All of these method types are shown in this example:

```
class transceiver
  method
    transceiver (frequency_type tf,
      frequency_type rf,
      transmission_type rq,
      transmission_type tq )
    destroy_transceiver()
    int send(transmission_type transmission )
    transmission_type receive()
    set_transmission_frequency( frequency_type frequency )
    set_receiving_frequency( frequency_type frequency )
  attributes
    frequency_type transmission_frequency
    frequency_type receiving_frequency
    transmission_type receive_queue
    transmission_type transmit_queue
end
```

There are six functions or methods in this example, five of which access specific data elements. The data elements are private, and can only be manipulated through methods defined on the object interface. The sixth function, "destroy_transceiver," is there to clean up the loose ends when a transceiver object goes out of scope. The constructor in this class is called "transceiver." It sets values for all of the data elements in the class.

There are four modifiers in this class. The "send" function changes the "transmit queue" by adding a new element to it, and "receive" removes a message from the receive queue, thus modifying the queue. The remaining two functions modify the transmission and receiving frequencies.

Often objects and methods are obvious from the requirements. An object is sometimes described as something that can be kicked (a sign or a gate, for example) and a method is the thing that performs the kicking (open gate, reset sign). If the nouns in the requirements translate into objects and classes, the verbs often become the basis for the methods or operations that define the object's behavior.

Requirements can give us a good idea of the object and methods we need, but we need to determine additional methods through scenarios, use cases, and prototypes. Scenarios and prototypes also aid in determining which methods belong in which classes. As the functional requirements of a system change, there's a temptation to add new methods into an existing object, even if the new method doesn't quite fit. If you're constantly adding methods and data elements to a class, it could be a sign you need to redesign some classes.

3.3.1.6 Encapsulation. The whole OO process drives toward one main goal: intelligently partitioned source code. All OO languages facilitate the partitioning of a system into classes, and no matter how elegant a set of object diagrams and descriptions is, it has to be modeled using the class representation provided by a language. The primary reason for drawing diagrams and writing descriptions is to partition the source code into classes, packages, objects, or some other language-supported OO construct.

Constructs such as classes offer a level of organized information hiding unknown in structured languages. The implementation of all the methods in the class, as well as all data structures are hidden within the class. Only those methods which are needed by other classes are visible through a strictly enforced interface.

A well-designed object in an OO system is a self-contained, autonomous entity. Objects contain data and all the routines necessary for operating on those data. This means a well-designed object can easily execute within its own process. If objects are well designed, their work is well coordinated, and one object does not attempt to directly access data in another object (some languages enforce this at compilation). Objects interact with each other by activating routines within other objects to access and transform data.

3.3.1.7 Inheritance. Once a useful class of objects has been created, the first thing the average developer would like to do is make changes to it. The usual copy-and-change-the-code paradigm leads to both reinventing the wheel and a proliferation of hard-to-maintain objects. What most developers would like is the ability to derive a new class directly from an old one, gaining some new functionality and keeping the capabilities of the old class.

This capability is implemented in OO languages largely by means of inheritance. Inheritance allows a developer to take one or more existing library classes and derive a new class without copying or modifying the library. The new class has the functionality of the parent classes, plus any additional functionality a programmer wants to add. The new class can then be added to the library, and additional new classes can be derived from it. Here's an example of inheritance:

```
class basic_sign
  base_function_A()
```

```
    base_function_B()
    ...
end
class parking_sign : public basic_sign
  added_function_A()
  added_function_B()
  ...
end
```

In some cases it is advantageous to create a new class by inheriting from two or more existing classes. Called "multiple inheritance," this approach is not supported in all OO languages, and can cause considerable confusion when it comes to determining the origin of any given member function. It can be difficult to understand and maintain two or more inheritance hierarchies, or to figure out where a function is defined. This is especially true if a function can take several different forms during execution (polymorphism). There are, however, a few circumstances in which multiple inheritance is appropriate. The following example illustrates the basics of multiple inheritance:

```
class telephone
  —telephone functions
  ...
end
class television
  —television functions
  ...
end
class videoconference : public television, public telephone
  —other functions not covered by television or telephone
  ...
end
```

3.3.1.8 Polymorphism. Polymorphism is a fancy way of saying many forms and is almost as old as computer science itself. The first time a programmer used a branch and a jump to a label, the concept that a program could run differently depending on the data it fetched was born. Later, "if-then-else" and "case" statements made it possible to create programs that made complex decisions during run time.

Programmers seemed to have all the tools necessary to create flexible software. Enter object technology. The object paradigm brings an entirely new meaning to adaptable, reusable software through polymorphism. In the OO world, polymorphism can mean a name or object ID that denotes, or refers to, different objects defined by different classes that all share a common superclass. When an object is referred to, say by sending it a message to "drawYourself," any object with the name can execute the drawYourelf method which, for different subclasses or subtypes of object, can be very different routines. (For example, a circle might use a radius and a square might use 4 x, y points, and the algorithm would be quite different.)

One of the beauties of inheritance is that a "hierarchy" of components can be created. Within this hierarchy, all components share or conform to the capabilities of those from which they inherited. This is, of course, roughly analogous to the role of genetics in biology. This means that *all* classes of the same parentage have abilities to respond to the same messages—in whatever way is necessary. For example, if child components C1 and C2 have the same parent P1, both C1 and C2 can have the same methods as P1, except that these methods can, as necessary, execute differently, and additional methods can be added to the children.

The I/O requirements for a system can offer a more concrete example of this feature. Suppose we define a class called `"broadcaster."` Its role in the system is to send data to the outside world. The data can be sent via different media, such as `radio frequency, telephone, e_mail, or print`. The pseudocode for the parent class might look something like this:

```
broadcaster class
   send (data_block)
   ...
end
```

A child class can now be added for each medium required:

```
radio class      inherits from broadcaster
   send (data_block)
   ...
end
telephone class inherits from broadcaster
   send (data_block)
   ...
end
e_mail class     inherits from broadcaster
   send (data_block)
   ...
end
print class      inherits from broadcaster
   send (data_block)
   ...
end
```

Each of these media implements "send" differently and overwrites the "send" in the parent class.

Now suppose you had a list of broadcasters. Each broadcaster could be any one of the subclasses defined above. To make all of the broadcasters in the list send out their specific type of information or broadcast all that is necessary is to traverse the list and send a "send" message to each broadcaster. Each broadcaster automatically figures out the appropriate "send" to use. This is polymorphism based not on control structures such as case statements, but on the very nature of an object. This makes it easier to write and debug code (providing a good OO debugger is used), reducing the amount of logic that must be developed and understood. Polymorphism is extremely useful in modeling real-world objects, such as I/O devices.

3.3.1.9 Persistence. One of the major characteristics of an OO system, or almost any nontrivial system, is the ability for entities created by the system to continue to exist on disk files after the program has finished execution. In an OO system, this feature, called "persistence," means selected objects and classes of objects continue to exist until they're explicitly deleted, whether the process they're in is executing or not. In complex systems, persistence is often achieved through the use of a database management system.

There are, of course, two major flavors of database management systems available: OO and relational. Each has strong and weak points when it comes to OO systems. It doesn't take a lot of extra effort (beyond that usually required for third-party software) to fit an OO database management system into a data model, and the system can be effectively optimized for storage and access efficiency. Relational systems generally offer more features (especially when it comes to querying and reporting). Relational systems are also a more mature technology, offering better compatibility with other tools, support, and reliability.

3.3.2 OO analysis and design

The raw material for the object model is in the requirements specification. Digging through the requirements specification to find information about objects, attributes, methods, persistence, inheritance, polymorphism, relationships between objects, and filling in the missing details is what requirements analysis is all about. The end product of requirements analysis is a good object model. The complete object model really consists of a static definition of objects and relationships along with a dynamic model of how the objects interact to fulfill the requirements.

Responsibility-based analysis and design seek to define a system in terms of concrete objects and their behaviors by writing narratives of user-system interaction and constructing typical object-interaction scenarios. A related technique elicits and describes responsibilities in terms of use cases—typical instances of usage of objects as they interact to perform a task or tasks. (Data-driven approaches generally approach design by first considering the information content in terms of attributes and variables associated with a domain, together with the locations or owners of that content. This approach is a legacy from both earlier programming methodologies and relational database design.) We and others have found in practice that responsibility-driven design produces less complex (in terms of the metrics we'll discuss later) and more easily understandable designs, both hallmarks of a good design.

From these narratives and scenarios arise detailed object definitions in terms of what they must do to participate in the scenarios. Thus their responsibilities emerge. The responsibilities are assigned to the appropriate objects, and some of the details of implementation, such as method signatures, are explicated within the object diagrams that map to your scenarios.

Of course, there are several different (OOA/D) object-oriented analysis and design techniques, each with its own way of representing the various elements of the analysis, such as graphical icons or arrowed lines. Most of them focus on analyzing and describing objects in terms of their relationships and interactions. One recent development in this field has been the combination of two heretofore separate methods promulgated by Booch (1994) and Rumbaugh et al. (1991), together with contributions by Jacobson (1993) and others under the auspices of Rational Software Corporation.* This unification of sorts has a graphical OOA/D language called the Unified Modeling Language (UML). It is being submitted to the Object Management Group (OMG) as a possible industrywide standard. Wirfs-Brock et al. (1990) is an excellent reference on a relatively pure approach to responsibility-driven OOA/D methodology. Information on use-case analysis is found in Jacobson (1993).

3.3.2.1 Static model. The static model is a detailed graphical and textual representation of all the classes and objects in the system. It partitions the requirements into classes of objects which (ideally) have a one-to-one mapping to the customer's world. The object model is roughly the equivalent of structure charts and data-flow diagrams in a structured methodology. It is also similar (but certainly not identical) in concept to the entity relationship diagrams that relational database architects use to partition a system into tables of records.

The initial key to success in the OO paradigm is to identify the objects and classes of objects in the system. In a complex system it can be extremely difficult to correctly divide the system into entities which both map to the real-world elements in the customer's domain, yet support the solution developers are creating in software. This generally involves many iterations of definition and redefinition of the object model. The objects identified form the basis of encapsulation, inheritance, and polymorphism in the model, which eventually leads to defining the relationships between objects. Initially every noun, implied or explicit, in the requirements document is a candidate object.

3.3.2.2 Dynamic model. After an initial set of objects has been identified, and while the relationships between objects are being determined, the object interactions needed to accomplish the tasks the customer has specified must be determined. This is often done via a set of use cases or scenarios which map the customer's "problem space" to the developer's "solution space."

---

*http//www.rational.com

Our primary tools in developing the object model are a good set of requirements and some detailed scenarios which portray accurately how the system is to be used. A scenario is a method-by-method representation and description of what happens when an external event, also known as a "message," triggers a reaction in the system.

Through requirements and scenarios we can flesh out the system with functions, or methods, for each object. A complete set of scenarios is often called an "object message analysis" or a "domain model," since it illustrates the behavior of objects in the user's domain. The scenarios illustrate this behavior by showing when and how each method must be invoked in every object.

As the model is being created and refined, the elements of implementation must be considered, such as prototyping, concurrent execution, network access, and reusable-component libraries. How well these elements work with the object model and how well the object model maps the customer's world are often what determine the success or failure of a project.

3.3.2.3 Making a model work. It's important to stop iterating on the static and dynamic model alone, and bring other aspects of the design into the process as soon as it is reasonable to do so. It's very easy to get stuck in the loop of constant model redefinition. Design and implementation considerations are the driving forces in bringing the model "down to earth."

Completion of the static and dynamic models proceeds by including variable definitions and pseudocode for each method. These are added as the design merges into implementation.

An object model is a kind of map, and as explorers in the sixteenth century found out, a map is only the beginning. To make a map truly accurate, it is necessary to diligently explore and survey the area being mapped. We must find out not only how a system looks (static models consisting of class and object diagrams) but how it works as well (user scenarios).

Scenarios can be a powerful tool in defining the behavior of a system. However, unless exhaustive, scenarios are not sufficient to provide a complete picture of what a system really does. What you really need to do is provide an example of each type of scenario that your system is being designed to accommodate. Some *case* tools offer a limited simulation capability, but to take a truly dynamic look at how a system behaves, prototyping is still the best tool.

One of the best ways to explore and improve a system is through a "vertical sampling" approach. This involves taking a scenario and prototyping it from user interface to the lowest-level utilities, making a "core sample" of what the system really does. After taking several such cores into the cyber strata, you'll have a much needed "reality check" to balance out the theoretical view of the object model. You can also get a pretty good idea of what must be done to make the various pieces of the system work together, and what, if any, objects and methods are missing from the model.

Another part of prototyping is performance benchmarking. From the best object models it is impossible to predict accurately whether the system defined fits into the customer's timing and memory restrictions. This can be a major sticking point in the acceptance of a mission-critical system. One way to prevent major design overhauls late in the game is to write programs to stress-test the parts of the system you suspect. Some of the usual suspects are network access, number crunching, data sampling, and database. But study your model carefully; memory and processing hogs can be hiding in unlikely places.

Once you know the performance risks, a number of effective actions can be taken short of redesigning the object model. In the best of cases, simply using the automatic optimization options available on a modern compiler solves the problem. Beyond this, there are many more aggressive approaches you can take. For example, if the language and compiler you're using support in-line routines, you can use this feature to improve execution time where the overhead for a call is roughly the size of the routine. Selective linkers and shareable libraries such as Microsoft's dynamically linked libraries (DLLs) can help relieve bloated source code, and storing the data and applications locally can prevent expensive server accesses across the network. Since copying blocks of data can be expensive in both memory and processing, using pointers and references to blocks of data can be very efficient.

Be aware of time and memory hogs, and how they may be optimized, but proceed with caution before undertaking ambitious rewriting to gain efficiency. Without executable software and thorough timing and memory analysis, you may end up spending several weeks optimizing a routine that is executed so rarely that you save only a few microseconds over the useful life of the program.

3.3.2.4 Detailed design and implementation.
Like prototyping, OO development is tightly iterative. Developers begin with an initial set of classes, then refine them, add to them, merge, split, and delete them throughout the life of the project. Because of this relentless updating, a successful OO design must be extremely flexible and resilient.

In the OO paradigm there is often significant overlap among the requirements, analysis, design, and implementation activities. Since the development of an OO system is a highly iterative process, a team is often involved simultaneously in the design, analysis, prototyping, and requirements gathering.

3.3.2.5 Tasks and concurrency.
Object technology and structured technology are quite different when it comes to flow of control, since each object (and thence each agents) can be autonomous and could be running in its own process. This model lends itself quite easily to the client-server paradigm. An OO system can easily be comprised of objects executing in parallel with each other, passing messages back and forth. This can be both a curse and a blessing. Once you progress from the standard structured call-and-return model, tracing program execution may be more complex, but errors may be more easily isolated. In addition, you'll have to start worrying about sharing data between tasks, deadlock, starvation, and so on.

Developers who are new to the object paradigm are often pleasantly surprised at how easily objects fit into a multitasking model. Unless specified as synchronous, object methods are assumed to operate asynchronously. The details of concurrency, such as what goes on what server and how to implement the needed scheduling model, are often addressed in the coding stages.

Because the object paradigm lends itself so well to multitasking, several OO languages, including Smalltalk and Ada, offer built-in support for concurrency. For C++, concurrency is available through the AT&T task library, among others. On many projects, however, developers must rely only on the target operating environment, such as Microsoft Windows, IBM OS/2, or UNIX. Operating systems support a variety of multiprocessing facilities, including semaphores, event flags, process waits, and critical sections. Since some operating systems offer only a limited tasking model, it is extremely wise to investigate how multitasking works on your target system very early on in the life of the project.

3.3.2.6 Network and interobject communication considerations. The OO model often seems to progress naturally into a client-server or peer-to-peer paradigm, but that doesn't do much to lessen the complexity of communication between objects and agents across a network. Detailed design is often the time when networking interprocess communication (IPC) details are addressed.

As with most aspects of a mission-critical system, the devil is truly in the details. Some of the details you'll need to work out are whether to use remote procedure calls, CORBA-compliant object request brokers (ORBs), OpenDoc, OLE-Active/X, or some other means of communication across the network. To support object or agent distribution and concurrency effectively, the infrastructure, language, and operating system must break down address and process spaces to allow location-transparent message passing. Operating system or CORBA-like broker services use location services to route messages between agents in different address spaces.

A globally accessible agent is known by a unique ID or name; name spaces across domains are managed by domain servers so name collisions are avoided (such as some agent in domain A having the same ID or name as an agent in domain B). Domain name management is accomplished via a registration process that uses global data store procedures of the operating system to publish the global name, or handle, of the agent. Domain names, as well as any containment hierarchy within which the agent lives, can be addressed by hierarchical naming structures. For example,

myNet.myNode.myApplication.fredFinancialAgent.myStockWatcherAgent

At run time, the handles are the means by which agents are addressed by other agents; local operating systems or ORB-like dispatching facilities hide message routing details from the message sender. Of course, when you design your agent-collaboration scenarios, remember to specify whether each message is synchronous (wait for reply, or block other tasks within this thread or process until reply) or asynchronous (continue with other stuff without waiting for reply). Notice that although concurrency certainly is a concern that goes to the core of your agent-interaction model, in most agent-based applications, distribution of agents is a configuration issue that affects performance, but should not affect your agent-object model or scenario diagrams. Especially with mobile agents, even efficiency may not be an issue if agents can move to take advantage of idle processing resources, or go where the needed information is located.

For more information, refer to Section 4, which discusses some of these infrastructures, Section 6, where we discuss agent mobility, and Section 7 on tools and environments.

3.3.2.7 Dynamic binding. In many applications of OO technology, late binding is necessary due to cross platform compatibility, debugging, and the requirements of the language. Late or dynamic binding means that variable declarations are not type-checked by the compiler (or interpreter) until the variable is fetched or stored at run time. The same applies to the argument list and, indeed, the full method signature check. It is not verified to exist on a particular instance until run time, when a lookup is performed.

Dynamic binding is certainly not required for all OO programming tasks. But many environments, particularly those with incremental compilers (interpreters), use this mechanism to increase productivity and reuse via polymorphism.

3.3.2.8 Classic programming. In the object paradigm, requirements analysis and design are very closely coupled, often overlapping to some extent. Even starting with a solid analysis, however, there are usually plenty of details to be worked out in design. Each of the classes in the design diagrams now becomes an encapsulated component.

The variable values are hidden and private and can only be read or written via methods. These classes, sometimes called abstract data types, represent the real-world entities in the problem domain. You may need to define some implementation-oriented classes beyond those in the design, and certainly you may use low-level classes (stacks, queues, strings, and the like), which are not explicitly diagramed in the design. Utility classes are used for database access and screen I/O.

In some cases templates are instantiated.* In other instances one of the defined subclasses of a more general class, such as last in, first out (LIFO), first in, first out (FIFO), and priority queues, which inherit from a general queue class, is used. The three subclasses are related to the parent class, but are template-instantiated since the range of types or subclasses that apply (that is, the set of all integers) is too great to be practical for inheritance. In other cases we

---

*Depending on the run-time implementation of templates, there may be performance penalties incurred by their use. Research your development environment and language carefully.

must write a new class which inherits from an existing library class, then add new methods and variables.

On the bumpy road from requirements analysis to the implementation of an OO software product, a number of tasks must be completed as we advance from one iteration of the product to the next. Among these are defining new classes, deriving, instantiating, and otherwise using existing classes to create new classes to build a hierarchical library of reusable components.

A well-designed class inheritance hierarchy is the best thing that ever happened to software reuse. Classes build on each other, getting increasingly complex as the hierarchy progresses, yet the simpler classes are still available for further derivation. These advantages are important, but the biggest advantage is that new classes can be derived and added to the hierarchy without making changes to existing classes. This all sounds great, but it is not without risks. A mistake in a basic class definition can mushroom into a systemwide problem, especially if you factor in multiple inheritance and polymorphism.

One common problem is placing too many attributes in a class that is near the top of the class hierarchy. This can increase memory usage (all objects instantiated from this or any child class contain the attribute, whether actually needed or not). This problem can also affect performance. Instantiation of any child object requires instantiation of the attribute.

As classes are identified in the requirements specification, wise software developers first look for existing library classes that fit the requirements. Classes such as those used for GUIs or data storage and retrieval are excellent candidates. Once a class is found that might be used in the design, a decision must be made about how to make it fit. A class may be instantiated directly as in this C++ example:

```
sign *parking_lot_status_sign = new sign(a,b,c...);
```

Similarly, a parameterized component may be specialized as in this Ada example:

```
package parking_lot_status_sign is new sign(a,b,c...);
```

Or a class may be derived from one or more other classes, with additional methods added or nonuseful methods ignored, as in the previous inheritance examples.

Ideally, a new system consists entirely of classes inherited and specialized from an existing class hierarchy. Usually, however, it is necessary to resort to more mundane approaches, such as including several reusable components from a library into an otherwise newly written class. A class that meets some of the required functionality, but has too much excess baggage to make it useful, may simply be copied and modified into a new class. In the worst of cases, a new class can be written completely from scratch, but this should always be a last resort.

3.3.2.9 Object orientation is not the holy grail or silver bullet.

OO technology is not a cure-all, and a poorly designed OO system can be as big a mess as any other poorly designed system. Debugging and testing can be more difficult than in the structured paradigm, and there is a lack of good development tools, although these continue to mature. Depending on what kinds of libraries are available in the development framework, OO development can be extremely difficult. Success in OO technology requires not only formal training but a paradigm to shift away from procedure definition and toward an object or class-centric view. This shift is being made easier with the influx of client-server technology and GUIs.

It can be much more difficult to trace the flow of control in an OO system. The complexity is in the interactions between objects, not in the methods themselves, as is usually the case in systems designed using the structured paradigm. Thanks to encapsulation, however, problems can be more easily isolated to the responsible class.

The key to making an OO project work is a commitment at all levels of management to a new way of thinking about software. Without this, an OO approach, or any new technology for that matter, is doomed to failure.

3.3.2.10 Object orientation and team-based development.

The heavy emphasis on reuse, and the fact that the complexity is in the interactions between classes, as well as the iterative nature of the object paradigm necessitate more and better communication between team members. In some ways, closer communication is needed. Class interfaces aid in this, but where reuse is concerned, there must be a meeting of the minds.

There are two, sometimes parallel, efforts that take place in the early stages of design work: architectural work and specifying component interface contracts and standards. During architectural work, system components and compositions of components (composition hierarchies) are identified, "natural" layers of functionality are formed, policies are established, and rough interaction diagrams and scenarios are explored. As architectural details develop, the component interactions can be detailed with interface specifications, build and run-time connection and references explicated, and detailed object diagrams created from more detailed scenarios.

Early and well-defined object interfaces are the key to teamwork on an OO system. It is not necessary for those working on one set of classes to know how the other parts of the system work; only the interface definitions that apply to them have to be known. In fact, the users of a class often help define the methods needed in external classes without knowing any implementation details. This is one of the major advantages of well-encapsulated objects. Of course, nothing ever goes exactly as planned, and a poorly specified interface can cause serious problems.

Interface contracts and standards are the key to coordinating the multiteam efforts. Very early on in the project, work must begin on solidifying the interfaces between subsystems. Standards for defining abstract data types, error handling, and parameter passing must be codified in a widely available document. This requires that interface architects have experience with the development tools and third-party libraries being used on the project. Since an interface is essentially a contract between two software entities, the persons involved in developing each software component or partition must commit to providing needed data and functionality to the other side. This commitment sometimes takes the form of a formal document detailing the interface, which is signed by all parties involved. Whether or not signatures are involved, each side must be informed early and often of the needs of the other side until the interface is solidly defined.

3.3.2.11 OO metrics. No discussion of OO technology would be complete without giving some idea of what it means to produce "good" OO designs and implementations. The de facto standard method for measuring programmer productivity is lines of code produced. In the best of cases, this is a silly measurement, but it is absolutely ridiculous when applied to the object paradigm. A more reasonable measurement is the number of reusable components used or created.

In the best of cases, software developers on an OO project don't spend much time writing new code. The best use of developers' time is finding and understanding existing classes and integrating them into the solution. This is done by instantiating a template class from a library (framework), adding another level of inheritance (expanding an abstraction), or, in some cases, copying and modifying the source code of an existing class. Finding a code rather than writing it from scratch can reduce the amount of debugging and testing needed.

The following metrics measure the complexity of your OO agents or other components.* This information should convince you to observe good OO development principles. If you see what can cause your design to go awry when it runs in the real world, you'll be more apt to consider these constraints and ideas while in the (much less expensive) design stage. We offer this information and the references so that, during your implementation, you are not caught flat-footed—so to speak—by a system that is slow or unwieldy due to a lack of consideration of the factors affecting complexity (and performance, maintenance, and so on). Following each of the metrics, we give you some idea of how they can affect the design and implementation of your system.

When you attempt to determine the complexity of your design, remember that your executing agents will encounter conditions different from those con-

---

*Most of these metrics come out quite acceptably if, all other things being equal, you do your analysis and design using a responsibility-driven approach, in contrast to a data-driven approach, as summarized in Subsection 3.3.2. Of course, there are several approaches to OO analysis and design. Certain heuristics apply more to some approaches and not to others, and there are always tradeoffs. Seek out and learn various OOA/D approaches and apply those metrics most applicable and important to your efforts. Also, refer to (1993) for a more complete review of OO complexity.

sidered when measuring your class-based architecture. For example, one agent class can represent several actual agent instances that are potentially executing on different platforms and encounter conditions requiring methods that don't apply to other agents of the same class. In addition, many agents share the storage of methods represented by one class. This means that your instance-based measurements would not account for redundancy. (Each agent would appear as though it had all its class or superclass variables and methods located with the agent, if the agent instance were separate from the class.)

Here are the metrics:

- *Weighted methods per class (WMC)*: The complexity of a class is given by the complexity of its attributes and its methods. WMC is the sum of the complexities of the methods of a class. The complexity of individual methods can be measured by cyclomatic complexity or lines of code or some other measure.

- *Depth of inheritance tree (DIT)*: The deeper a class is in an inheritance hierarchy, the greater the number of methods it inherits, making it more complex. DIT for a class is defined as the number of its ancestor classes. Try to keep your inheritance trees relatively short, about 3 to 5 deep. Though this is not a hard and fast range—for example, in a development environment with a large class library, you'll find deeper inheritance trees—but you are starting out with, presumably, a RootAgent class and a root domain class, for example. And you should not have to go more than a few subclasses down to get the level of specificity that is operational for your agent system. If you have a tree that is 20 deep, something is wrong with your design, and some aspect of your functioning system will pay the price, quite possibly when you try to add a new feature to an existing agent or to add new agents to a working agent-based system.

- *Number of children (NOC)*: For a class, NOC is the number of its immediate subclasses. This is an indication of the potential influence a class can have on a system. When you design your agent and domain classes, keep them as simple as possible. Put frequently used routines in some utility class for all to share, for example. Follow good software-engineering practices with respect to modularization, and you should be safe here.

- *Coupling between objects (CBO)*: Coupling can exist between classes that are not related through inheritance. One class is coupled to another if its methods use the methods or the attributes of the other class. CBO for a class is the total number of classes to which such couples exist. For example, different types of agents collaborating on a task would query each other as to each other's available resources and capabilities. Agents would also communicate with their domain's objects to affect changes and with the base or run-time environment (such as the operating system) to discover the available resources. All these involve invocation of methods and should be carefully planned out via scenarios before implementing.

- *Response for a class (RFC)*: RFC is the sum of the number of its methods and the total of all other methods that they directly invoke. This measures a combination of the complexity of a class through the number of its methods, and the amount of communications with other classes. Since a distributed (or even a single-node-based) multiagent system would presumably involve a great deal of communications, this is an important measure for you to get a handle on during your design decisions.

- *Lack of cohesion in methods (LCOM)*: The cohesion of the methods in a class increases with the degree of their similarity. Methods are more similar if they operate on the same attributes. The metric attempts to measure the degree of similarity by counting the number of disjoint sets produced from the intersection of the sets of attributes that are used by the methods. Keep your methods singular in purpose; and don't duplicate the method's behavior elsewhere.

- *Weighted attributes per class (WAC)*: Attributes and methods are both properties of a class, and both contribute to its complexity. Whereas WMC measures the contribution of the methods, this metric measures the contribution of the attributes. WAC is defined as the number of attributes weighted by their size. If you don't think you'll need an attribute operationally, then don't include it just for sake of completeness. Also, ensure that you type your attributes properly—don't define unlimited arrays when properly sized static arrays will do.

- *Number of tramps (NOT)*: The signature of a method, including the types and number of its parameters (arguments), gives an indication of its function. Extraneous parameters, or tramps, that is, those not referred to by the body of the method, both increase complexity by increasing the number of parameters, and give a misleading indication of the processing done by the method. This metric is defined as the total number of extraneous parameters in the signatures of the methods of a class.

- *Violations of the law of Demeter (VOD)*: The law of Demeter attempts to minimize the coupling between classes. If a class follows this law, then its methods can only invoke the methods of a limited set of other classes. This set is composed of the classes of the attributes of the object, the classes of the parameters of the method, or the classes of objects created locally during the execution of the method. In practice, this law is difficult to follow without exception.

3.3.2.12 OO and maintenance. Due to the iterative nature of OO development, all coding and design in an OO system is a form of "maintenance." Fortunately OO programming makes maintenance easier, since the modularity is much higher than in procedural or structured methods. The classes map more directly to the real world, so someone with a good understanding of the problem domain can easily understand the class hierarchies.

In a well-designed OO system it can be much easier to determine what needs to be changed, added, or deleted to enhance functionality or repair bugs. Testing newly added or changed code can also be easier since the effects of the changes can be isolated.

3.3.3 OO agents

Now that you have a grasp of OO fundamentals (if you didn't already), we will expose the benefits of agent development using OO technologies. We propose in this subsection that IA technology *is* distributed object technology coupled with AI.

IAs are expected to provide better support for enterprise computing and business models. In distributed environments, today's languages, infrastructures, and operating systems treat just about every resource as an object, even if the resource is not traditionally viewed or created as an object. In both client-server and peer-to-peer networks, access mechanisms [broker method calls, remote procedure calls (RPCs), even database access] will be addressing the resource as an independent entity, either directly or via proxy. So you may as well think of your agents and their environment as (or as composed of) objects.

The following describes how OO ideas and techniques can be applied to an agent-based system. Although many simple agents may be built using only portions of the OO ideas that follow, if you get involved with the development of a large-scale agent system, these ideas, if followed in detail, result in an elegant, internally consistent architecture.

Realize that you may not have the luxury of developing your agents using all of the concepts discussed. Perhaps your development system and the components with which you start do not have the same structure or the same capabilities that we discuss. Don't let that deter you. What we present in the following paragraphs reflects characteristics of an ideal system that may be built from scratch. Such purity is usually only possible by using fairly pure OO development environments like Smalltalk-based VisualWorks or VisualWave. But many of the ideas presented here are expressed within architectures discussed in Section 5.

Although some of these beneficial aspects discussed here may be available through structured or procedural programming methodologies, it is our contention, based on experience, that the OO paradigm facilitates these benefits to a greater degree via built-in language mechanisms and idioms of use. In addition, as we mentioned earlier, most mainstream programming has undergone the great conversion. Indeed, the most productive agent-development environments are exclusively OO, for example, Smalltalk-based, Java-based, and Telescript-based environments. Furthermore the infrastructure support so essential to distributed architectures, CORBA, for example, is also based on an object paradigm.

3.3.3.1 OO agents are reusable.
A system of agents is reusable if the individual agents or the system's overall design can be used to build a new system.

Reusability is enhanced when each agent or support component is built to serve a single, well-defined purpose. A reusable design should define generalized patterns for interactions between agents. An OO development environment fosters reusable design and implementation.

When you reuse and refine things that already exist, you jump a major hurdle in solution development, namely, reinventing the wheel. By starting with agents or other intelligent software solutions that are well tested or in use elsewhere, much of what you do is transformed from creation or programming from scratch to development by integration, composition, encapsulation, and refinement. OO development ensures that reusability is a primary characteristic of agents and other components that are provided by others in the form of class libraries, and that you develop.

3.3.3.2 Reduced agent-development costs. In general an OO system is based on reusable, refinable components. Although it is certainly possible to produce well-designed, modular code based on procedural or other language types, object orientation has some definite advantages. Considering the inherent support for code management, organization, and maintenance in modern OO development environments, the OO-based infrastructures available, and the host of paradigmatic enhancements that object orientation brings to the design table, we believe that, for distributed agent development, object orientation is the way to go.

If agents are developed as components in the OO paradigm, you do not have to start development from scratch. You can find many commercial classes and other software, which can be encapsulated as objects, that fulfill many of the functional requirements of many types of agents, or parts of the total functionality of an agent. For example, math packages, network communications protocols, user interfaces, and database access routines are all available as referenceable or encapsulable code. In addition, when a component must be changed and maintained, you do it in the *one* place where the agent is defined—the agent's class definition.

For example, say you have 15 types of agents all representing intelligent searchers for different databases that reside on different types of networks. If you defined a basic search agent once, in a Search Agent class, and then make a few subclasses that specialize the database access and network protocol methods, then all the changes and documentation for the different agent types can be made in one place.

Those changes can then be propagated throughout the various systems to the instantiated agents. (Or, more accurately, the instantiated agents can call upon the changed methods or the definitional changes in their respective classes, which could exist in a central "agent server.") You or your users may instantiate hundreds or even thousands of individual agents. Changing your code in one place and seeing the results in many places via the built-in class-instance relationship mechanisms lowers development costs.

3.3.3.3 OO development enables flexible agent structuring. The OO approach to agent development provides a wide spectrum of flexibility in the way you structure an agent-based application. From coarse to fine grained modularity, OO development presents the agent developer with the flexibility to define an agent system in a way that makes sense, without the system imposing arbitrary constraints.

Especially with systems consisting of many agents, it makes sense to factor the degree of functionality, autonomy, mobility, intelligence, and other application aspects over your agents as appropriate. Since there is a wide spectrum associated with each of these aspects, a responsibility-based OO analysis and design of each agent and the whole system inherently helps you make the correct factorings.

Consider the relatively trivial example of an IA-based librarian that gets user directives, searches, retrieves, analyzes or collates, and presents customized information based on the directives. The responsibilities are fairly well delineated from the basic description. You could assign each distinct phase of the librarian's overall task to a separate agent. The application aspects discussed earlier could be factored into each agent as appropriate. For example, the search and retrieval agents might be highly mobile, possessing network protocols enabling them to traverse networks. They would also have database interfaces enabling them to interrogate information sources, gain access to sources relevant to the user's requirements, and carry or download that information back to the other librarian agents. The analysis and collation agents could possess more intelligence in the form of lexical and semantic analysis capabilities, as well as having algorithms for sorting information. They would not have to be mobile. The presentation agent would interface to the native platform's user interface library and would format and present data per user preference.

3.3.3.4 OO agents are maintainable. Individual agents and agent-based systems are maintainable if they can be changed to accommodate changes in hardware and operating systems, or to correct deficiencies and improve performance. OO agent systems are inherently maintainable for two reasons.

First, OO agents are usually built within a larger framework or infrastructure, which provides for a degree of agent isolation from underlying hardware and operating systems. For example, VisualWorks or Microsoft's Foundation Class Library abstracts platform-related issues into classes that are part of the overall OO environment, and are thus immediately usable and extendable. Or OO agents using a CORBA-based infrastructure to communicate across network nodes are isolated from the underlying communications details. You therefore have much less code to maintain.

Second, OO development encourages and inherently facilitates designs and implementations that exhibit a weak coupling between agents and other components, and allows interactions to occur only through an agent's external interface. This is the definition of modularity. It also comports with several good design concepts as measured with the metrics discussed earlier.

3.3.3.5 OO agents are extensible. A system of agents or a single agent is extensible if you can change it to support changes to its original purpose. As agents become more ubiquitous, they will be used in environments that the original designers did not consider. This makes the ability to modify and extend the agents to take on new responsibilities and to function rationally in new surrounds very important. Indeed, it may be the case that the end user or power user is the person to whom the task of extending the agent falls.

OO development enables extensible agent systems. It does this by allowing new agent types and new kinds of relationships to be defined by the original developer and, if desired, by the end user via the configuration of certain aspects of the agent. Object orientation has inherent modularity; and together with inheritance and polymorphism, adding and changing agent features and behavior are greatly simplified and less fraught with side effects than traditional programming methodologies.

Adding new kinds of agents and relationships to existing agent systems is independent of the underlying system support mechanisms (such as an agent browser or agent-to-agent communications mechanisms). Also, it is important that existing relationships be easily navigated by the user and updated to reflect changing requirements. You don't need a major system release to take advantage of new capabilities. You simply express the desired capability in terms of a new or updated agent or support component or relationship.

3.3.3.6 OO agent systems are understandable. OO agent development allows you to map the individual types of operational things in your domain (search, human interface, assistance, control) directly to individual types (classes) of agents or protocols within agents. This is a natural way to design and build agent applications.

Subsequent users of agents designed and built using OO techniques will find the design easy to understand because your users will understand what each type of agent does in your application. Self-documenting agents (that is, agents with documentation or help components embedded within them) can guide users in their custom configuration (if any) and use.

3.3.3.7 OO agent development supports interconnected hierarchies of agents and domains. OO development enables a relationship called composition, which can be used to make more complex agents by building relationships between simpler agents and other primitive components. Composition hierarchies are useful architectural structures. They represent a natural approach to organizing agents according to behaviors. Hierarchies also represent a natural way to organize the domains (or models thereof) and environments in which your agents function; you'll see an example of agent hierarchies in Section 5.

Most other development paradigms to date have enabled composition only at a single level of hierarchy, that is, combining functions or procedures into an application. More complex relationships, such as multiple levels of composition and other hierarchies, are more difficult to fashion, and do not map as naturally using nonmodular, nonreusable languages like C, FORTRAN, COBOL, and Pascal. Now through object orientation's formal support of a uniform interface between objects and the relationship known as composition, you can develop agent-based solutions that support as many levels of hierarchy as your domain demands.

3.3.3.8 OO agents and intrinsic system knowledge. An agent system or architecture is a set of agents *and* their interrelationships. The relationships comprise a great deal of knowledge about the system. This knowledge is intrinsic when it is explicitly represented. We believe that to develop a robust, rich system of collaborating agents and other components, explicit representation of relationships as well as the ability to form and break relationships at run time are essential. In so doing, relationship knowledge is available to the system itself and to individual agents to support system functions such as built-in change management (agents that automatically update themselves), security, user-interface help facilities, navigation (both by agents and through the agent's interfaces), modeling, and simulation.

Making changes to agent-based systems can cause headaches. The problem is one of managing existing agents: who made what changes, when were the changes made, where are agents and other components now located, and how does a change in one agent's definition affect other agents in the system. Because a system potentially consists of hundreds if not thousands of agents and other support components, change management becomes an onerous task.

OO development techniques support an improvement in these problems by endowing the agents themselves with knowledge of their own status and, very importantly, how they relate to other agents in the system.

In addition, agents can represent integral or support components which are encapsulated to a developer-determined degree. Encapsulation means that a developer can hide the implementation of the agent from the users of that agent. The only thing the user knows is what the developer made visible. Therefore that part of the implementation that is invisible to others can be changed without affecting other agents in the system, as long as what is visible (the public interfaces) does not change. If interfaces do change, OO development environments have change management tools that can navigate relationships to see how changes to one agent affect other agents and components. You are aware of the change impact before you commit to the change, because the system is aware of it and can report its knowledge to you.

Objects can represent not only the agents of an agent-based system, but also the relationships among the agents. If agents know about their relationships with other agents, they can more readily query them without the aid of an intermediary. They would also know what the overall capabilities of the agent system are. This would help in refining resource negotiations, task and subtask division of responsibilities, and other coordination activities.

3.3.3.9   OO agents plus OO domains enable modeling and simulations.   OO agents representing some or all of the operational aspects of your system can run independently of the actual system hardware. In other words, agents that model some or all of the activities within a particular domain can be run in a domain simulator object. Agents running in this context are simulating what they would actually do in the real domain. This helps you to test your agents in a real environment before "commissioning" them to your users and the real world. It also lets you perform what-ifs and other experiments without fear of harming or annoying systems, platforms, networks, or people.

Depending on the type of development environment and language you decide to use, your development platform could support a pseudo run time or interpretive mechanism. Using such a mechanism, you can test your agents to some level before releasing the agent's definition (that is, the agent class) and instances thereof to other developers (that is, into a class library) for reuse in other agent-based solutions and systems, or to the real world for direct use. For example, all of the virtual machine (VM)-based interpretive environments, such as Smalltalk, allow you to run your agent within the development environment.

3.3.3.10   What is an OO agent?   Since the object provides a natural way to represent anything of interest within a system, objects can represent the components or agents in a system or application, such as scheduling agents, human interface agents, search agents, and so on. An object also represents its relationships to other objects, a way of capturing the compositional and other knowledge of the system.

Agents and their interrelationships, and the appropriate support components and objects, are the software of an OO-based agent system. An OO agent's definition is a class—the unit of OO reuse and maintenance. An agent developer defines agents of the same type with a class. For example, all agents have certain things in common, such as a name or ID and other attributes, and a basic set of communications and error-handling protocols, for example. These most basic attributes and behaviors can be defined in a RootAgent class, with more specific types of agents for specific domains defined in subclasses of the RootAgent.

For example, after defining the RootAgent, say, you want a subclass of agent that includes all the attributes and protocols that would be found in a generic text search agent. Rather than defining all these things each time you want a search agent, you can specify these "attributes" and behavior once and save it in a class called TextSearchAgent for reuse within other agents.

3.3.3.11   What are OO agents made of?   In a pure OO architecture, agents can be *composed* of other agents and objects. Agents are *defined in terms of* attributes and operations. Attributes, also referred to as, or modeled and implemented with, variables or parameters, have values that define or reflect the state of an agent. Operations (also known as methods, procedures, or scripts) define the agent's behavior. Agents can also have other agents, as well as other kinds of component objects, "inside" of them. This is known as a composition relationship. We'll talk more of composition later.

3.3.3.12 Agent attributes. An agent's attributes are modeled, or implemented with, variables (parameters) that have values reflecting an agent's state. Variables themselves can be specified with attributes that define the variables: name, type, default value, security access, comments, and other attributes. Variable types can be simple (integer or real) or complex (arrays or records).

A variable has default or user-defined read and write methods. A new variable value could be directly written as it is presented to the agent's interface, or some manipulation can be done to, for example, translate the variable value into a different format before storing it. Postprocessing can also be defined. If a variable is stored as a certain value, some postprocessing method could examine it and decide if other processing is required.

3.3.3.13 Agent operations. An agent's operation (also referred to as method, script, or handler) is a specification of behavior. It is composed of an operation signature and a body of code that runs to produce the behavior. The operation signature is the interface other agents use to invoke the behavior. Depending on the implementation language used, an operation signature is typically composed of an operation name, any parameters (arguments), and any return value. The code body can express an algorithm that includes delegating subtasks via message sends to other agents.

3.3.3.14 Encapsulation—the agent boundary. You can make any of the agent's variables, operations, and component agents or objects (actually, the names of these entities) public, and therefore on the boundary, or private, or encapsulated. The public entities on an agent's boundary are the uniform specification of an object's interface to other software (agents) in the system. For the purpose of further discussion, we represent an agent object graphically as a roundtangle box with the agent name inside, as shown in Fig. 3.2. Of course, any graph-

Financial Info. Search Agent 1

Figure 3.2 Agent icon.

ical development environment could represent classes and objects differently, and even give you the ability to create a graphic the user might see when using the agent.

3.3.3.15 Agent classes and agent instantiation.

An agent class defines an agent—how it behaves and what information it carries. The agent class specifies the types of an agent's variables, the code within an agent's operations or methods, any relationships with other agents, and any component agents or other objects. In addition, the class can contain components which specify the agent's user interface, which defines how the agent is viewed and manipulated. A class can be shown graphically as a square boundary, as illustrated in Fig. 3.3.

The class is a common unit of work allocation, reuse, and revision control for all components of an OO agent system. A class is a software structure built by an agent developer to define a class, or kind, of agent. Agent classes that define what agents are to do in specific domains or scenarios are subclassed from the RootAgent class. For example, Fig. 3.4 shows a FinancialAgent subclassed from the RootAgent; it would have attributes and methods tailored for the financial domain.

In Fig. 3.4 the solid line between the classes, with the arrow pointing to the superclass, indicates a relationship between classes. Specifically, the pointed-to class is the more abstract class. For example, whereas RootAgent might have generic communications protocols and attributes common to all agents in your architecture or system, FinancialAgent would have various methods having to do with specific financial calculations like returnOnInvestment, annualizedReturn, percentReturn, and so forth. Similarly, FinancialInfoSearchAgent is subclassed from FinancialAgent. It would have even more specific attributes and methods that enable it to perform searches on financial databases and return results to a user.

A particular agent is instantiated from its class. For example, FinancialInfo SearchAgent1 has been instantiated from FinanciaInfoSearchAgent. One or more agents can be created (instantiated) from a single class. Figure 3.5 shows a simple example of both a search agent class and its many individual instantiated agents. The dashed lines between the class and its instances indicates a relationship. The pointed-to object is the class; objects at the nonarrowed end of the relationship are the instances of the class. Notice that there are three

Figure 3.3 Class icon.

Technologies that Facilitate and Enable Agents   79

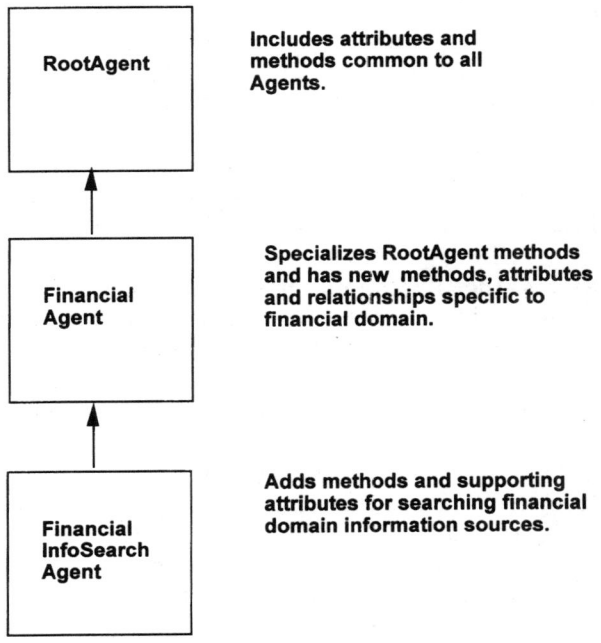

Figure 3.4   Agent subclassing.

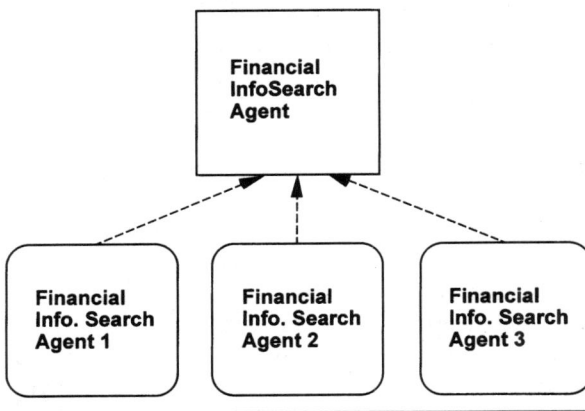

Figure 3.5   Agents instantiated from agent class.

search agents. There could be as many as necessary, each defined by the common FinancialInfoSearchAgent class.

The instantiated agents (and the class) in Fig. 3.5 have different names. The class is named and unique; the individual agents are named and unique. When a class is instantiated to make an agent (or at other times), variables can be initialized with values specific to that agent but not to the agent's class. Other variables might be configured by the user (for example, which financial database to search and for what stock to search for information).

3.3.3.16 Agent classes—The RootAgent class. All classes, including the RootAgent class, are derived from some class that is given as part of your development environment. For example, in ParcPlace-Digitalk's VisualWorks, the root class is also the root object and is simply called "Object." To build a RootAgent class in VisualWorks, you would probably go further on down the inheritance hierarchy, perhaps to "Model," to gain advantage of all the support for other aspects of the agent application inherent in the up-line superclasses, such as user interfaces and various relationships.

All agents within a particular system share a common set of variables and operations. Therefore these common components are defined in the RootAgent class. They include an agent "name" variable, default "variable read/write" operations, which allow other agents to access their variables, operations whose methods return information about the class definition itself, and default interagent communications methods (for example, all methods using an infrastructure like CORBA or a TCP/IP socket).

All agents could have an operation or method called "initialize," which readies the agent for execution in a particular domain or a platform or operating system. All agents could have an operation shutDown which temporarily stops their execution. The operation signature for this operation is defined in the RootAgent class. Subsequent subclasses derived from the RootAgent class would supply a particular method for this operation, depending on the domain and the platform operating system.

All agents could assume that they will be loaded in a node containing a scheduler or other facilitator-type agent or object (refer to Section 5) and therefore specify a reference or "uses" relationship to that scheduler or facilitator agent.

Further all agent classes would share common operations and variables that could be very different from the common operations and variables present in other aspects of a complete application. For example, whether you encapsulate the user-interface aspect literally into the RootAgent, or rely on a reference relationship to an object or set of code routines completely outside of your agent architecture, is up to you. Some examples of RootAgent specifications, at a high level, are presented in Sections 6 and 7.

3.3.3.17 Domain-specific agent classes. Within any class hierarchy there are classes that represent particular solutions to particular domains or aspects of a domain. For example, consider the problem domain addressing search strategies. We can imagine some variables and operations which are common to all agents used in search strategies: a variable representing the input string or pattern of the user, various search methods employing search algorithms appropriate for the target platform and the databases or domains being searched, and so on. In addition, the search agent would include user interfaces specific to search criteria gathering and results presentation, such as resultSortOrder or searchTermHilighting.

In a similar fashion, we can imagine other families of classes, each defining the variables and operations common to some other problem domain (a particular expert system, product scheduling, authoring, user interface, work-flow automation, and so on).

3.3.4 Relationships among agents

The relationships among agents and objects make a system. The richness of the system and its architecture depends on its ability to interconnect things. One goal of OO agent development is to create an agent architecture that enables other agent developers (or you, in subsequent releases) to add new kinds of relationships to meet existing and emerging needs. Indeed, it is our claim that if agents are to become more sophisticated and intelligent, they must be able to establish their own relationships dynamically, as required. Meeting this goal requires a clear understanding of what relationships are and how they can be represented in an agent system.

This section presents a summary of a comprehensive model for relationships so as to allow agent developers to create new ones, and to provide an infrastructure wherein agents themselves can negotiate with other agents and objects with the purpose of establishing a relationship on the fly, and dissolving it when no longer pertinent to the agent's goal.

Of course, you may decide to forego formal representation of relationships in your agent development efforts. We advise that, as you try to extend your agent architecture, having at least a clear model of (if not plans to explicitly and consistently implement) the relationships well thought and mapped out in the analysis and design stages, will provide many benefits over the life of your agent system.

Some specific examples of agent architectures with relationships embedded as either discrete objects or components, or as a consequence of using a specific agent communications language, are discussed in Section 5.

3.3.4.1 Relationship defined.
A relationship is a connection between two things. In OO agent development, relations exist between agents, between agents and other software entities, between classes of agent, and between a class and its instances (objects). (You saw an example of the latter relationship in Fig. 3.5).

A relationship can be one-way or two-way. In a one-way relationship only one end of the pair knows about the other. In a two-way relationship both ends know about each other.

3.3.4.2 Composition relationship. One important kind of organizational knowledge is about collections of things that make up a more complex thing, which can be a component of a still more complex thing. This is called compositional knowledge and is represented as a "has-a" relationship.

One example of composition is shown in Fig. 3.6, where a FinancialAdvisorAgent (FAA) is the overall coordinator of the other agents and functions in the financial domain for this advisory application. It processes information obtained from other agents and reasons about the economic factors affecting markets as a whole vis-à-vis the user's financial goals. It prepares and presents advice using the user interface. It is composed of (has-a):

- *PortfolioAnalyzerAgent* (PAA): Manages a portfolio by collaborating with its contained agents. It may have a genetic algorithm–based facility that launches multiple PAAs, which perform discrete what-if scenarios. The best performing PAA (best predicted return) over several "evolutionary" trials is kept until specified financial conditions change, where a new set of PAAs can be spawned and evolved. The PAA, in turn, is composed of:
  - *TransactionAgent* (TA): May have hooks to the broker to place buys and sells automatically. This agent could have a fuzzy-logic-based facility that helps determine the appropriate buy or sell conditions. It would also execute the buy or sell commands by collaborating with the NetworkAccessAgent (NAA).
  - *PatternWatcherAgent* (PWA): May reference or encapsulate a neural network that "watches" for economic or financial patterns it is trained to recognize as signaling a buy or sell situation.

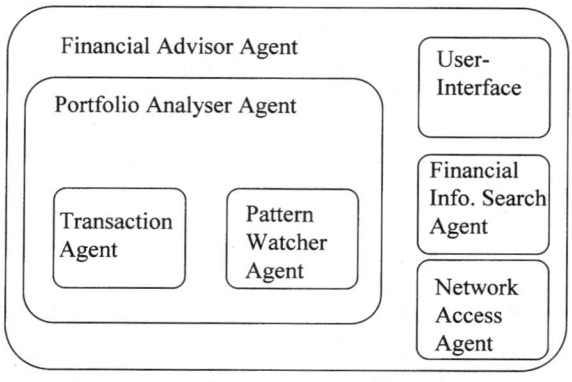

Figure 3.6 Example of composition.

- *UserInterface (UI)*: An object that would have methods that make calls to interface objects created from, for example, Microsoft's Foundation Class Library or Apple's Toolbox/Shared Library. The user interface object could receive messages from all the other components and agents in the FinancialAdvisorAgent to present their data as instructed. Or the user interface code could be dispersed into each agent as appropriate.
- *FinancialInfoSearchAgent (FISA)*: May have search algorithms, or has access to search engines to gather and filter information from financial databases over the network.

In Fig. 3.6, we represent the composition relationship as agents visually contained within other agent's boundaries; that is, agents are composed of agents and other components (such as user interface aspects). This way of representing composition is a very natural way. You can easily see what agent is the composing agent and what agents are the components.

In Fig. 3.7 we represent the same composition hierarchy using perhaps a more formal method, more closely aligned with Booch's notation. In this figure a line with an open circle at one end means that the agent with the circle attached to its boundary uses the agent at the other end of the relationship line. A line with a filled circle at one end means that the agent with the circle attached to its boundary is composed of the agent at the other end of the relationship line.

In the foregoing example FinancialAdvisorAgent participates in three composition or "has" relationships and one "uses" relationship. Notice that the UserInterface is not designated as an agent. This would be typical in a system composed of agents that display their activities using host-platform/OS–derived facilities. Continuing, PortfolioAnalyzerAgent has two components, one of which, TransactionAgent, participates in a "uses" relationship with NetworkAccessAgent.

There are many kinds of composition, which differ according to the specific degree of dependence between the composing object and its components. Some of these subtle subtypes are described next.

Physical composition—sometimes referred to as containment—can represent a physical relationship between two parts or objects. For example, a pail contains water but is not composed of the water. In another example, a car contains passengers but is not composed of them. A forest is composed of the trees, but you can take away a single tree and the forest still exists. A car is composed of a transmission (and other things), but if you take away the transmission, the object car is no longer, truly a car (if the object is behaviorally defined).

Some OO practitioners use containment in place of composition when, for example, there is an interaction between two objects that is controlled by a third class. The controlling class is the containing class for the other two classes, which are in a "uses" relationship to each other. In either composition or containment relationships, agent naming becomes important. As mentioned

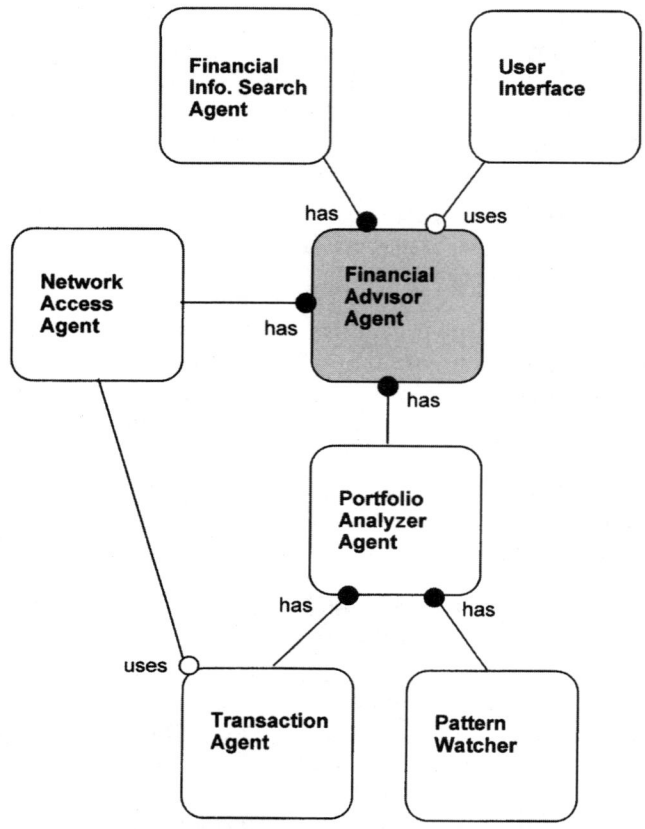

Figure 3.7 Agent relationships in Booch notation.

earlier, hierarchical naming can be used to identify agents uniquely at any level within a multilevel composition or containment structure. Your chosen language, operating system, or infrastructure should have the ability to accommodate hierarchical naming and referencing across process, address, node, and network boundaries.

There are several other kinds of relationships. Many are specializations of the "uses" and "has-a" (composition) relationships. For example, a generic relationship can be specified between two objects. The relationship can further define:

- *Cardinality:* Specifies how many objects are in relation at one end or the other; for example, 1:1, 1:n.
- *Constraint:* For example, does this relationship always exist, or does it come and go depending on the state of the related objects?

Composition or containment can be specialized. A "by-value" composition means the actual containee (or copy thereof) exists inside the composing object. A "by-reference" composition means that the composing object only has a reference inside of it, to the contained objects and so on.

Whatever your decision concerning the relationships to use in modeling your system, use them consistently. For a rigorous explanation and usage guidelines with regard to modeling relationships, refer to Booch (1994). For a good treatment of responsibility-based design, we recommend Wirfs-Brock et al. (1990).

3.3.4.3 Relationships can be modeled as connector objects. Earlier we pointed out that explicitly representing a system's relationships provides dividends in terms of system richness, maintenance, and so forth. In Fig. 3.7 notice that the relationships are represented graphically as lines and circles on the boundaries of the objects participating in each relationship. These graphics can also be modeled and implemented as objects in your actual architecture. If connectors are objects, we would expect connectors that enable a certain kind of relationship between agents to be defined with a connector class. Each connector class defines operations and variables that support the creation and deletion of a relationship with another agent or object, as well as information about a specific relationship. For example, a connector class, when instantiated to represent a specific connector, could have a variable whose value contains both the name of the other agents in the relationship and any dependence information.

This connector-based representational technique, when implemented as objects, provides a powerful mechanism for dynamically creating and deleting relationships between agents, as required, while they are running. Remember, interagent communication is a fundamental aspect of any agent architecture, as you will see in later sections. If you can model the communications aspect and other relationships of an architecture in the same way as you model the agents themselves, you are well on the way to an elegant, robust, and maintainable agent system.

The RootConnector class assures that all connector objects share a common way to express the fundamental information and operations essential to all connectors (for example, {connectTo agentList} and {disconnectFrom agentList} operations, as well as operations to support navigation of relationships). Specific connector types are then derived from the RootConnector class and extended to support specific types of relationships (such as scheduling relationships). To reiterate the importance of connectors as objects, the connector (and connection) objects are the places to put, or encapsulate, your interagent communication mechanisms, whether hand-crafted or reused components of purchased tool kits.

3.3.4.4 Dynamic relationships—scenarios revisited. In Subsection 3.3.2.2 we briefly discussed the need for a dynamic model that expresses how your agents collaborate or interact with each other and with nonagent aspects of the application or system. Here we give you a simple example of a scenario based on the following classes of financial agents and other components we talked about in Subsection 3.3.4.2:

- FinancialAdvisorAgent (FAA)
- PortfolioAnalyzerAgent (PAA)
- TransactionAgent (TA)
- NetworkAccessAgent (NAA)
- PatternWatcherAgent (PWA)
- UserInterface (UI)
- FinancialInfoSearchAgent (FISA)

It may be helpful to refresh your memory with a short narrative of each agent's functions as described there. Figure 3.8* illustrates a simplified scenario (simplified because there are lots of assumptions hidden within the inner workings of most of these agents that reflect, for example, the various goals of the user), called manageMyStockPortfolio, whose narrative might go something like this:

> The user requests that the FAA carry out periodic stock portfolio analyses and seek and carry out those financial transactions that optimize the user's portfolio based on explicit but high-level criteria. The FAA tells the PAA to generate and return a list of existing holdings, including cash or equivalents, and categorize the holdings by industry exposure. The FAA also instructs the PWA, and through it the FISA, to work together to see if any stocks meeting certain criteria can be purchased for the resources the user has available, and that meet the FAA's knowledge of the user's preferences and financial goals. The FISA either uses the NAA to move to other locations over the net to investigate economic data and company financial information or, again with the NAA, acts as a client to search engines across the net to obtain relevant information. After some period of time, configured by the user, the FAA, using the UI, presents the user with a list of stocks meeting discrete, preestablished criteria, as well as stocks the FAA, by virtue of its ability to learn and weight relevant factors, suggests may also be positioned to meet the overall goals of the user. The user selects those stocks he or she wants to purchase and instructs the FAA to do so at the best price obtainable within 48 hours. The FAA passes these instructions to the TA, who then works with the FISA and the NAA to consummate the transactions with a broker's on-line system.

Notice that in the narrative, as in the diagram, there is no direct indication of *how* these activities are carried out, that is, there are no detailed algorithms, whether neural networks are used in the pattern watcher, or some MEU-based optimization is embedded within the FAA. The only relevant details at this stage of analysis and design are the high-level collaborations between agents and other objects to get the specific jobs done. Also note that messages of the type `get Foo Bar()` assume a return message to the requestor that contains the requested information. (In other words, there is no arrow going back to the request or in this diagram.)

---

*Note that in Fig. 3.8 the names of the agents participating in a particular scenario, have a "1" at the end. This indicates that in a real system there are probably several instantiations of the different agent types, each requiring a unique name. In Figs. 3.6 and 3.7, and associated text, on the other hand, the discussion is centered on generic relationships between types (classes) of agents.

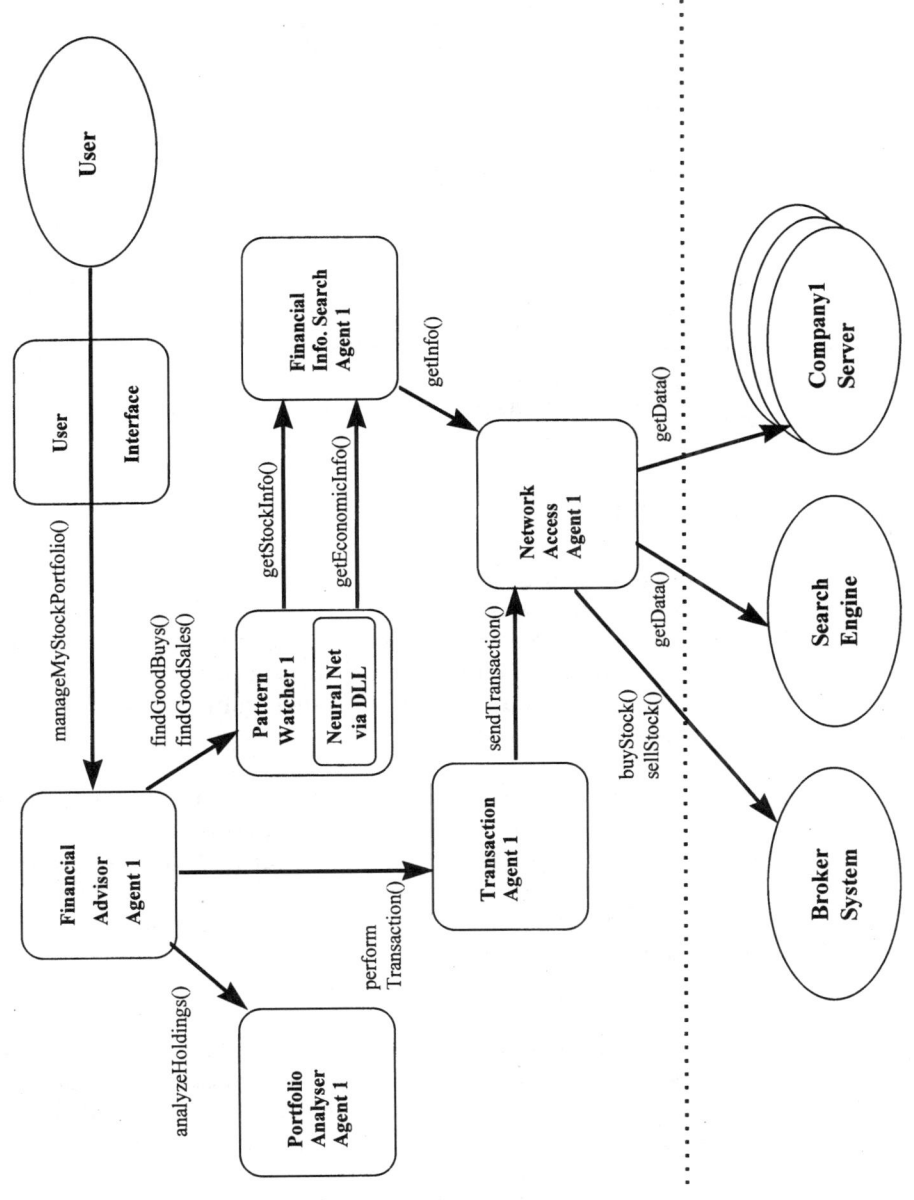

Figure 3.8 Agent-based manageMyStockPortfolio scenario diagram.

We also want to point out that besides the objects named "___agent" there is no indication from this level of scenario diagram that this is indeed an IA-based system, as opposed to a function- or ordinary-object-based application. What would make this a true IA system is the design and implementation details, encapsulated within each agent object, that imbue them with goals, decision-making authority, autonomy, mobility (if necessary), and the reasoning capabilities needed to *represent* (that is, act as an *intelligent agent* for) a well-informed stock buyer and the other entities a buyer would employ to get the job done. The next step is the design stage, where you provide the agents with just those attributes and capabilities.

3.3.5 OO agent architecture benefits from uniformity of representation

We can now see that everything of interest in an agent system is uniformly expressible as classes and objects. All entities in an agent system share a common foundation of representation and communications. This uniformity, in turn, leads to greatly decreased development and maintenance efforts, and provides a basis for modeling any domain and its attendant agents.

3.3.6 Agent development by extension and refinement

An individual agent or a complete agent architecture is developed by reusing and refining or extending existing classes, if suitable ones exist. In other words, you don't necessarily have to construct any agent or support component from scratch. You can browse through existing libraries of agents and objects to find relevant behaviors that you want in your own solution. You can reuse those behaviors in several ways, as follows.

3.3.6.1 Reuse by specialization and generalization. Specialization is the process of deriving an agent subclass from an existing agent class and extending or refining its behavior by modifying or adding to its operations and variables to perform more specialized functions. This is what we indicated earlier in the example collection of financial agents. When you specialize an agent by subclassing, the derived class inherits the behavior of its superclass or classes. When a new class is derived, the methods or operations inherited from its parent agent can be overridden simply by reimplementing them to suit the need. However, it is very important to point out that the form of interface to the operations—their signatures, which define what is accessible at the class's boundary (to other agents)—must remain the same.

Inheriting the behavior of a parent class is reusing that behavior. Behavior can be inherited from more than one class. This is called multiple inheritance.

Another way to create a new class of agents is to abstract (that is, "move") common behavior out of many classes into the new class. This new class then becomes the superclass ("parent") of the original classes. This process is called generalization.

Specialization and generalization result in a hierarchy of class definitions. Not only does this OO architectural approach allow other developers to extend your agents by creating new classes of agents, it provides a way to organize these class definitions. Derivation produces a hierarchical organization of classes (also known as an inheritance hierarchy). An example agent class hierarchy was shown in Fig. 3.4.

3.3.6.2 Reuse by composition and encapsulation. When a new class is created, other classes can be included or referenced as components in it. For example, to create the FinancialAdvisorAgent class discussed previously, you could:

- Derive the FinancialAdvisorAgent from the FinancialAgent class. The effect of this is that the FinancialAdvisorAgent class inherits all the variables and operations common to all FinancialAgent classes.
- Include (create a composition relationship) into the FinancialAdvisorAgent class the PortfolioAnalyzerAgent, FinancialInfoSearchAgent, and NetworkAccessAgent classes representing the component agents of the FinancialAdvisorAgent.
- Specify other relationships as connections between the component agents and other agents or components to be used in the system.
- Add operations and variables appropriate to the subclassed agents.

Support for composition can be as simple as a language that enables one piece of code to reference or call another piece of code. However, in terms of object orientation, we would want more explicit support, especially when we need to maintain the code. We are talking about encapsulation mechanisms. Encapsulation is the process by which one object, or in our case agent, is made part of (contained within) another object or agent.

The reason we discussed the use of connector objects earlier is that, by creating connector objects as part of your agent system, you can provide both the definition and the behavioral aspects of encapsulation. A composition connector relating two agents would have the attributes necessary to express, for example, a one- or two-way reference. This attribute could be a simple pointer type.

Methods on the composition connector object would determine just how the encapsulation process proceeded. This can happen dynamically in a running system. For example, if you wanted to dynamically encapsulate or include the TransactionAgent into the PatternWatcherAgent, a message is sent to a newly instantiated connector to write the identity of the TransactionAgent into a list of pointers that is part of the PatternWatcherAgent's attributes. (Maybe this attribute is called "myComponentAgents.") The connector might also be responsible for breaking that composition relationship if requested by the connector's owner.

3.3.6.3 Reuse by instantiation. Because you can instantiate a class an indefinite number of times, you are, in effect, reusing one definition of an agent as many times as required. For example, a financial counselor or a broker could provide several instances of the FinancialAdvisorAgent—one on each client's PC, or they could all exist on the broker's server, accessible by clients over a network. Each would have a unique identification attribute (in other words, a name; for example FinancialAdvisorAgent 1, and FinancialAdvisorAgent 2. Each would belong to a particular client. But they would all be defined by the one Financial-AdvisorAgent class maintained by the broker.

3.4 Intelligent Agents via Soft Computing

The ability of machines and software to reason with uncertainty, recognize patterns, and evolve responses to ever-changing situations would "certainly" bring them close to a realization of what we humans would call intelligent. Most computing over the last half-century has been based on implementations of specific, well-behaved algorithms acting on known ranges of input data and producing known output. But increasingly, machine intelligence, as defined earlier, is being developed with the help of several specific AI technologies.

Over the years AI research, as well as insights into how naturally intelligent systems, such as we humans, process information, have brought forth at least three important computing technologies: fuzzy systems, genetic algorithms, and neural networks. We center our look at these particular "intelligence-producing" technologies not because they are the only ones worth considering when designing your IAs. Rather, it is because these technologies are fairly mature, are available commercially, and can be readily incorporated into, or accessed from, agents coded with the languages we talk about in this book.

For example, in the process and production-control domains:

> ...advanced control technologies include expert systems, fuzzy logic and neural networks. These technologies can be used to generate an optimized model of the process. Expert systems are sets of decision-making rules, or algorithms, that embody the combined knowledge of many human experts. Neural networks can be trained from historical data and can detect patterns not apparent to a human observer. As a result, they can be used to generate setpoints for controllers that take into account interactions among process variables such as pressure and flow. Fuzzy logic controllers enable more rapid recovery from process upsets by more closely approximating setpoint values. All these technologies can be used to increase efficiency by reducing feedstock requirements, improving yield, and minimizing waste. (Honeywell, 1996)

The average computer program knows how to deal with concepts such as "false" and "true" or "hot" and "cold," but is less adept at using concepts such as "maybe" or "somewhat warm." Human agents have the ability to deal with shades of gray, but their cyberspace counterparts are stuck in a world of crisp black and white landscapes. The addition of technologies such as genetic algorithms, neural networks, and fuzzy logic can give agents a fighting chance at succeeding in a world where they must compete with other agents for scarce resources, and work to benefit a human user who lives in a relativistic world. Combined with agents in dynamic agent development and execution environments, they form more shrewd and savvy servants.

Often an agent's main task is to monitor a data stream. A "dumb" agent simply checks for values above, below, or within a specific threshold, or for a match between data and specific canned criteria. Access to intelligent subsystems can make a huge difference. An agent with access to AI functionality such as a neural network can, for example, look for complex patterns in multiple data streams, regenerate goals as its environment changes, or combine forces with other agents to achieve new goals.

Of course a good programmer could write a genetic algorithm or neural network that would be more efficient in memory and execution resources needed, but the reasons for using an off-the-shelf tool such as an ANN development environment is that it makes a standard interface available. Of course, using an off-the shelf tool kit means that agent developers need not spend huge amounts of time developing technology not directly related to the agents they are designing, possibly finding out that they had taken the wrong approach in the first place.

One agent can spawn other agents, which can then act as experts in achieving specific parallel subgoals. Agents can then meet and decide which move to make based on current conditions.

A simple-minded bidding agent would attend an on-line auction with a ranked list of products a user wished to purchase and the maximum prices the user could afford. An IA would use a neural network to figure out competing agent's bidding patterns, and constantly evaluate bid scenerios to determine the best choices at the moment, considering how much its user desired certain items up for bid, money on hand, and the bidding patterns of other agents. Thus the user would get the most for his or her money.

Fuzzy logic is good for dealing with uncertainty. Genetic algorithms and ANNs are great for pattern recognition and determining the best of good choices. A neural network works well at picking the best performing mutual fund. Applying a neural network to historical data patterns, combined with current statistics, can predict the best choices in anything from mutual funds to the winners of a professional golf tournament.

In addition, combinations of these technologies have produced useful solutions to difficult problems. For example, genetic algorithms have been used to "train" fuzzy systems, and a neural network is used to model a process very accurately. The model is then used to generate fuzzy-logic rules that are used to run the process. The next few subsections look at these soft computing techniques in more detail.

3.4.1 Fuzzy systems and fuzzy logic

In this subsection we discuss fuzzy systems and applications of fuzzy techniques. Imbuing your agents with fuzzy reasoning capabilities can give them the ability to exist within domains where knowledge may not be quite as crisp as we are used to dealing with in traditional computing approaches.

3.4.1.1 Introduction. Fuzzy logic and, more generally, fuzzy-based systems are, despite the moniker, mathematically based systems that enable computers to deal with imprecise, ambiguous, or uncertain information and situations—in other words, the real world. Since your agents will exist in the real world, fuzzy logic is one of the converging technologies with which you should be familiar. You may find equipping your agent with fuzzy reasoning capabilities enhances its ability to cope with imprecise users, uncertain and new domains, and other agents.

Traditionally the computer sciences have involved only absolutes based on precise mathematical formalisms, including the predicate calculus, two-valued logic, information theory, and graph and group theory. Indeed, computer languages, language compilers, and language execution environments (operating systems and hardware) are all based on crisp, exacting syntax and semantics. In contrast, our natural human language is full of ambiguities (unless spoken by the likes of a Spock or, more recently, Data).

This is one reason why many programmers are attracted to so-called higher-level languages. These languages, while being further away from the exacting machine level, have expressive ability closer to our own language. One of the first things many programmers do is cast a natural-language narrative of the programming problem to be solved, then proceed to refine the narrative into pseudocode (p-code) and thence to whatever computer language the programmer is using. Just as software engineering (despite those "cowboy artists" out there) is expressly precise, computer hardware is likewise designed and constructed using a plethora of rigorous engineering techniques and principles.

It is due to this adherence to traditional scientific teachings that most engineers and scientists over the years have looked askance at the propositions, put forth by fuzzy-logic proponents, that uncertainty should be embraced by systems design, rather than engineered (or ignored) away. The only uncertainty allowed was usually restricted to circumstances where the laws of large numbers were valid and randomness was an accepted attribute of certain systems' characteristics. But even as more complex, computer-based systems become commonplace in industry and our personal lives, there remains an acute inability for the average, nontechnical human to easily interact with these systems in all but the most trivial scenarios (such as ATM machines).

3.4.1.2 Need for fuzzy reasoning techniques in IA systems. This disconnect between computers (and computerized devices) and humans seems to be of the most fundamental kind. Humans are used to interacting with a real world that, at our level of consciousness, is primarily analog and ambiguous. Our very language is. This is why it has been notoriously difficult to get computers to truly understand humans' natural language.

This is not to say that brute-force-based or clever-pattern-recognition-based speech or word recognition systems haven't made great strides recently. But the computer still doesn't "understand" what is being said. For any semblance of understanding, natural-language processing techniques include reference to context to extract meaning. And to reason about what is being said—for example, to determine truth or falsity of a statement—computers rely on formal logic reasoning techniques and access to large databases of stored knowledge.

If computers are to function one day as adequate assistants, then it seems that we must either conform to the rigors of precise communications or produce computers that can mold their processes and understanding to the humans' primary modes of communications: natural language and mathematics. While we are very good at expressing mathematical relationships succinctly and unambiguously, natural language has proved problematical. Enter fuzzy logic.* It may seem that information regarding a task or problem domain, if available at all, can be precise, and that only when dealing with future contingencies would imprecision, or fuzziness, creep into the equation. In fact, current multistate logical formalisms apply degrees of truth to available data. Fuzzy logic continues where multistate logic leaves off.

In 1965, Zadeh proposed fuzzy set theory to help computers reason with uncertain and ambiguous information. His work was based on attempts to understand complex nonlinear systems. Zadeh proposed fuzzy technology as a way to model the uncertainty of natural language. Indeed, he saw that many difficult problems can be expressed much more easily in terms of linguistic variables.

Linguistic variables are simply the words (usually adjectives) we use to describe certain aspects of the real world. Adjectives are used to describe or enhance the description of properties of things. But one of the most important features of a linguistic variable, as used in fuzzy systems, is the notion of its utility as an expression of data compression. Consider Zadeh's explanation: "Consider a linguistic variable such as Age whose linguistic values are young, middle-aged and old...Clearly a numerical value such as 25 is simpler than the function young. But young represents a choice of one out of three possible values, whereas 25 is a choice of one out of, say 100 values..." (Zadeh, 1994).

Zadeh goes on to call this kind of compression granulation. He maintains that it is very important because it is more general than quantized or discrete values. It mimics humans' use of words and their meanings, and transitions between linguistic variables are gradual . This last feature means that your use of linguistic variables may enable your agents to deal with more continuous and robust descriptions of reality and problem spaces.

---

*Although not discussed directly in this book, other areas of language processing and understanding are also important to an IA designer, such as relationship analyses that produce lexical relationships (synonymies and antonymies, taxonomies, or part-whole relations) and semantic relationships. Cyc is one system designed to enable the representation of such relationships explicitly.

Fuzzy logic originally extended boolean logic to handle truth values between the extremes of completely true and completely false. Fuzzy technologies now encompass much more than originally designed. For example, there is now a subset of the technology called fuzzy differential equations.

Fuzzy systems technologies have several benefits, including higher, parallel computational capabilities than symbolic processing, a higher level of abstraction—the set-theoretic level ability to deal with imprecise or conflicting information—and denser or more compact encoding of knowledge than models using crisp state transitions.

3.4.1.3 Overview of fuzzy technologies. As Cox (1995) analyses fuzzy technology, it encompasses three related disciplines:

- *Fuzzy set theory:* "Describes the mechanics of how fuzzy sets are organized and what operations are allowed."
- *Fuzzy logic:* "The process of making logical inferences from a collection of fuzzy sets."
- *Approximate reasoning:* "A combination of mathematically precise logic and powerful heuristics...such as fuzzy set hedges, rule contribution weights, and alternate sets of operations."

As Zadeh (1994) has said: "In a broad sense, fuzzy logic is almost synonymous with fuzzy set theory. Fuzzy set theory, as its name suggests, is basically a theory of classes with unsharp boundaries." But Zadeh also says that when fuzzy logic is considered more narrowly, fuzzy set theory includes it and other branches of it or ideas, including fuzzy arithmetic, fuzzy programming, fuzzy topology, fuzzy graph theory, and fuzzy data analysis. These branches hint at the wide application areas of fuzzy ideas.*

A fuzzy set is a collection of things that belong to the set to varying degrees. For example, consider the fuzzy set representing the linguistic variable "short." The fuzzy set of short people could have people under 4 feet belonging 100% to the set, people 5 feet 6 inches belonging 50% to the set, while people over 7 feet would belong perhaps 1% or maybe not at all. In other words, the short fuzzy set does not give a crisp delineation between those people that are short and those that are not short. Rather, a short fuzzy set contains everyone, but to varying degrees or percentages. The way this is represented in fuzzy language is to define the degree to which a statement of the form:

$$y \text{ is in } G$$

is true. This is done by examining a mapping of ordered pairs $[y, q]$ that express the membership functions of all such (relevant) things denoted by y. The value

---

*Note that boolean logic constructs are completely expressible using fuzzy logic.

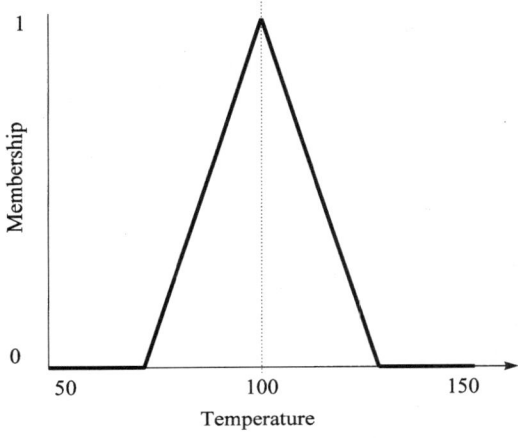

Figure 3.9 Hot-temperature membership function.

denoted by q is the degree of membership, or truth, of the statement "y is in G." Membership functions can be graphed to give you a visual idea of what fuzzy is all about. An example membership function for a "hot" temperature could be plotted as shown in Fig. 3.9.

In Fig. 3.9 the range of hot temperatures is 50 to 150 degrees, with 100 degrees at the peak. An input temperature variable value of 100 degrees, when fuzzified, or plotted on the graph, would give 100% hot, whereas a temperature of 70 or 130 degrees might be classified as 60% hot. Of course, the 70-degree value might find itself at 100% within a fuzzy set "warm," and the 130-degree value might be at 90% within a "very hot" fuzzy set.

From this example you can see that fuzzy sets can be made to model the real world in a manner more in line with how we approach categorization—with some common sense and flexibility. Fuzzy sets can be combined to produce composite, multidimensional fuzzy sets. In this way, complex entities with many descriptive attributes can be accurately modeled. Composition of fuzzy sets can be summarized as follows:

> In the composition subprocess, all of the fuzzy subsets assigned to each output variable are combined together to form a single fuzzy subset for each output variable. In MAX composition, the combined output fuzzy subset is constructed by taking the pointwise maximum over all the fuzzy subsets assigned to the output variable by the inference rule. In SUM composition, the combined output fuzzy subset is constructed by taking the pointwise sum over all the fuzzy subsets assigned to the output variable by the inference rule. (Kantrowitz, Horstkotte, and Joslyn, 1996)

Since fuzzy sets are an extension of classical sets, set-theoretic operators can be applied to fuzzy sets. For example, intersection (logical, algebraic, and bounded product), union (logical, algebraic, and bounded sum), and complement operators can be applied to fuzzy sets.

A fuzzy system uses a set of rules that determine to what degree a proposed set of solution variables fits or solves a problem. In classical logic, modus ponens is an important logical rule having the form:

Premise: A is true

Implication: If A then B

Conclusion: B is true

In this classical form, A and B are crisply defined propositions, or values, such as 5.0 or yellow. In fuzzy logic, A can be a fuzzy proposition (alternately called a fuzzy set or fuzzy subset) such as: current temperature is approximately 55°F. Fuzzy modus ponens then transforms into a inferencing form that more closely approximates how we humans naturally converse about the world.

Premise: Current temperature is approximately 65°F.

Implication: If current temperature is approximately 55°F then higher thermostat setting a little

Conclusion: Higher thermostat setting a little

Using fuzzy logic, you can see that using different premises, there can be several different conclusions using the same implication.

The input values of solution variables are fuzzified to determine their degree of membership in the fuzzy set modeling that particular variable. Each input value then contributes to a final solution to the degree its rule fits the value into its related fuzzy set.

Fuzzy logic operations can be performed as follows. Consider a rule such as:

$$\text{if } P \text{ is short and } Q \text{ is tall then } R \text{ is medium}$$

where P and Q are input variables having known values and R is a data value that needs to be discovered via computation. Short is a membership function (that is a fuzzy subset) defined on P, tall is a membership function defined on Q, and medium is a membership function defined on R. The first part of the clause: "if P is short and Q is tall," called the antecedent, defines the degree to which the rule applies. The second part of the clause: "then R is medium," the rule's consequent, determines the membership function for R. Where there are many fuzzy rules, composition is used to come up with a single fuzzy set for each output variable of interest.

Fuzzy systems process rules using certain inferencing methods, two of which are explained as follows:

> MIN and PRODUCT are two inference methods or inference rules. In MIN inferencing, the output membership function is clipped off at a height corresponding to

the rule premise's computed degree of truth. This corresponds to the traditional interpretation of the fuzzy logic AND operation. In PRODUCT inferencing, the output membership function is scaled by the rule premise's computed degree of truth. (Kantrowitz, Horstkotte, and Joslyn, 1996)

Finally, defuzzification of an output fuzzy subset is done to arrive at a crisp value when needed to drive or input to a conventional, nonfuzzy system. Kantrowitz et al. say that two of the more common methods are CENTROID, where the output value is found by finding the center of gravity of the output membership function, and MAXIMUM, where the output value is simply the membership function's value at its maximum.

3.4.1.4 Applications of fuzzy systems technologies.
Fuzzy systems technologies have been used in a gamut of application areas, including:

- Expert systems
- Device, process, and production control
- Market prediction and trading, asset allocation
- Inventory control
- Loan analysis
- Fraud detection, detection of patterns
- Production scheduling
- Route planning
- Econometric modeling

Expert systems are some of the biggest beneficiaries of fuzzy-logic technologies. Fuzzy expert systems are used for problem diagnosis, planning, prediction, natural-language processing, and within intelligent robots and user interfaces. A traditional expert system is a set of rules that constitutes particular reasoning processes that capture what a so-called expert would do in the search for a solution to a problem. One of the most difficult aspects of building an expert system is collecting the expertise, or knowledge, which is often encapsulated within one or more humans.

This process, called knowledge engineering, is often fraught with imprecision and combinatorial explosions. This isn't because the experts don't know their subject—it is because the way in which knowledge engineers elicit the knowledge, and the manner in which the knowledge is expressed, is usually via human language. Because human language is not especially crisp, except in the most trivial of domains or contexts, it is a tedious and potentially error-prone process to convert a human's expertise into something a computer can deal with using traditional methods and logic. This is especially true when the logic is full of exceptions and conditionals. For example, a process operator, when asked how he controls the process, might give the answer in the form:

"If the pressure is too low, I increase the reactor's fuel flow a little and also inject some more catalyst, except when the pressure ramp-up is very fast, then I only increase the fuel flow slowly."

A fuzzy-based expert system also consists of a knowledge base composed of rules. But the rules read like the expert spoke; for example, "if the temperature is about 20°C, and the pressure is high, then close the hydrogen feed valve a little." For each input variable value, like pressure and temperature in the previous example, fuzzification takes place, producing membership functions that determine the degree of truth for each rule's premise. When the expert system starts an inferencing process, a fuzzy subset is produced for each output variable of each rule.

Our own work at Honeywell deals with such nonlinear systems. Device, process, and production control has historically depended on the formal, mathematical model called the proportional, integral, derivative (PID) form of control equation. Given various feedback mechanisms (feedback, feedforward, and so on), a set point is used to alter an output such that the process variable of interest (the monitored and controlled variable) comes as close to the set point as practical. Many times, though, the characterization of the process is such that PID forms of control are not adequate to describe highly nonlinear or chaotic processes.

This is because PID equations depend on specific transfer functions that define what is to happen to a system, given a certain perturbation (such as a set-point change). The difficulty is expressing the ranges of actions to take given the ranges of measured process variable values. Scaling up from a system of a few variables to a great many variables is quite difficult. The system being modeled with PID forms of control is assumed to be fairly linear, or can be described with monotonic functions. Unfortunately many processes are nonlinear, and the intereffects of interacting portions of the process as a system are unpredictable from an analysis of the parts.

Enter fuzzy-based control. Fuzzy logic is good at expressing ranges or shades of values within a range. In addition, fuzzy rules can model portions of a system very accurately, and when combined via inferencing, produce desired outputs (those minimizing the error between set point and input values) across the whole system. The fuzzy system becomes an expression of the process in terms of its continuous behavior rather than discrete mathematical equations.

Examples of fuzzy-based control abound. They include the following:

- Blue Circle Cement in Denmark implemented a cement kiln control system based on a fuzzy expert system. It uses fuzzy rules gleaned from operators and mixes and grinds ingredients more efficiently and with smoother control than the mathematical model of the kiln.
- Sendai subway train control in Japan. Hitachi generated a fuzzy rule–based system to predictively control the cruising and braking of trains in order to minimize energy usage while maximizing passenger comfort and safety.

- Discrete devices that incorporate fuzzy-logic–based control, including washing machines, dryers, air conditioners, and video cameras (compensating for users' unsteady hands, automatic exposure, and light balancing), and automobile-based devices such as braking, cruise controllers, and transmission systems are all on the market.
- Pattern-recognition applications include the Sony PalmTop, which uses a fuzzy-logic decision tree algorithm to perform Kanji character recognition.

Another application of fuzzy logic is as the basis of a search engine (or search agent). Fuzzy-logic–based searching can, for example, retrieve items containing some string or other pattern, much as search engines based on regular programming. But fuzzy rules also make it much easier to search for items that are semantically related to the desired string or pattern. Because the matches are members related to the target by their percentage of "belonging" to a fuzzy set representing the target pattern, the presentation of the search results can include some indication of how closely they are related to the target pattern.

From the foregoing discussion you can see that IAs incorporating fuzzy-logic capabilities would be well suited to environments where they need to maintain autonomy in the face of conditions that would break more rigidly coded solutions with precise logical constructs.

3.4.1.5 Making fuzzy agents.

One of the first things to consider about incorporating fuzzy technologies into your agents is: under what agent-based scenarios would fuzzy be better than other possible solutions? In other words, why use fuzzy and under what circumstances? What we can say here is that, in general, use fuzzy technologies in or with your agents for the same reasons fuzzy is used in stand-alone applications. If an agent needs to reason with imprecise or incomplete information, or the domain variables are expressible using linguistic variables, fuzzy logic is a candidate to make your (and your agents') job easier.

For example, suppose your domain is a control system and you are monitoring a process unit for a special kind of alarm condition using a fuzzy inferencing system. A typical nonfuzzy alarm system might compare expected or desired values of process variables, or rates of change of those variables with alarm limits. When a limit is violated, an alarm is annunciated and propagated to those other parts of the control system with a need to know, as configured by the process engineer. Typically this would include sending an alarm message to an operator's station. Moreover, alarming functions are already present in all control systems. What we are looking for, though, is something more autonomous—an agent that could go from node to node in a distributed process-control system checking up on conditions and other resident agents, collating the information, and reporting back to a central operator. This mimics a real process operator's duties. The operator looks for trouble by walking both the plant and the cyberplant; a mobile agent could certainly serve as operator proxy in this situation.

An alarm agent that incorporates fuzzy inferencing might move itself from controller to controller in the unit to see if alarm conditions are *about to be exceeded* by fuzzifying process variables and applying the appropriate fuzzy logic to combinations of the variables. These fuzzy variables can be examined as to their fitness, or degree of belonging to fuzzy sets defined by linguistic variables. In other words, the situation can be expressed by the phrase: "if the temperature of the reactor is almost at maximum, and the coolant valve is just about closed, doWhatever."

There are many ways you can incorporate fuzzy systems technologies into your agents. The applicability of any method depends heavily on your selection of the agent development language and delivery platforms as well as your overall agent architecture (see Section 5) and, to a lesser extent, the infrastructure (such as CORBA, OLE, and so on; see Section 4) over or within which your agent system exists.

That being said, the most obvious method would be to first produce a fuzzy inferencing system, a fuzzy-logic–based expert system, or a fuzzy rule base that deals with uncertainties your agent may have to deal with. For example, one way to incorporate fuzzy technologies into your agent system might involve the following. A stand-alone fuzzy system can be produced using a commercial package that has a callable API.

After your fuzzy system is constructed and debugged, a dedicated agent can be produced that interfaces with the fuzzy system. Other agents can then access fuzzy capabilities via the FuzzyInterface agent.

A slightly different approach would be to place the agent-to-fuzzy system communications protocol in an agent class from which all appropriate agents can inherit, or make use of, the fuzzy system as though it were their own interface. An agent requesting fuzzy processing can send the fuzzy system interface agent the inputs that represent information in its environment. The inputs could be the premises of various fuzzy system rules. In addition, the agent can request that the fuzzy system execute some set of its rules using those inputs. Information sent back to the requesting agent could include conclusions or recommendations reached by the fuzzy system. These conclusions could help the requesting agent clarify a nebulous situation, environment, or event that it has been exposed to.

Developers can embed fuzzy logic directly into agents using a package called FuzzyExpert from Indigo Software Ltd. of Cheltenham, England. FuzzyExpert is a C++ based class hierarchy of membership functions, fuzzy sets, and rule-based chaining components. It contains over 25 classes representing more than 200 fuzzy functions. FuzzyExpert provides developers with components such as multidimensional fuzzy collections and relations.

Another company, Modico, Inc.,* makes a fuzzy system development environment called Fuzzle for Windows. It is a particularly easy to use tool, employing

---

*e_mail: modico@aol.com

menus and icons to define the fuzzy system. It checks your system for consistency: input validation, completeness of logic, and an indication of the soundness of the design. When the fuzzy system is finished, you can generate a stand-alone C or FORTRAN program, or a subroutine that can be called by an agent.

A group called European Research in Uncertainty coordinates deployment of "soft" computing technologies such as fuzzy logic.* The group is producing a common reference platform for the development of an infrastructure that facilitates the sharing of knowledge about several technologies that can be used within IAs, including fuzzy control, decision support, data fusion from simultaneous sensor streams, and mining of databases for "hidden" information, meanings, and patterns. Two participating companies produce fuzzy products that you may want to investigate further for use within your agents. SireneF by UCI Microelectronique of France is a neurofuzzy development environment on Windows PCs. ZeTec GmbH from Germany produces Fuzzy-Box for making hybrid software and hardware systems for fuzzy-based industrial control.

MathWorks, Inc., is another company that proffers add-ons to enable nonlinear aspects of systems to be modeled with neurofuzzy and fuzzy techniques. Matlab supplies a toolbox of routines to use these technologies. The Fuzzy Systems Toolbox for use with Matlab provides the ability to design membership functions and fuzzy rules for a given application. The toolbox provides an environment within Matlab to test fuzzy systems as well as hooks to link in neural networks and genetic algorithms.†

Yet another company, Exsys, Inc., of Albuquerque, New Mexico, has developed a rule-based expert system (Exsys Professional) which can provide relatively broad fuzzy-logic rules in lieu of complex rules based on boolean logic. This fuzzy expert system can then be used by your agents.

As you can see, the avenues for incorporating fuzzy logic into your agents, to give them some semblance of those qualities we consider intelligent, are manifold. And we have only scratched the surface with the companies and techniques discussed. Do some research on your own (for example, do an AltaVista search on Fuzzy Logic), and you will find a plethora of fuzzy system offerings on virtually any platform. Your agents will thank you (well, maybe not).

3.4.2 Evolutionary computing

In this subsection we discuss evolutionary computing and the application of evolutionary computing techniques to agents. Imbuing your agents with the ability to evolve their behavior and reasoning capabilities can give them the ability to exist within domains where conditions are dynamic, where you or your users may have time to let a cadre of agents fight it out among themselves to see who serves the end user best.

---

*Navigate to http://ftp.pws.com/pws/engr/fuzzy for free demonstration routines.

†http://www.haley.com

3.4.2.1 Introduction.
Evolutionary computing consists of system-design and implementation approaches that use biologically based evolutionary processes as a model. In addition to the popularized genetic algorithms, classifier systems, evolutionary programming, and evolution strategies are some of the other techniques roughly categorized as evolutionary processing. Evolutionary computational techniques seek to evolve computational entities via simulation of the observed natural evolutionary processes of selection, mutation, and reproduction. Evolution of computational entities according to these processes proceeds to develop entities successively more suited to their tasks than previous incarnations. Specifically,

> ...Evolutionary Algorithms maintain a population of structures, that evolve according to the rules of SELECTION and other operators, that are referred to as `search operators,' (or GENETIC OPERATORs), such as RECOMBINATION and MUTATION. Each INDIVIDUAL in the population receives a measure of FITNESS in the ENVIRONMENT. REPRODUCTION focuses attention on high fitness individuals, thus exploiting the available fitness information. Recombination and mutation perturb those individuals, providing general heuristics for EXPLORATION. (Heitkoetter and Beasley, 1996)

Genetic algorithms work by maintaining a gene pool of possible parameter settings (the complete parameter set being somewhat analogous to a chromosome) and a performance metric or goal. The parameters can belong to, or form all or part of, the characteristics of the software entities of interest—in our case agents. Successive evaluations of the performance of individual parameter sets from the gene pool results in the unfit sets (agents or algorithms within them) being eliminated. Then mutations and crossbreeding (crossover) produce new parameter sets. Then the genetic algorithm repeats the evaluation against the goals. After many "generations" or repetitions, the approach ensures that the fittest parts of each parameter set end up in a single optimally tuned parameter set. This means that the agent best suited to the task or goal is selected to survive the evolutionary pruning.

To simulate evolutionary processes, sometime parameters from different parameter sets are exchanged in what is called a recombination operation or crossover. Determining which parameter sets recombine is usually done by choosing more than one of the fittest individuals. Their characteristics are combined in a successive generation, which is even better than previous generations of fit individuals. Genetic algorithms usually use fixed-length encodings of the salient characteristics or parameters of software entities, such as agents.

Other evolutionary computing approaches, such as genetic programming, do not need to represent the parameters as fixed-length structures. You have more freedom to represent the entities being evolved as appropriate to the problem. Further, genetic programming techniques focus on the relationship of parent solutions to offspring solutions. An offspring solution is created via replication of a parent solution, then mutated to some predetermined degree. This is done several times to produce several offspring along a spectrum of mutation (assessed by determining how much the parent was changed to produce the mutated offspring). Each new offspring solution is then evaluated as to its fitness as a solution. Notice that evolutionary programming does not use recombination or crossover of parameters between different individuals to produce a new individual. The sole mechanism for change is reproductive mutation.

A further branch of evolutionary computing techniques involves using a combination of approaches used in genetic algorithms and genetic programming. It is called evolutionary strategies. Here individual software entities can be mutated individually and combined with each other, then subjected to fitness evaluations or selection mechanisms, as appropriate to the problem.

3.4.2.2 Need for evolutionary computing techniques in IA systems.

So what are evolutionary computing techniques used for? Many applications deal with optimization problems—trying to find an optimal or best solution to a problem. Because you have flexible control over the selection mechanism or fitness function that evaluates solutions against some criteria, evolutionary computing techniques are sometimes easier to implement than alternative methods such as constraint-based programming, where you have to have foreknowledge of all the constraints and the interactions between software entities. Constraints can be implemented within evolutionary computing techniques by simply making them weighted components of a single fitness function which represents the desired optimization strategy.

Optimization problems well suited to evolutionary computing approaches include various forms of scheduling, such as timetabling (scheduling exams or classes), resource or job scheduling, and maintenance scheduling common in process and production control. One of the problems we encounter in our own work at Honeywell is scheduling the optimal use of our customers' resources (such as feed stocks, reactors, blenders, valve manifolds, pipes, holding tanks) in multiple batch processes occurring simultaneously.

Evolutionary computing techniques hold promise in solving these types of problems because, in an ongoing production scenario, it is often the case that any high-quality solution is preferable, if reached quickly, to the best solution that is found too late. In other words, time is a resource that must also be taken into consideration when applying computing resources to finding solutions. In this case, evolutionary computing techniques very often find "good enough" solutions, in some cases very close to the optimal solution, much more quickly than brute-force techniques, such as branch and bound searches through large search spaces (Fang, Ross, and Corne, 1993).

Evolutionary computing techniques are good in any situation where your agent must deal with many interacting variables that can result in many possible solutions to a problem. The agent's job, in such situations, is to find the optimal mix of values of those variables that produce an optimal solution. Some other examples of agent applications that might benefit from evolutionary computing include the following:

- Management agents such as:
  - Distribution agent, which helps figure out how to best get finished goods from a storage facility to retail outlets, given various constraints.
  - Task assignment agent, which, within a workflow process, matches workers or machines to tasks.
  - Project management agent, which prioritizes tasks based on optimal resource utilization.
- Financial agents such as:
  - Portfolio balancing agent, which monitors and dynamically rebalances the dollar amounts over a variety of securities.
  - Forecasting agent, which finds the best performing set of financial agents (in our previous example).
  - Cash-flow optimizing agent, which analyzes cash inflow and outflow and staggers payments to maximize cash on hand, while avoiding late payments.
- Engineering agents such as:
  - Design assistant agents in a variety of engineering disciplines, which find optimal sets of components to meet a specified set of constraints.
  - Process control agent, which optimizes production rates based on unitwide and enterprisewide goals.
  - Network configuration agent, which could dynamically balance the movement of software, including other agents, based on resource utilization, network topologies (which could be changing), and bandwidth considerations.
- "Front end" agents such as:
  - Fuzzy-logic training agent, which sets values for membership rules automatically, evaluating the rules to see how good the results meet expectations.
  - Neural net training agent, which finds the best set of weights in which to focus learning.
  - Curve-fitting, statistical evaluations, and so on.

For example, genetic algorithms can be used for designing an optimum hardware circuit starting with a simple circuit and a set of rules. A design assistant agent can use such an approach in combination with an ability to surf the web for circuit components and their specifications. The agent would construct a circuit and evaluate its efficacy against some criteria. Then a genetic algorithm could mutate the circuit-generation rules, or the components used, and the agent would compose another circuit. As this process proceeds, the best circuit (within some time constraint) would evolve.

A relatively simple agent can make flight and hotel reservations and put together a vacation itinerary, but an agent that incorporates a genetic algorithm can iterate over many possible vacation plans and evaluate them with respect to criteria based on the desires of a human user to plan a better vacation. Though an agent cannot understand quality in the subjective sense, or know the meaning behind the symbols it manipulates, it can determine, just as a chess-playing program can, that one move is better than another based on how any move can get the system closer to a goal the user has set for it (checkmate the opponent or plan a fun and frugal vacation). A genetic-algorithm-equipped agent does this as it produces successively better generations of solutions by mutating previous solutions.

Where a nonintelligent agent may compare airfares to find the lowest fare available to a specific destination, an agent that provides successive generations of better and better vacation solutions would be much more adept at pleasing the user. You could use a genetic algorithm tool to encapsulate the process, which lets a vacation-planning agent not only formulate a vacation plan, but find the best one for its user's interests, available funds, time, and so on. This would involve dynamically modeling many possible vacations and ranking them based on current criteria. The agent may find, for example, that there is new snow at a user's favorite ski resort, but that lift-ticket prices just went up. It may also detect additional funds available in the user's vacation account, so a little fancier vacation might be possible. As a wide range of vacations are constantly modeled and judged for maximum expected utility, a short list of winners eventually emerges, and can be evaluated by the user. If none on the short list is satisfactory, the user can kick the whole process off again, whereby the parameters would be combined or mutated by the genetic algorithm to arrive at a new set of vacation solutions.

3.4.2.3 How developers and agents themselves can make better agents: incorporating evolutionary computing technology into agent systems. One of the first things to consider about incorporating evolutionary computing technologies into your agents is: under what agent-based scenarios would evolutionary computing be better than other possible solutions? In other words, why use evolutionary computing and under what circumstances? What we can say here is that, in general, use evolutionary computing technologies in or with your agents for the same reasons evolutionary computing is used in stand-alone applications. If there are many possible agents, each incorporating a unique algorithm or solution to a problem, and you want to find the best agent for the job, evolutionary computing is a candidate to make your (and your agent's) job easier. Or if you just have a general idea of how an agent might achieve a goal, you could code that approach, and progressively mutate and combine solutions until a more detailed, specific, optimized solution is found.

There are many ways you can incorporate evolutionary computing technologies into your agents. The applicability of any method depends heavily on your selection of the agent—development language and delivery platforms as well as your overall agent architecture (see Section 5) and, to a lesser extent, the infrastructure (such as CORBA, OLE, and so on; see Section 4) over or within which your agent system exists.

It also depends on whether the agent evolution is done during development, to send out the best agent for a job, or whether you (or a user) will unleash many agents, each with a less than optimal solution. In the latter case you would have to provide a way for those agents to evolve toward an optimal agent via combination, mutation, and selection. This means that another entity (an "Evolver" agent) would have to supply the mutation, recombination, and selection mechanisms on the fly. It would be an integral part of your agent system. This is an attractive option for those situations where an agent's environment is changing so much that any hard-coded solution or algorithm, or even a switching mechanism among several embedded algorithms, might not be flexible or optimal enough.

That being said, the most obvious method would be to produce first an evolutionary computing agent that could interface with the other agents actually doing the work. A stand-alone evolutionary computing agent can be produced using a commercial evolutionary computing package that has a callable API. The "worker" agents could be created with appropriate solutions or algorithms and interfaces with the evolutionary computing agent. The evolutionary computing agent would need a representation of each of the starting solutions or algorithms used in the worker agents.

One example of this approach is illustrated by a product called Evolver from AXCELIS, Inc. Although Evolver works with Excel directly, its engine is written as dynamically linked libraries (DLLs) that can be called from any Windows application or from an agent written in any language that can call DLLs (such as VisualBasic). If your agent models those variables that affect the solution and can specify its goal, then it could feed the Evolver engine those variables and goal and the engine finds the optimal variable values and hands them back to the agent. The agent would then set its variables to those values, which could be used to determine what behavior the agent exhibits next. There are several solution generating methods (crossover, mutation, and so on); each is represented by a DLL. You can even produce your own methods and incorporate them into the set of DLLs that is referenced by the Evolver API.

One of the most popular uses of genetic algorithms is as a front end to neural networks (discussed in the next subsection) and fuzzy systems. With respect to fuzzy, genetic algorithms can be used to train fuzzy systems. For example, the FlexTool genetic algorithm is an add-on to MatLab by MathWorks, Inc. FlexTool is a tool that lets you build software that learns the optimal parameters for a problem using survival of the fittest. Combining a FlexTool genetic algorithm with the FlexTool Fuzzy Systems tool, the environment, called FlexTool Evolutionary Fuzzy Modeling by Flexible Intelligence Group LLC, permits up to four performance goals to be set simultaneously and evaluated for Darwinian fitness. It builds fuzzy systems by using accelerated genetic algorithms to optimally adjust fuzzy membership parameters and rule structures. In this case, then, the evolutionary computing technique is used to produce the best fuzzy rule set for your agent.

As with other soft computing technologies, incorporating evolutionary computing into your agents, either directly, letting them evolve on their own at run time, or indirectly during development, can ensure that your agents are the very best they can be.

3.4.3 Artificial neural networks—computers that learn while they compute

Perhaps no computing technology has elicited as much excitement, and stimulated the imagination from layman and professional alike, as artificial neural networks (ANNs). In this subsection we discuss ANNs and the application of ANN techniques to agents. Incorporating ANNs into your agent systems is another step toward making truly intelligent agents. Imbuing your agents with the ability to learn and thereby increase their chances for dealing with their environment and perform their tasks better over time certainly increases their acceptance. Indeed, lots of ANN-based agents could lead to that emergent intelligence that seems to have occurred within natural systems like us.

Imbuing your agents with neural processing capabilities can give them the ability to exist within domains where pattern recognition and deciphering ambiguity are important to function adequately. Of course this limited subsection cannot do justice to the huge and exploding base of information and applications of ANNs. However, the summary presented here may serve to pique your interest in ANNs and convince you of our position that ANNs' use as or within agent systems serves as a complementary, possibly even a driving, technology in the quest to make computing systems and agents "intelligent."

3.4.3.1 Introduction: the brain as model.
The brain performs remarkable feats: memory, three-dimensional visual processing, language skills, generation and enjoyment of music, motor skills, analytical and reasoning skills. Many neuroscientists believe that one of the primary reasons the brain can perform such phenomenal feats is via functionally dedicated architectural structures and an overall parallel-processing mechanism within functional units, such as the visual cortex and the motor cortex. In Section 2 we discussed speculation that the brain's many interacting functional processes can be described as analogous to many collaborating agents on several hierarchical layers of increasing complexity. In addition, for certain memory and reasoning capabilities the brain employs possibly large-scale parallelism across the cerebral cortex and other large substructures.

Smaller-scale processing at some level occurs when individual units called neurons "fire," sending a signal down a dendrite, causing a chemical release across a synapse, or gap, to another neuron which, if it receives enough such input stimuli, fires itself, and so on. Neurons are specialized and grouped together in various functional units throughout the brain, giving rise to a partitioning of functionality, vision systems, speech center, motor centers, and so on. But processing at the lowest biochemical level within each neuron is identical. In addition, each neuron is relatively autonomous and independent, processing its inputs and producing an output asynchronously with respect to other brain events.

The brain is a very large, natural neural network, composed of smaller, dedicated neural networks, which are composed of individual neurons. This simple yet powerful architecture belies the tremendous complexity and capability that inheres within the brain and its most important (some would say emergent) artifact, the human mind. We can deal with the world in large part because our neural network evolved to recognize the patterns that are inherent within universe's structures. This pattern-recognition prowess is obviously extremely important to our survival. Virtually every interface with the universe and our codification thereof is via pattern recognition, including:

- Visual processing
- Speech and language skills inherent in pattern-based communications protocol based on formal lexica with a specific syntax and semantics
- Science—the methodology of recognition, recording, and testing of hypotheses regarding patterns
- Motor skills, being nervous-system habituation to repeated physical movement as complementary to natural forces such as gravity and acceleration
- Auditory processing, especially music composition and enjoyment, based on sound-wave-front translations into recognizable patterns that repeat

Indeed, much of what we do consciously and subconsciously depends on our ability to recognize patterns and relationships. One important aspect of patterns and our brains is that the brain seems (via evolution) particularly adept at learning new patterns. With ANNs we have imbued computers with the ability to learn and store new patterns, and to solve pattern-recognition problems heretofore claimed intractable (or very nearly so). This promises to be very helpful to the IA developer as well as to computer science in general.

3.4.3.2 Need for neural networks. Neural networks have several benefits as computational systems. Some of these benefits are shared with fuzzy systems and evolutionary computation:

- Higher, parallel computational capabilities than serial, symbolic processing
- Higher level of abstraction—representations as raw-data input patterns without precodification
- Ability to learn; neural networks being trained to recognize patterns and do optimization based on actual or desired output
- Capability of implementing solutions without complete knowledge of the algorithms or data transformations required to solve problems using conventional approaches
- Ability to work with very noisy or incomplete data

In the quest for higher productivity and efficiency over the decades, business has tried to make use of the wealth of data that they collected about their enterprises. As the data were reduced and filtered, they became information. As the information was further analyzed, it was transformed into knowledge. Now business needs the ability to use that knowledge wisely—they seek wisdom or expertise. This often takes the form of guidance or the prediction of trends. That is the last and most important transformation as we look to a future of increased worldwide competition.

ANNs offer the ability to help perform these transformations. As raw data and information proliferate, it is becoming increasingly difficult to determine their relevance to a problem or decision. ANNs can help filter unmanageable quantities of data and can find hidden relationships and patterns.

But it is not only at the front end of data reduction and interpretation that ANNs shine. In may agent-based systems, agents require a degree on autonomy by virtue of their isolation from other sources of "guidance." Mechanisms fostering built-in autonomy are also important when a lack of a priori knowledge about a situation makes it difficult to program proper responses to every situation an agent may encounter. ANNs offer such a mechanism—learning. The ANNs' ability to learn the proper behavior (that is, the proper output given a particular input) without explicit algorithmic specification makes them well suited to perform tasks otherwise much too cumbersome to "brute-force" program. This relative lack of brittleness becomes an important characteristic of computing systems, as such systems engage the real world with anthropomorphic expectations; Turing testers watch out.

As part of an agent-based decision-support system, for example, ANN-enabled agents could contribute directly to a company's bottom line by predicting important micro- or macroeconomic trends. The trends could help company management in managing production schedules better or determining optimized pricing structures. As part of a complex agent-based medical diagnosis and advisory system, ANNs could filter vast amounts of patient data and information about related cases, and recognize significant similarities or differences in treatment modalities, allowing doctors to make more efficacious treatment recommendations.

The need for ANNs in many existing and potential computer systems is exemplified by another characteristic of said computers, noticed by quite a number of members of the populace who have dealt with computers over the last few decades, namely, that computers only do what some programmer has "explicitly" constrained them to do. This has made the exclamation: "do what I want, not what you've been told!" to echo forth from many a human-computer interaction. As mentioned, this brittleness in the face of either uncertain input or the classic "operator error" has dampened the acceptance of computers by a large segment of the population. Remember, despite their reported ubiquity, the so-called personal computer is only in about 20% of American households, and it has been estimated that despite the much-heralded information age, not only do most workers *not* work directly with computers, but also they would *rather not* work with them. This defines an issue of both trust and comfort.

ANNs promise to humanize computers. Together with the use of some of the other converging technologies summarized in this section (evolutionary computing, fuzzy logic, expert systems and commonsense databases), ANNs are needed to bring computing to the people. Because agents are supposed to be, by definition, a proxy for a human, or as a friendly human front on an otherwise unfathomable computing system, the use of ANNs within agent systems should bring these suppositions closer to actuality.

3.4.3.3 Overview of artificial neural networks. ANN and, more generally, neural-based processing systems are attempts to mimic certain aspects of the architecture and proposed mechanisms of biological neural systems such as the human brain. ANNs also enable an approach to programming more akin to training the computer, letting it learn how to do a task, rather than codifying detailed specifications of the task with algorithms leading to implementation with various types of programming language.

ANNs enable computers to deal with information and situations where patterns are important characteristics, namely, the real world. ANNs can extract and codify patterns from complex and noisy data and predict trends that would be every difficult using other techniques. Since your agents exist in the real world, the ANN is one of the converging technologies with which you should be familiar.

An ANN is composed of groups or layers of interconnected individual processing elements. These elements are roughly analogous to the brain's neurons, and are referred to as artificial neurons (we'll just call them neurons). Figure 3.10 shows a typical simple neural network architecture. Just like the brain's neurons, each of the ANN's neurons has one or more inputs and one output. A neuron typically works by forming a weighted sum of its inputs, which feeds an internal (nonlinear) function that determines whether the neuron "fires" by either producing a value at the output, or not. This function implements a real neuron's activation level. The neurons whose inputs interface to the outside world (that is, values representing a pattern to be learned and recognized) are called the input layer. The group of neurons whose outputs compose the final output of the whole network are called the output layer. Any groups of neurons between the input and output layers are called hidden layers.

The hidden layer or layers perform a lot of the work that an ANN does. Each neuron in the hidden layer is connected to all the neurons in the input layer. Each of the inputs into a neuron is typically multiplied by a weight (a number). The weights on the interlayer connections are crucial to the ability of the ANN to learn and, together with the interconnection topology, determine the architectural family or class of the ANN (for example, the Back Propagation of Errors, Hopfield Kohonen). An ANN is trained by:

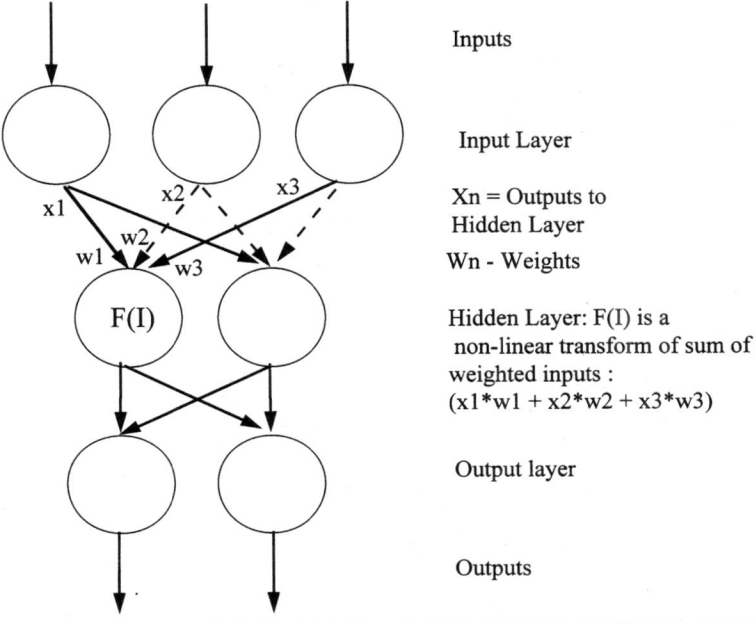

Figure 3.10 Typical artificial neural network architecture.

1. Setting the weights to some initial values
2. "Running the network" by applying some values at the inputs of the input layer neurons
3. Comparing the output to what you want it to be—the difference being the error
4. Resetting the weights and iterating this process until the output "goal" is reached

What is important here is that the process of resetting the weights, the learning process, is carried out by some set of rules (sometimes even with the aid of a genetic algorithm or fuzzy logic). An important part of the mechanism in a large class of ANN architectures, called back propagation of errors (referred to simply as BackProp), usually involves using this error value, suitably modified, as a feedback into one of the previous layers.

One of the important distinctions between training (or learning, depending on your point of view) methodologies is based on whether the ANN is supervised or unsupervised. In supervised training "...external prototypes are used as target outputs for specific inputs and the network is given a learning algorithm to follow and calculate the correct connection weights. Unsupervised learning is the sort of learning that takes place without a teacher...a learning algorithm may be given, but target outputs are not given" (Rao and Rao, 1993). In other words, the ANN organizes itself.

3.4.3.4 Applications of neural systems technologies. Neural systems technologies have been used in several application areas, including the following:

- Defense systems—recognition of military hardware for friend or foe identification
- Sales forecasting
- Optical character recognition—handwriting recognition
- Medical diagnosis by comparison to case histories
- Law enforcement; bomb detection in airports, residue identification, matching suspects characteristics (faces, fingerprints) to known perpetrator databases
- Process and production control—especially where anomaly identification and alarm-condition detection as important
- Market analysis and prediction
  fraud detection
- Loan analysis
- Validation and verification of data
- Sporting outcome prediction

If your agents will be working in any of those domains or application areas, or if the agent must learn its task, rather than being discretely programmed ahead of time, then ANNs should certainly be considered.

3.4.3.5 Making brainy agents. One of the first things to consider about incorporating neural technologies into your agents is: under what agent-based scenarios would neural-based processing be better than other possible solutions? In other words, why use neural networks and under what circumstances? What we can say here is, in general, to use neural technologies in or with your agents for the same reasons neural networks are used in stand-alone applications. If an agent recognizes patterns within input data, or your agent must be capable of learning, then neural processing is a candidate to make your job easier.

For example, if your domain is a control system and you are monitoring a process to detect a pattern of process-variable changes, or rates of change, that indicates a process upset is about to occur, a typical, nonneural system would have to be programmed from a priori knowledge of expected anomalous conditions and how those conditions are reached. A neural network approach could use historical or real-time data to train itself to learn the patterns that lead to the alarm conditions. Note that in this approach you need not know how the neural network represents the patterns or what algorithms are used internally (in the so-called hidden layers) in order to use the network.

There are several ways you can incorporate ANN technologies into your agents. The applicability of any method depends heavily on your selection of the agent development language and delivery platforms as well as your overall agent architecture (see Section 5) and, to a lesser extent, the infrastructure (such as CORBA, OLE, and so on; see Section 4) over or within which your agent system exists.

That being said, the most obvious method would be to produce first a neural network that deals with the pattern-recognition or learning situations your agents may be confronted with. For example, one way to incorporate an ANN into an agent system might involve the following. A stand-alone neural network can be produced using a commercial package that has a callable API.

After your neural network is constructed and debugged, a dedicated agent can be produced that interfaces with the neural network. Other agents can then access the network via the neural agent's interface. Input data can be passed to the neural agent with go/nogo pattern detection as the output. For example, IBM's Neural Network Utility (NNU) gives the agent developer the ability to create neural networks (and fuzzy-logic systems) graphically and embed them into agent applications. With NNU you can also create custom data filters and data translation utilities, which would also be useful within an agent.

A slightly different approach would be to place the agent to neural network system communications protocol in an agent class from which all appropriate agents can inherit, or make use of, the neural network as though it were their own interface.

Another example of using ANNs within agents are those produced with Agentware from Autonomy, Inc. They are being used for searching the WWW. Agentware spawns web agents to search through millions of web sites for data that would interest a user. These agents figure out a user's interests based on selections he or she has made previously.

Some packages use genetic algorithms to sort through many different ANN architectures to see which are best suited to a particular type of problem. For example, NeuroGenesis from BioComp in Redmond, Washington, uses a genetic algorithm to try different combinations of inputs, connections, layers, and activation functions. First relevant data sets are loaded into the environment and assigned to input layer neurons. Then the NeuroGeneticOptimizer builds and trains the neural network, evaluating it against others. As the best ANNs are kept, they are crossbred with each other using selected genetic algorithms. The result is a network that best fits the problem. The finished network can be embedded into a spreadsheet or examined with a tool. The genetic algorithms in NeuroGenesis are available as a separate API composed of C-based DLLs, so you can use the genetic algorithm feature by itself.

Section 4

Agent-Enabling Infrastructures

4.1 Introduction

From Section 3 you learned of certain technologies that we believe are very important in the quest to develop and deliver truly intelligent agents. By adopting an OO approach to agent development at the start, you can benefit from all the advantages object orientation bestows on various software-engineering tasks. By learning about and incorporating commercialized AI technologies such as neural-network- and fuzzy-logic-based reasoning, you can offer, within your agents, the ability to diagnose certain types of problems and solve them faster and with more accuracy than with traditional, brute-force algorithmic means. In addition you read how evolution's theory of natural selection has led to the commercialization of genetic algorithms. These algorithms can be applied to your agent-development efforts in two ways. First, during simulation and testing of a population of agents that differ in their approach to a problem, genetic algorithms can help determine which agent will be the best for a certain task. Second, you can actually use genetic algorithms at run time, where your agents can "mutate" their way through challenging situations and thereby gain more autonomy. And, perhaps with most impact, you learned to take advantage of expert system knowledge bases, or ontologies, within the domain in which your agents exist and carry out their tasks.

Of course, there are many other such technologies, both in academia and in the commercial world; we hope we have given you a taste for the possible. But now you need to progress to the next level of thinking about your agent-development efforts. If you are to benefit from technologies like those mentioned in Section 3, not to mention building the agents themselves for particular application domains, you certainly shouldn't also be bogged down in the minutiae of building what, in this section, we refer to as infrastructure.

Agents depend on well-behaved, structured, low-level systems to interoperate with other software and themselves. Bernstein (1996) describes what we refer to as infrastructure as middleware services: "To help solve customers' heterogeneity and distribution problems, and thereby enable the implementation of an information utility, vendors are offering distributed system services that have standard programming interfaces and protocols. These services are called *middleware services,* because they sit 'in the middle,' in a layer above the OS and networking software and below industry-specific applications." In this section we take a look at some of these middleware offerings. Note, however, that the dividing line between infrastructure and architecture, and between infrastructure and modern operating system or networking functions is a fine one.

Of course, in addition to middleware-type infrastructure capabilities, we assume, but do not explicitly cover in this book, many other lower-level elements of a completely functioning system. In other words, we assume the platform consists of an operating system and any standard networking protocols such as TCP/IP or an implementation of appropriate layers of the ISO (International Standards Organization) (OSI) seven-layer protocol suite. Here, then, are the categories of technologies, protocols, or mechanisms we consider to be infrastructural in nature:

1. Interapplication communications, or interoperability mechanisms and frameworks, including proprietary mechanisms such as Microsoft's OLE2 and Active/X, or "open" mechanisms such as DCE's RPC and CORBA-based and CORBA-compliant products such as IONA Technologies' Orbix, Expersoft's PowerBroker, and Apple's OpenDoc.

2. Features, capabilities, *and idioms* inherent in standard language implementations, including, if applicable, any standard class libraries. In particular, ParcPlace Digitalk's VisualWorks Smalltalk development environment inherently supports as part of its framework a powerful idiom or approach to an application's architecture called model-view controller (MVC). Computer programs have always dealt with data, of course, and as programs have become OO, the data consist of what is stored in objects as variables and information on how they relate to each other. The idea of basing a system on a model of the data in the system is not new. Today OO systems may be built upon not only a data model, but the views (or GUIs) associated with the data, and a "controller," which is the logic to make the model and the views work together to fulfill a customer's requirements.

3. Stand-alone environments that, while not strictly infrastructural or standard in any way, nevertheless could be used to provide some capabilities similar to those needed by some agent systems. This would be the case if, for example, the agents you had in mind were very focused and contained such that typical distribution concerns or heterogeneous situations were

not a primary consideration. An example of this would be an OO database management system (OO DBMS). Another example might be self-contained expert system environments like Gensym's G2 or Neuron Data's Nexpert OO expert system's development environment. Of course, any such subsystem could be part of the overall system in which your agents exist.
4. Higher-level capabilities now being delivered as part of advanced operating systems that significantly aid the developer. These include communication policy mechanisms such as Apple's Publish/Subscribe and AppleEvents.

In this section we intend to focus on the first of these categories—interapplication and interoperability mechanisms. For information on the other types of infrastructure-like support mechanisms and environments, pick up the latest documentation on the operating system, language, or expert system you are interested in.

Infrastructures, as used in this book, enable agent architectures and agents themselves to be built on top of a bedrock of functionality. This enabling functionality gives you, the agent developer, a head start in your quest to forge robust agent-based systems. We advise that you familiarize yourself with this material (if you haven't already) and follow up on the references given.

As mentioned earlier there is a fine line between infrastructure and architecture. Our definition of infrastructural software is as follows: software that, while not having functions produced directly for the purpose of supporting agent development, can nevertheless be directly useful to the implementation of certain agent-specific mechanisms—especially interagent communications.

Yet another fine line is that which exists between typical development tool kits and a more comprehensive development environment that seems to offer the best of all worlds. Two examples come to mind—Telescript and VisualWorks. Both not only offer a language in which to couch your agent's specification, but also provide underlying architectures and support for certain infrastructures discussed in this section, as well as cross-platform support. You therefore are relieved of many of the decisions as to what infrastructure will serve as the scaffolding on which to hang your agent architecture.

Picking the platform or operating system on which you intend to deliver agents is half the battle. Indeed, many of the infrastructural standards presented in this section were developed to cross the "barrierus heterogenous," all too common in today's diversified computing universe. Using these standards will help your agents become (ideally) platform independent. By picking the right language and development environment that complements, or can be used with, your chosen infrastructure, your agent system may approach to within an agent's whisker of that ideal.

The review of the various infrastructures we present in this section is not meant to be all-inclusive. Such an attempt would overwhelm this book's focus on agent-design concepts and development ideas. We briefly delineate the major infrastructures that are out there now, and working: object linking and embedding(OLE)/ActiveX, OpenDoc, the common object request broker architecture (CORBA), and remote procedure call (RPC) mechanisms as part of the distributed computing environment (DCE) from the Open Software Foundation. The usefulness of these standards to agent architectures and development will become apparent in this and later sections, especially in Section 5, where we talk about agent architectures.

Because the world of infrastructures is a rapidly evolving one, some of the details presented here may be out of date when you read this. (Indeed, a particular infrastructure morphed into or become subsumed into something completely different.) Before committing to any infrastructure, research the latest specifications for the infrastructure you are interested in. For a more detailed presentation on infrastructures conducive to agent development, such as OLE, CORBA, and OpenDoc, we recommend Orfali, Harkey, and Edwards (1995).

In addition, this section contains brief summary information on other types of infrastructural support in the form of operating systems, networking standards, unique software such as IBM/Lotus Notes, and emerging software like portable distributed objects (PDO). Again, the coverage is not intended to be complete; we want to give you a flavor for what kind of stuff is out there. You have to do the extra research and learning necessary to find the best fit for your planned agent system. As an adjunct to information in this section, Section 6 discusses many design considerations related to agent development. Some of those considerations further elaborate or relate to agents interoperating over the Internet, intranets, or the web.

Any of these software infrastructures is a base upon which you can build your agents and, thence, your agent architecture. Figure 4.1 shows the way your agents and agent architecture fit within an overall system—an architecture that *uses* infrastructure and other associated high-level components, as well as networking and operating-system related functionality. Figure 4.1 also graphically shows two other aspects of an overall agent system.

1. The relative connection strengths (indicated by the boldness of the interconnections) show what we believe to be the relative amount of use or dependence an agent or agent architecture will have on related system components. For example, we feel any sufficiently complex (Section 5 gives some definition of what we mean by "complexity") distributed agent system will make a great deal of use of interoperability offerings that are, for example, CORBA-based or CORBA-compliant.

2. The shaded system components represent those items discussed in this section. The darkest component is the one we give the most emphasis.

As you read Section 5 on agent architectures, you will see how some of the ideas and capabilities presented there can be built on top of the functions available within the commercial versions of these infrastructures. Serious agents do not necessarily have to execute *directly* on top of any of these infrastructures—instead, these infrastructures can provide the basis for a rich agent-execution environment.

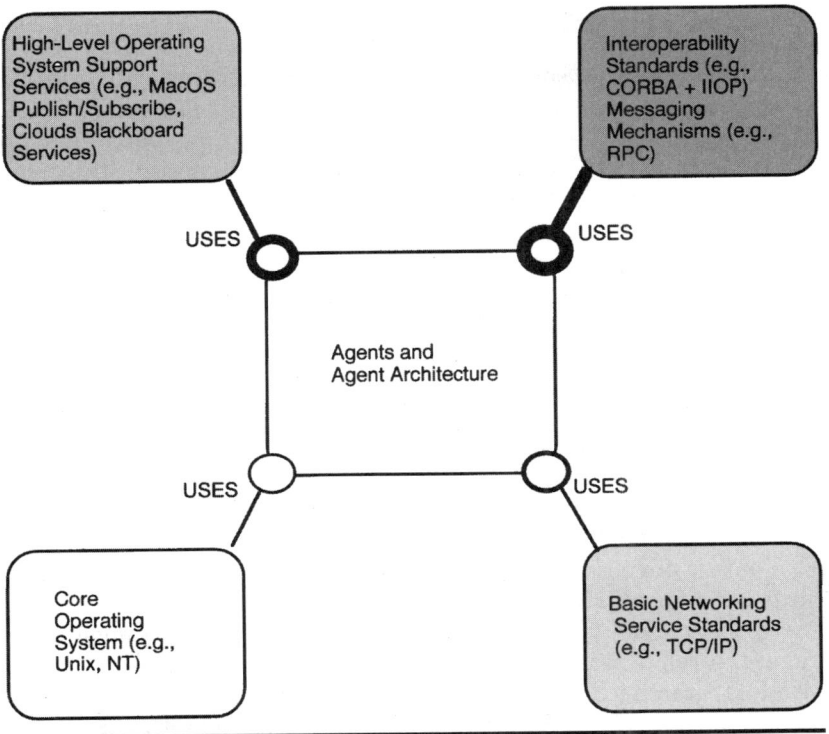

Figure 4.1 Relationship among agent system components.

4.1.1 A word about interoperability "standards"

There are many standards currently available to aid your agent-development efforts. Some are only partial solutions to infrastructural support for interoperability, especially considering our predilection in this book toward OO development. For example, X/Open's portability guide (XPGn, where n is the latest revision), NIST's APP, and IEEE's POSIX offer guidelines for infrastructure developers and applications developers.

But in practice, these guidelines "…are only partial solutions to the problem of interoperability. The goal of true interoperability has yet to be reached" (Mowbray and Brando, 1993). Even the Open Software Foundation's distributed computing environment, which we summarize later, may not fit the requirements of collaborative agent systems, given the desirability, in our opinion, of a full OO implementation of such systems. We'll focus on more complete solutions to the interoperability problem, solutions that are modular and are currently available as implemented, commercial frameworks.

As indicated in Fig. 4.1, the interoperability component supports interapplication or interobject communications. These include, for example, pipes, sockets, RPCs, and other high-level message-passing mechanisms. Within this component, middleware products are built conforming to certain interoperability standards.

What standards you adhere to and your architectural philosophy depend on the focus of the agent system. For example, in the figure a combination of an ORB together with functions that might make up a distributed object management system (DOMS) constitute an infrastructure best represented by CORBA. Several CORBA-compliant products are available—we'll mention some in this book. In CORBA's case, the architectural focus is supplying support for distributed applications, especially interobject (or interagent) communications management. This is of great importance to distributed agent development.

On the other hand, the focus of other middleware-type products, such as an object database management system (ODBMS), is representing a schema of objects within a central repository, and enabling access to that repository of objects and their interrelationships. ODBMSs (and their precursors, DBMSs) have a great deal of capability in the areas traditionally served by DBMSs, including backup and restore, centralized management, replication, and transaction control.

For our purposes, namely, agent development, we focus on the CORBA-type infrastructures. However, ODBMSs, because of their ability to represent a rich architecture of objects (and, therefore, agents), bear consideration by you as agent developers. As you research these infrastructures' capabilities, both here and in referenced material, you'll find another fine line between the capabilities of a CORBA-based product and the advanced ODBMSs.

4.1.2 A word about client/server

Before we go on, let's clear up what may be some confusion as to the meaning of client/server. First, CORBA and other object broker architectures (as well as within the OO community) refer to client/server in a different way than, for example, the database world. In this book a client means the requester of a service; a server is the provider of a service. A software entity, whether it is a true object or a legacy program, can be both a client and a server, depending on what it is doing at the time.

In some software architectures, like X-Window, designers set up specific subsystems to be nothing more than a client *or* a server—not both. For example, in X-Window the server is the hardware or software complement that "serves" up the user-interface displays—the windowing system. The client is the application (which could be on the same hardware or software, or not) that drives the server with application-specific content. The server's client is the application. In a similar fashion, in the RDBMS world, the data repository and associated functionality is termed the server; the application is the client.

In this context, however, client/server, as it applies to agents, means that you, the developer, can set up your agents to be exclusively a client, exclusively a server, or both, depending on your design requirements. When an agent is requesting a service of another agent or component, it is acting like a client. On the other hand, when it is providing a service or data for a requesting agent or component, it can be termed a server. As you can see, these designations are time-variant. Therefore, unless you wish to split up your agents according to the old lines of client and server, you can ignore the distinction. We discuss the client/server and peer-to-peer computing paradigms throughout this book, including Subsection 6.3.2.

4.1.3 Evaluating infrastructures for agents

Incorporating commercial infrastructures (like those presented in this section) and architectural ideas (like those in Section 5), "...requires a firm understanding of the [agent] application's present and future software requirements, coupled with a thorough analysis of how candidate ORB [-based infrastructural and architectural] solutions satisfy those needs" (Abowd et al., 1996). In addition to what we discuss here, Subsection 5.1.3 contains other guidelines that, while in the context of architecture, apply equally for choosing infrastructural mechanisms and middleware.

4.1.3.1 Interoperability infrastructures—agent communications and agent application models. As a precursor to looking at some common infrastructures and the services they provide, it may be helpful to summarize the kinds of things to look for in an infrastructure with respect to concepts, which you (and your agents) could find useful. Tables 4.1 and 4.2 categorize and delineate these communications and general capabilities. In Table 4.1 we see the kinds of abstract communications services an infrastructure can (but may not) provide for your agents. Table 4.2 lists the types of activities, or scenarios, that your agents could participate in (remember your scenario-based OO agent analysis?).

In support of agent applications, look for infrastructures to support the following typical, platform-independent (if possible) services*:

- *Presentation management:* Forms manager, graphics manager, hypermedia linker, printing manager
- *Computation:* Sorting, math services, internationalization services, data converters, time services
- *Information management:* Directory services, log manager, file manager, record manager, RDBMS, OO DBMS, repository services
- *Communications:* Peer-to-peer messaging, RPC, message queuing, E_mail, Electronic Data Interchange support

---

*Note that some of these services may also be provided by the operating system or a robust developer's kit via class or function libraries.

TABLE 4.1 Infrastructure Communications Models

Message protocol type	Definition	Network-dependent agent example	Network-independent agent example	Agent applicability notes
Datagram	Single outbound message	Network device status notification; agent sends a command to another agent and doesn't require any answer	Paging system	Used for agents that control devices, other agents, and systems
One shot	Single outbound message and single response	Network-based calculation server; agent expects an answer	File transfer	Agent A wants agent B's available resources
Query	Single outbound message; chained responses	LAN-based database query	Web search	Agent requests stock price updates at regular intervals
Asymmetric	Multiple outbound messages and responses; single session	Customer service call center	Web-based publishing	Facilitator or coordinator; agents broadcast tasks
Symmetric	Multiple outbound messages and responses; 2 dedicated channels	High-throughput reservation system	High-throughput web publishing	Intensive agent collaboration

- *Control:* Thread manager, transaction manager, resource broker, fine-grained request scheduler, coarse-grained job scheduler
- *System management:* Event notification services, accounting service, configuration manager, software installation manager, fault detector, recovery coordinator, authentication services, auditing services, encryption services, access controller.

Notice that the infrastructure-provided services needed to support agents are very similar, if not identical, to those needed for any complex application. However, an agent application may require capabilities beyond those available in the infrastructure. Agents primarily need support for messaging and remote operation invocation, tasking, scheduling, persistence, resource allocation, and deallocation. Most importantly, mobile agents need the infrastructure (or other facilities) to provide mobility or transportability, security and authentication, and support for handling erroneous conditions occurring during execution.

A good agent-execution environment is built on top of robust middleware. Indeed, some CORBA-compliant ORB providers, for example, Orbix + Isis,

TABLE 4.2 Infrastructure Application Models

Application type	Definition	Agent applicability notes
Store and forward or publish and subscribe	Directs message to recipient by pushing data or commands out (store and forward and publish) or by pulling data in (subscribe); subscription-closely related to polling	E_mail agents (S&F) software or news gatherer autoupdate (P&S)
Work flow	Directs message to recipients according to either conditions or policies	Information routing and agents in work-flow-automation applications; groupware agents, database triggered agents
Distributed transactions	Manages simultaneous messages to multiple targets under transaction control; i.e., database updates with two-phase commits	Financial transaction agents
Remote file access	Redirects file requests across a network	Network management and troubleshooting agents
Remote database access	Transmits SQL or similar requests across network to servers	Information search and gather agents
Distributed object interaction	Supports messaging between objects and agents across a network	OO version of RPC
Access to remote functions	Redirects procedure call to an application or object across a network	RPC-, ORB-based coordination agents
Distributed database management	Maintains a single local logical database across multiple physical databases	Reconciliation agents, status update agents
Database replication	Synchronizes copies of a single database	Replication error, status, and new or updated information notifiers
Display/user interface distribution	Client-server type user interface split; such as X Window	Workspace management agents

produce a fault-tolerant infrastructure for distributed applications. Using a product like this, agent developers can gain a foothold in those industries, such as banking and stock trading, not to mention the process-control industries, that rely on their applications running in the presence of node failures, telecommunications bottlenecks, and so on.

4.1.3.2 Criteria for choosing an infrastructure. So how do you make a choice between two major competing approaches to infrastructure—OLE/ActiveX, based on DCOM, and CORBA-based products such as OpenDoc and other CORBA-centric ORBs? There are several criteria, as pointed out by Roy and Ewald (1995), including the following:

- *Platform:* What platforms are supported? CORBA is currently supported on many platforms whereas OLE is only on Windows NT 4.0. But other Microsoft operating systems and others are sure to follow. However, CORBA-based platform support may be more mature. And remember, Microsoft will probably get the enhancements and updates out on their platforms first—so you may have a lag in your development efforts if you go with OLE. On the other hand, if you are developing exclusively for Microsoft platforms, then OLE is certainly a viable, and possibly preferred, choice.

- *Language:* What languages are supported? Direct support for several languages is available with CORBA-based products, including C/C++, Java, Smalltalk, Ada 95, and COBOL. OLE/DCOM supports C/C++ and Java (via Microsoft's Jakarta).

- *Cost:* The relative costs of CORBA and OLE/DCOM depend on whether you are a Microsoft-only platform developer, or will need these products on other platforms. Since OLE DCOM comes with the Microsoft tool kits, you pay no up-front cost. Costs on other platforms should be commensurate with CORBA products.

- *Complexity:* Several developers have reported in the trade press that OLE/DCOM is more complex to understand and implement than CORBA-based applications.

- *Integration capabilities:* Even if you are not a Microsoft-only developer, chances are that your agents will need to interoperate with OLE/DCOM-based entities. Several CORBA vendors supply integration with OLE, including IONA and Expersoft. With ActiveX entering the picture, ActiveX controls could integrate with other applications over the network. Both solutions have the capability of integrating with existing, or legacy, applications, but CORBA being multiplatform, more opportunities exist if your agents need to take advantage of legacy functionality. In addition, both infrastructures enable interaction with Java-based applets and servelets.

- *Robustness:* Both Microsoft and the Object Management Group are addressing the lack of robustness in their offerings. Microsoft is preparing a product called Viper, which will be a transaction coordination service for those applications requiring scalable, fault-tolerant, and secure transaction-oriented functions, like those encountered in financial transactions from ATMs. The Object Management Group is preparing the object transaction service (OTS), which will provide similar functionality within CORBA-compliant products.

- *Mobility support:* Either the infrastructure or some aspect of your development and delivery environments need to provide your agent with mobility. Most infrastructures are based on standard network protocols that, for example, allow pipes, RPCs, and other mechanisms for transporting software entities and data across a network. But your application's agent mobility requirements could be much more sophisticated and complex.

4.2 OpenDoc

"OpenDoc is the emerging open standard for compound documents, providing a cross-platform, object-oriented architecture based on interoperable software components" (Kelly-Bootle, 1995). Why is this important to developers of IAs? In order to perform software engineering a little more scientifically and with more of a traditional engineering discipline—making something complex out of simpler, already working parts—you need to find the already working parts to incorporate into your agent schema.

Kelly-Bootle (1995) states: "...at the heart of OpenDoc, is the advantage of software components, chunks of trustworthy code and data with infallible public interfaces." This is just what you want when considering the construction of simple or intelligent agents. In fact, constructing a truly intelligent agent by composition, that is, orchestrating the collaboration of many simpler agents to perform a difficult or complex task, is facilitated by architectures such as OpenDoc.

Sponsored by Component Integration Labs, the open forum for OpenDoc technology, the architecture is supported by Apple, IBM, Novell Applications Group, Adobe, IBM/Lotus, Metaware, the Object Management Group, and the X Consortium. The alliance with the X Consortium will promote further interoperability with Fresco, X's compound document architecture.*

"OpenDoc extends the 'suite' concept to include the selection of components from different suppliers and the transparent sharing of data across heterogeneous platforms" (Kelly-Bootle, 1995). An advantage OpenDoc has over many other architectures is its support of the open scripting architecture. This enables developers to customize existing interfaces, which is needed when developing agents that make use of old and new applications and data. The ability of an infrastructure to support scripting should not be overlooked. Indeed:

> Scripting technology provides the basis for *roaming agents*. These are interpreted programs that carry their own environment (and state) with them....OpenDoc documents are almost the perfect carriers for these roaming agents. This is because an OpenDoc document can store—in addition to part data—scripts that can be associated with the various parts or with the entire document. In the world of OpenDoc, a roaming agent is a document with its data and scripts. (Orfali, Harkey, and Edwards, 1996)

Figure 4.2 presents a high-level view of the layered OpenDoc architecture.

Kelly-Bootle states that OpenDoc is much more that just an applications framework: "OpenDoc itself is based on SOM (System Object Model), a language neutral, CORBA-compliant mechanism for creating cooperative objects"

---

*For more information on Component Integration Labs, contact them via their net site: cilabs @ cil.org. For more information on OpenDoc itself, contact Apple at opendoc@applelink.apple.com.

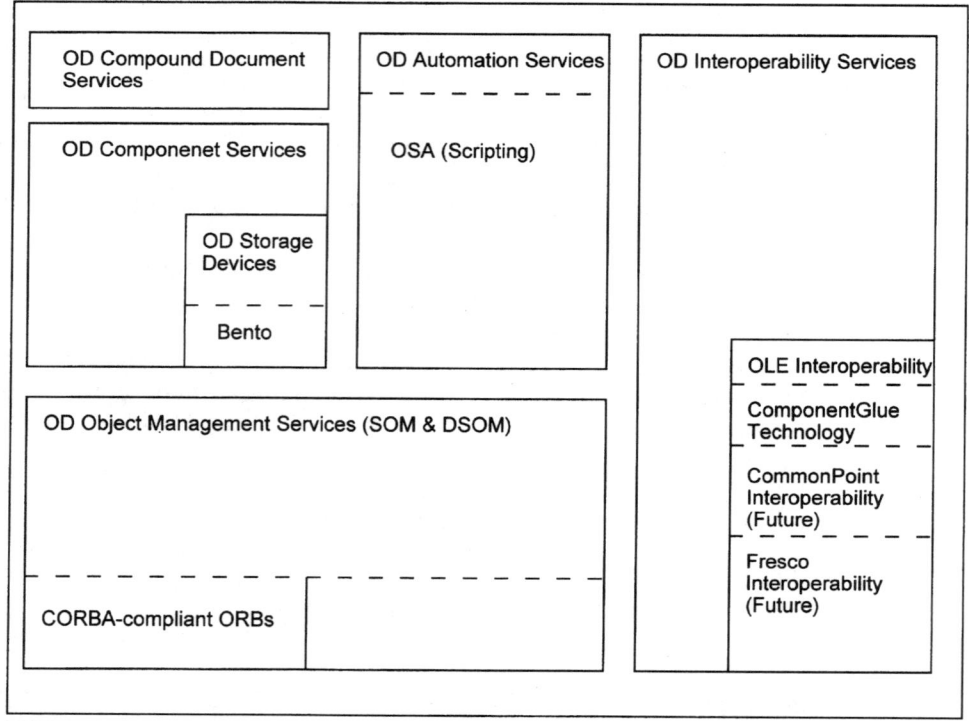

Figure 4.2 OpenDoc high-level architecture.

(Kelly-Bootle, 1995). Read that last word in the quote as "agents," and you see its potential import to your agent-development efforts. He continues: "...Open Doc documents transcend both the conventional 'pieces of paper' and the conventional 'content' stored in data files....OpenDoc can access compound documents (text, graphics, multimedia) without invoking separate applications" (Kelly-Bootle, 1995). Since agents can be OpenDoc users as well as humans, agents could take advantage of these access mechanisms.

What do documents have to do with developing IAs? Indirectly, agents can be thought of as intelligent documents. In other words, you could model an agent as a document—one that is composed of several parts reflecting the different aspects of an agent. Documents that we usually encounter are passive; but with new technologies, like embedded codes [such as Standard Graphic Markup Language (SGML) or Hypertext Markup Language (HTML)], attached scripts, and active parts, documents take on a whole new, active meaning. A "live" document could act as its own update facility. For example, say you have an on-line encyclopedia with embedded agents that monitor new developments in various domains and prepare and format the new information into "their" document. Directly, certain types of agents not necessarily made of OpenDoc parts still may have tightly integrated relationships with documents. And remember from Section 3, objects and agents and their interrelationships *are* a system. For example, filtering agents, work-flow agents, retrieval agents—all have something to do with documents. Now consider that the OpenDoc documents are really "parts" that consist of code and data, just like the objects you learned about in Section 3.

The essence of the OpenDoc document is that it is made of different kinds of parts (and the parts can contain parts...), each part having an editor, viewer, or services applet instead of a monolithic application that does everything (which is usually too much). Again, this is the essence of creating software by composition, as elaborated in the section on OOA/D/P (Analysis/Design/Programming).

OpenDoc supports Microsoft's OLE by directly enabling the incorporation of OLE objects via the ComponentGlue and OLE interoperability pieces of the architecture, that is, OpenDoc components can be mixed with OLE 2.0 objects in the same document, and information can be exchanged between OLE and OpenDoc using the clipboard, linking, and drag and drop mechanisms.

As of this writing OpenDoc implementations include IBM (OS/2, AS400, MVS, AIX, and HP-UX), Apple (MacOS and OpenDoc development Framework), and Novell (Windows 3.1, Windows 95, and Windows NT).*

On technical grounds, OpenDoc is a more OO (with true inheritance) and network-oriented architecture than OLE. A part can use its own methods for viewing and editing or for performing services, or it can ask its container part ("up" the composition or containment hierarchy) to do some or all of those things for it.

The requirement placed on infrastructures by agent developers is summarized as follows: "The distributed object infrastructure must make it easier for components to be more autonomous, self-managing and collaborative" (Orfali, Harkey, and Edwards, 1995). Intelligent components, that is, IAs, aren't going to be able to get the whole job of "intelligence" done by themselves. Even if infused with soft computing technologies, expert systems, and clever algorithms and code, other aspects of intelligence emerge as a result of:

- Interoperation à la "society of mind" (Section 2)
- Inclusion of specific AI-based technologies, including soft computing techniques (Section 3)
- Their architecture, that is, how they are put together (Section 5)
- The support systems they can rely on, that is, the infrastructure upon which they function

---

*As of this writing, Novell is planning to move all their OpenDoc activity to Canopy Technology, Inc.

For all CORBA-based (refer to Section 4.4) infrastructures, however, most of the currently implemented services concentrate on the server side. OpenDoc is an exception.*

In many cases CORBA and other interoperability standards really support peer-to-peer computing, which is just what a distributed-agent developer wants. Certainly, true peer-to-peer requires an always open pipe between the communicating agents, especially for heavily collaborative situations. You don't want the overhead of opening and closing channels while your agents try to get lightweight messages to and from. However, sending a request to an agent to do some task, closing the channel, and later having the agent reopen a channel to return a result is perfectly congruous with many types of agent-based applications.

With that in mind, let's look at what is available to you, the agent developer, on the client side with OpenDoc. "OpenDoc is a good choice because it uses CORBA as its underlying object bus" (Orfali, Harkey, and Edwards, 1995). In fact, CORBA and OpenDoc may become even more intertwined because the backers of OpenDoc (Xerox, IBM/Lotus, Novell, SunSoft, Xerox, Oracle, Apple, and others) have submitted OpenDoc to serve as the compound document portion of CORBA's common facilities reviewed earlier.

An OpenDoc part is a CORBA object with intelligent desktop behavior. If your agents take the form of OpenDoc parts, they can collaborate on the same desktop *and,* if you also use CORBA, your agents can be distributed over many desktops. In fact, some vendors provide direct interoperability between CORBA-based ORBs and OpenDoc. For example, Iona Technologies' product Orbix 2.0 for the Mac OS 7.5 provides connectivity between Orbix and OpenDoc.

In addition to the compound document architecture, OpenDoc enhances the capabilities of CORBA components (and thus your agents) as follows:

- A memory-management technique tracks references to components and automatically releases memory when a component is no longer referenced (needed).
- A way for an agent to find out what kinds of things other agents or components can do. This is similar to OLE's IUnknown call.
- An agent's properties (that is, attributes) can be updated by the user by interacting with dialogs when your agent is running.
- A rich library of about 60 classes includes methods that can be overridden by you when you create an agent as a subclass of the existing classes.

OpenDoc-based parts-cum-agents can make use of a rich set of features, as follows:

---

*To reiterate, who's the client and who's the server is a matter of point of view. It is dynamic, as an agent at one point in time can be a provider of a service and thus is a "server." Soon thereafter the same agent could request some service, and therefore becomes a "client."

- Use of the CORBA-compliant system object model (SOM)-based facilities, including:
  - Local and remote interoperability for OpenDoc-based parts (that is, agents)
  - DLL packaging
  - OO support, including multiple inheritance
  - Version control
  - Dynamic addition of developer-created methods (no recompile required)
- Bento is a container structure (remember composition and containment?) that enables:
  - Platform-neutral storage and retrieval of collections of objects
  - A persistent ID (for agents that hang around a long time)
  - Relationships (via references) to other objects and agents in other Bento structures
  - Incremental versioning and changes
- Uniform data transfer is a single set of APIs that allow you to store and move data and parts.
- Compound document architecture defines the containment hierarchy of parts containing data and other parts, and also defines the human-interface characteristics of the parts via frames. Each part has an operational aspect, usually an editor, for those parts that are literally documents such as text files, video, and graphics. As you can see, this part's architecture is analogous to OO components in that a part has data and methods (the "editors") that act on the data.
- Open scripting architecture is essentially a language allowing developers to coordinate the activities of parts. If you don't like the AppleScript-based language, OpenDoc allows you to record parts' activities and save them in your own scripting language.

If your agent needs to take advantage of multiple sources of knowledge from different "servers," then OpenDoc is the ticket. An agent *as* an OpenDoc part (or incorporating an OpenDoc part or parts) can get to information from E_mail, text files, images, databases, spreadsheets, or any other source accessible via ORBs. That agent then also has use of the native application's functions when the information needs to be edited. Section 6 has more on intelligent documents as agents.

Another important feature of OpenDoc is its compatibility with Microsoft's OLE. Methods within OpenDoc-based parts (agents) can be invoked from within an OLE container. And OLE objects can be activated from within an OpenDoc part. In fact, "OpenDoc parts are OLE compatible because they are OLE applications....So this may be a case where you can have your cake and eat it too" (Orfali, Harkey, and Edwards, 1995).

LiveObjects is OpenDoc as delivered to the end user. Apple's CyberDog, recently delivered on the Mac OS, consists of several LiveObjects, or component parts, that agents can interact with, including Mail/News, Notebook, Log, and an AppleShare server selector that integrates the desktop, the local LAN, and the Internet.

An important aspect of OpenDoc/LiveObjects and CyberDog and other components becoming available is that Java applets can run in any LiveObject-friendly document, thus opening the way for Java-based agents to interact directly with, and be based within, the OpenDoc infrastructure. For agent developers, if you use Java to develop LiveObjects, they will work on every major GUI platform, in that both OpenDoc and Java are both multiplatform environments.

4.3 Object Linking and Embedding and ActiveX

OLE, based on Microsoft's common object model (COM), can provide the capability to share resources and communicate in a quasi-OO way between applications (clients and servers) on a Microsoft desktop. Perhaps this was sufficient before the distributed computing model exploded first with the Internet/intranet paradigm. Now Microsoft has developed ActiveX™, a technology based on the distributed common object Model (DCOM). As this book is going to press, Microsoft has announced plans for fulfilling its vision of openness for ActiveX by transitioning specifications and appropriate technology to an industry standards body, called the Active Group. This group will be formed under the auspices of the Open Group, which consists of the Open Software Foundation (the DCE body) and the X/Open Company Ltd. The material that follows covers both OLE and ActiveX. For agents based on Microsoft technology we suggest you concentrate your efforts on ActiveX.*

4.3.1 OLE—an overview

OLE, although treated as an infrastructure in this book, is an integral part of Windows 95 and other Microsoft operating systems. If you develop in a Microsoft-only environment, then OLE supplies the mechanisms whereby a developer can fashion some basic interobject communications.

Within the OLE paradigm, processes on a single platform can relate to each other in a common object-oriented format, the COM. Each COM object can communicate with others via standard interfaces. Through the interfaces IUnknown and IQueryInterface and the OLE automation interface, OLE provides means for one COM object to find out what methods other COM objects support. Through OLE automation, COM objects can determine potential interfaces and alter behavior at run time. This could provide a foundation for agent interaction in the future. OLE's overall architecture is shown in Fig. 4.3.

---

*For more detailed and more up-to-date information, especially on ActiveX, navigate to http://www.activex.org/.

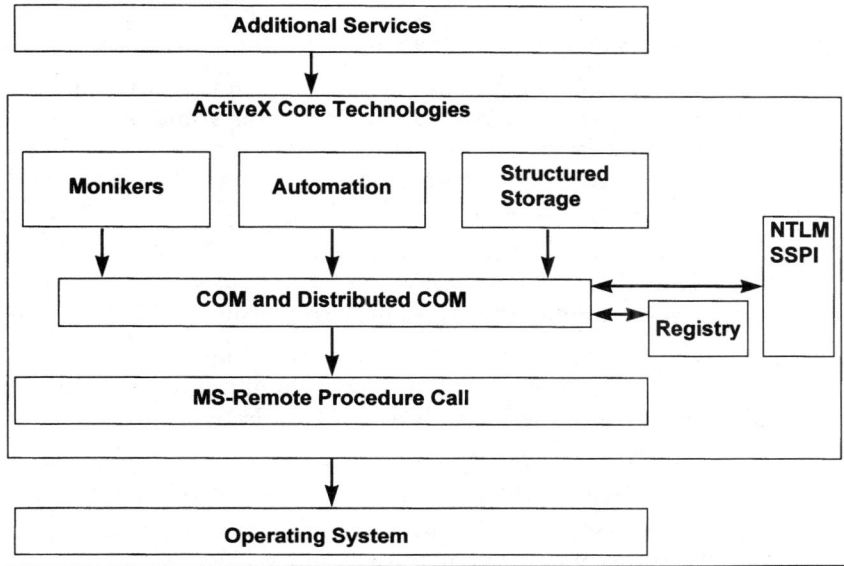

Figure 4.3 OLE architecture.

4.3.1.1 OLE common object model. OLE is based on, or built on top of, a COM. COM, and therefore OLE, is not based on a pure OO implementation. For example, OLE-based "objects," referred to as components, do not support inheritance—limiting the extensibility of any OLE-based development effort. OLE components can be aggregated or composed from other OLE components. In addition, extension of OLE components can be accomplished by building them from true OO development environments. COM:

- Defines facilities that enable OLE-based components, such as agents, to locate and communicate in a consistent fashion
- Defines the structure of objects and agents
- Provides a standard binary format for object and agent implementation

COM has the following major features:

- *COM functions:* A library of routines that you can call to perform common object procedures such as initialization and creating an instance of an object.
- *Marshaling:* The process of packaging and unpackaging an object's interface to other nodes across a network, or to other processes within a single node. When objects communicate within a node, they use what is known as a lightweight remote procedure call. This is analogous to a standard RPC, but

takes less overhead processing and can therefore be used where fine-grained (that is, small) objects need to communicate rapidly and frequently.

- *Network OLE:* Supports intercomponent communications over a network. It uses a COM extension that supports the standard RPC mechanism as encapsulated in the DCE (refer to Subsection 4.5). (Microsoft plans to integrate Network OLE with its operating systems; for example, NT 4.0 in late 1996 and Win95 in 1997).

- *Structured storage:* A set of facilities that enable objects to be hierarchically stored and retrieved in a variety of formats like files, database records, or in memory—a compound document architecture.

- *Moniker:* The name of an object, which includes where the object is located. Providing a consistent way to access an object, developers can use built-in name types, such as a file, or develop their own.

- *Uniform data transfer:* Provides underlying mechanisms that allow objects to perform many typical operations using a consistent format.

- *Version management:* Maintains the connections, or interfaces, between objects as objects are themselves updated.

4.3.1.2 Structure of OLE. OLE, built on top of COM, is composed of a set of services, categorized into the kinds of things the services relate to, as follows:

- *OLE documents,* the core of Microsoft's compound document architecture, enable work product produced in one type of application, say Excel, to be linked to, or embedded within, a document created by another type of application like Word. An OLE document is called a container. It contains the document or object created by a server application. Objects within another object can be edited (or other types of operations or methods performed) in place.

- "Application programs that create compound documents are called OLE containers, and applications that furnish objects are called OLE servers. It's possible for an application to be both an OLE container and an OLE server" (Pleas, 1996).

- *Linking* an object to a compound document means that the object actually is still physically and logically located outside of the composing document or object. If other users or agents make changes to the source object, that object's "proxy," or representation, attached to the composing document will reflect the change.

- *Embedding* an object inside a compound document means that the object is physically and logically located in that composing object. Any changes made to the original object do not get made to the embedded object, and vice versa.

- With either linked or embedded objects, the behavior they exhibit emerges from the methods that are part of the object's "creator" application.

- *OLE automation* is the way a controlling entity or application works with other objects. This is done via OLE controls.

- *OLE controls* is an automation object that can process events (from users or from other objects or applications). Other OLE controls can gain access to a particular controls interface, and establish a communications link that lets one object know when another object has changed. But there is a (current) limitation when OLE controls are applied to a typical agent application. For example, suppose you have a document or container application in which you want to embed an OLE-based agent (control) that gets and displays live information, say stock prices, from some on-line service. (An IA would even act on these data, perhaps using some pattern analysis function to determine automatically a buy or sell signal and actually place the order...but we digress.) Although an OLE-based agent, container, server, or control, would support this, it does so while that application is active, that is the front or top window. If you were to click off that top window, the feed would stop.*

- *Drag-n-drop* allows users to drag objects onto resources such as printers or into other objects on the desktop.

4.3.2 ActiveX

ActiveX is essentially the next generation of OLE and is taking the COM paradigm onto the web. While OLE has been successful at linking certain types of applications together on the Microsoft desktop, the Internet can be hostile territory for COM-based objects. As discussed previously, OLE uses COM to provide high-level application-centric services such as linking and embedding. This enables users and developers to create compound documents on the Microsoft desktop.

The idea of ActiveX is to allow the COM architecture to execute on web browsers as ActiveX controls (buttons, list boxes, pull-down menus, animated graphics, and so on). These can be embedded in Java applets, on HTML pages, and in scripting languages to liven up web sites and provide interactivity. ActiveX is a lean, stripped-down version of OLE, optimized for size and speed so it can execute in browser space.

As mentioned, ActiveX has been turned over to a standards body, the Active Group, as part of the Open Software Foundation, ostensibly to get ActiveX acknowledged as an independent technology. The newly formed Active Group will be composed of companies that have agreed to participate as a steering

---

*An application window or object that does not have the current focus, yet can still interact with other entities, is called in-place activation and is something OpenDoc supports.

committee, including (as of this writing): Adobe Systems Inc., Computer Associates International Inc., Digital Equipment Corp., Hewlett-Packard Corp., Lotus Development Corp., Microsoft Corp., NCR Corp., the Powersoft Division of Sybase Inc., SAP AG, Siemens Nixdorf InformationSysteme AG, and Software AG.

4.3.2.1 ActiveX technologies. ActiveX is a set of technologies that integrate software components in a networked environment, regardless of the language in which they were created. This integration of components enables content and software developers to create interactive applications and web sites easily. As a leading commercial object model, ActiveX has been widely adopted by corporate management information system (MIS) and independent software vendor (ISV) communities and is used by millions of application and content developers today. Hundreds of ISVs currently market more than 1000 ActiveX controls.

Microsoft is providing the following technologies (see Fig. 4.4) in the ActiveX collaborative development process:

- *COM and DCOM:* Provide the underlying object model that all ActiveX components use.
- *Microsoft remote procedure call (MS-RPC), including Microsoft Interface Definition Language (MIDL), excluding transports:* Optimized version of DCE-RPC, which provides scalability, marshaling, privacy, and support for pluggable network transports (protocols).
- *Standard security provider interface (SSPI):* Allows objects to be invoked securely, with user authentication.

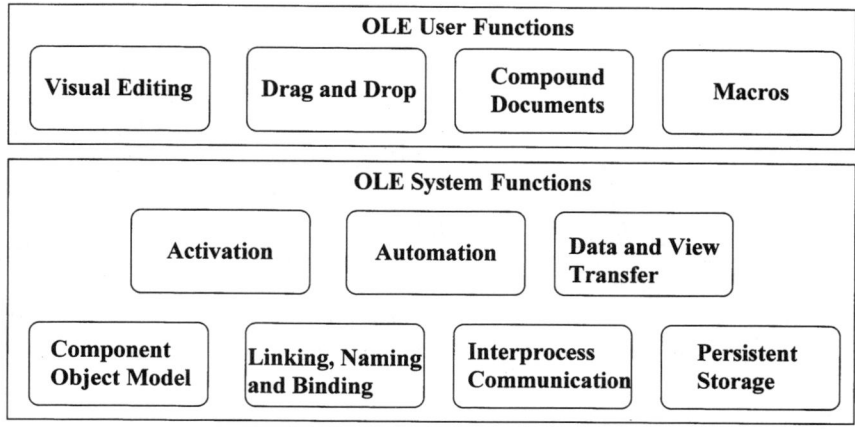

Figure 4.4 ActiveX architecture.

- *Structured storage:* Provides a structured file format, which can be implemented on multiple operating systems.
- *Registry:* Provides a database of ActiveX objects on a given system.
- *Moniker:* Allows objects to be invoked and communication between objects to be maintained asynchronously.
- *Automation:* Allows method invocation and "programmability" of objects.

ActiveX supports key industry and de facto marketplace standards, including HTML, TCP/IP, Java, COM, and others. ActiveX includes both client and server technologies.

- *ActiveX controls* are the interactive objects in a web page that provide interactive and user-controllable functions and hence enliven the experience of a web site.
- *ActiveX documents* enable users to view non-HTML documents, such as Microsoft Excel or Word files, through a web browser.
- *Active scripting* controls the integrated behavior of several ActiveX controls and/or Java applets from the browser or server.
- *Java™ VM* is the code that enables any ActiveX-supported browser such as Internet Explorer 3.0 to run Java applets and to integrate Java applets with ActiveX controls.
- *ActiveX server framework* provides a number of web server–based functions, such as security, database access, and others.

4.3.2.2 ActiveX and Java. ActiveX provides a standard mechanism to extend any programming language, including Java. ActiveX extends the capabilities of the Java language by allowing Java developers to integrate their applets with the richness of ActiveX. ActiveX ties Java applets together with objects created in other languages, so that Java programmers can link to ActiveX controls directly from their Java programs.

By the same token, objects written in other programming languages from multiple vendors can link to Java applets. ActiveX is the glue that ties them all together, delivering the most powerful web technologies in an open, integrated platform. By providing a common way to extend and link programming languages, including Java, ActiveX maximizes developers' resources for interactive web development.*

4.3.3 Developing agents based on OLE and ActiveX

How can you use OLE and ActiveX as an underpinning for agents? If you make your agent in the form of an OLE control, which is eventually compiled into a

---

*For more information on ActiveX and Fig. 4.4, navigate your browser to http://www.activex.org/.

DLL, other applications that can make use of OLE controls can use your agent. Of course, you'll have to provide user-interface code that another application can use to display any data or behavior that your agent produces.

Your OLE-based agent would have, as do all OLE-based components, controls, or container applications, a messaging interface. This interface includes a receptor, or sink, located within your OLE agent, which receives instructions or events from its containing application. The agent-container object communications links are set up dynamically and can include a protocol or language negotiation.

As Linthicum (1986) states: "To ease development, OLE provides standard events (called stock events) for each OLE Control." These stock events emanate from the root control class called COleControl. Other key OLE-support objects found in Microsoft's Foundation Class Library—part of the Microsoft SDK (Software Development Kit)—include COleDocument, COleDataSource, COleDropSource, COleDropTarget, COleMessageFilter, and COleDataSource.

Many developers have voiced concern about the overly complex nature of OLE and its intertwining within the different Microsoft operating systems. A possible, though not entirely satisfactory, solution involves using a Microsoft App Wizard and an appropriate C++-based template. OLE controls and other parts can be generated from the template without having to know all the complex implementation details. But to really make a reasonably sophisticated agent-based application, you probably need to plumb the intricate depths of OLE's interface.

As Microsoft and the Active Group deliver a distributed OLE, possibly via ActiveX or Network OLE, and if they address deficiencies related to a lack of OO support, then we believe these technologies could be suitable for some aspect of a full IA infrastructure. Inheritance is important not only for ease of extension, but in its ability to hide proprietary code that you may not want others to see. Also, because OLE is proprietary and, so far, limited to Microsoft operating systems, your interoperability options are limited. Transitioning ActiveX to the Open Software Foundation may herald true openness and independence for this technology—we'll wait and see.

In Cairo, the OO NT operating system forthcoming from Microsoft, all services are implemented as OLE2 objects, linked together by COM. And Cairo's OLE is to be distributed across machines. Microsoft has stated that they will not make OLE, even in its newest Cairo-based incarnation, CORBA-compliant. This means that Network OLE will not be interoperable with CORBA-based ORBs (unless some third party provides a bridge service of some sort). This will limit its usefulness in the increasingly heterogeneous universe that agents will inhabit.

In addition, as Dunbar (1996) remarks, "OLE is a collaborative API; it requires that two developers work together on specifications." This is opposed to CORBA-based interoperability standards such as OpenDoc, which limit the amount of developer collaboration required to enable agents from different developers to work together. Adding to the view of OLE's current inadequacy from a good agent infrastructure, is the status of support from the software industry. Although many developers are supporting OLE in some fashion, especially to garner the Windows 95 logo for their products, the actuality is a bit different. As Pleas (1996) concludes: "Clearly software vendors continue to add support for OLE. But they're doing it in inconsistent ways, preventing true integration and communication from becoming a reality."

4.4 The Common Object Request Broker Architecture

Enabling smart, distributed software components, such as IAs, to interoperate and collaborate requires efficient provision of a number of important services. CORBA 2.0, Novell's AppWare distributed bus, and the COM from the combination of Microsoft and Digital are all aimed at fulfilling this requirement. Since distributed objects, and hence most agents, are binary components that must be accessible to other "clients," namely, agents, with only a name and the appropriate message signature (interface), how does the message get through to the right component?

The answer is: "...a common interconnection bus that hosts client components, core services needed by all components (including naming, persistence, events and transactions) and common facilities for component collaborations" (Orfali and Harkey, 1995). Since agent developers need to either use an existing bus or make their own, we'll discuss an important existing approach. You'll notice the difference in capabilities between existing standard approaches, which we discuss in this section, and some of the other, more agent-oriented approaches covered as architectures in Section 5. For example, the facilitator discussed in Section 5 provides many functions of an infrastructure broker, and then some.

So you have to make up your mind whether to use commercially widespread "standard" mechanisms and possibly work around their weaknesses by building facilitatorlike functions on top of (or along side) them, or fashion the specific functions or services required by your agents from scratch. Of course, what you choose depends a lot on how complex your proposed agent-based system will be. For example:

- Does it span multiple networks and operating systems?
- Will it be composed of agents created with different languages?
- Will they need to interoperate with conventional software based on the "standards"?

Let's take a summary look at the Object Management Group's CORBA. Vendors of CORBA-compliant infrastructure implementations include: "DEC,

HP, ICL Sunsoft, IBM, Expersoft, IONA Technologies and Postmodern Computing" (Brando, 1996, p. 56).*

CORBA is both an architecture and a specification for producing CORBA-compliant ORBs. A broker is analogous to the facilitator presented in Section 5 as part of an agent-specific architecture. Brokers mediate communications between objects or agents and provide services that facilitate their collaboration. Sounds like a great fit with your needs? Let's take a closer look.

The current incarnation of CORBA, version 2.0 (as of this writing), aims at providing an object (and thus agent) intercommunications bus. The main elements of CORBA are:

- *Object request broker:* The software "bus" that interconnects objects (and therefore your agents). It includes an interface definition language (IDL) in which you specify your agents' capabilities. CORBA 2.0 also specifies how multiple buses from different vendors can "connect" to enable interoperation.

- *Object services:* Components that provide services such as event notification, naming, persistence, transactions, concurrency control, life-cycle management, relationships, externalization, query, licensing, properties, security, and time. A CORBA-compliant ORB would be able to provide these services to an object or agent.

- *Common facilities:* Components that define how application objects and agents interact. The components are defined using IDL. The kinds of services that the components are intended to address are information management (similar to OLE and OpenDoc's compound document storage and data interchange), user interface (similar to OLE and OpenDoc's in-place editing), systems management (that is, the object or agent life-cycle activities such as install, configure, operate, maintain), and task management (rules, scripting services, transactions, work flow).

In addition, the common facilities part of the CORBA architecture intends to provide another dimension of support—that of the vertical application domain. These IDL-defined interfaces will cover domains such as medical, financial, manufacturing, retailing, and other major domains. As these interfaces develop and become widespread, your agents could have a ready-made playing field, so to speak.

In conjunction with common facilities, CORBA has defined standards for specific object services. CORBA's implementation of these services will allow you to use them via "mix-in" single or multiple inheritance[†] with your own

---

*For more detailed information on CORBA, navigate to http://www.acl.lanl.gov/CORBA/.

[†]A mix-in is a class that augments the behavior of another class. Mix-in inheritance allows the use of aspects from one or several separate classes within the definition of another class—mixing-in attributes or methods, as required. The exact mechanisms are usually language dependent and are beyond the scope of this book. Refer to Booch (1994) for further details.

agents or objects. In this way you'll also gain access to some of the important root object services CORBA has to offer, such as persistence and transactional facilities. The following object services are standardized and ready for your agents to use:

- *Life-cycle services* define how agents can be created, copied, moved from place to place, and deleted.
- *Persistence services* provide a way for you to store your agents in a variety of containers, such as databases, files, or as components of other objects.
- *Naming services* will allow your agent to locate another agent by name using a variety of naming contexts, including International Standards Organization's (ISO) X.500, Open Software Foundation's DCE, and Sun's network information services (NIS).
- *Event services* let your agent register an interest in specific kinds of events. An event channel distributes events within the CORBA "bus."
- *Concurrency services* provide a lock manager that allows your agent to obtain a lock on a transaction or thread basis.
- *Transaction services* provide a two-phase commit coordination between agents.
- *Relationship services* allow agents to create associations among themselves (for example, containment relationship). These services also provide a way for your agents to traverse the resulting relationship graphs.
- *Externalization services* let your agent stream data into or out of a software entity, including itself.
- *Application objects* are the developer-made, and therefore end-user-visible, objects and agents. They must also be defined using IDL. Note that certain CORBA ORB packages like Orbix enable IDL–IDL translations and bridges, somewhat simplifying communications code between agents built with disparate languages. (Section 7 contains a couple of examples.)

Remember our discussion of encapsulation or "wrappers"? The Object Management Group enables these services by providing IDL "wrappers" for existing standard mechanisms that already provide these services. "The ambitious goal of CORBA is to turn everything into nails and give everyone a hammer. The nails are the IDLized services, and the hammer is the IDL interface to these services" (Orfali and Harkey, 1995). The essential parts of the CORBA 2.0 ORB with an agent "twist" are shown in Fig. 4.5.

Let's see how you would get your agents talking to each other using these CORBA components illustrated in Fig. 4.5. First, you would compile each of your agents' interfaces into *client IDL stubs* using an IDL compiler. These stubs tell other agents how to invoke an agent's services. This is a static definition of your agents' capabilities. When an agent makes a request (sends a message) to

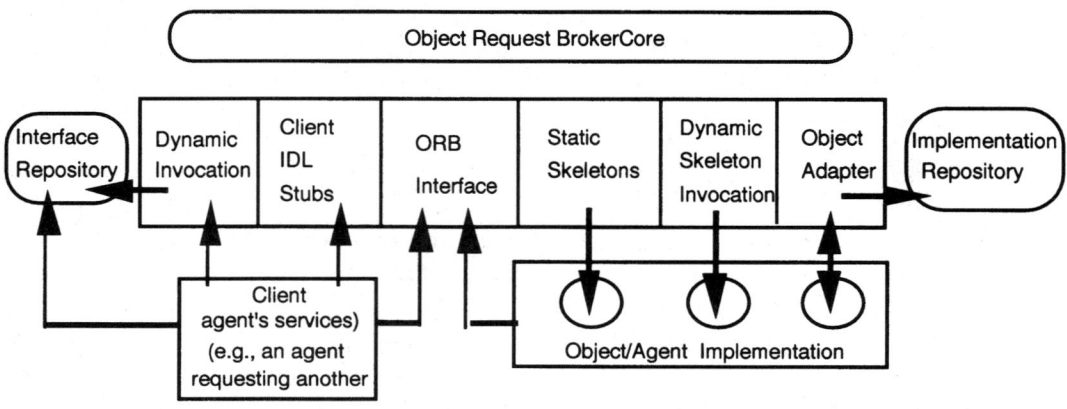

Figure 4.5 CORBA 2.0 architecture. (*Modified from Orfali and Harkey, 1995.*)

another agent, the CORBA server routes these requests to particular *agent implementations* by using the set of stubs defined within *static skeletons* or, at run time, using the *dynamic skeleton interface.*

If you want to use *dynamic invocation APIs,* you can arrange to have your agents find a particular service while it is running, find the definition, and call that service and get a reply.

The *interface repository* has APIs that let you modify the definitions of all registered agent interfaces, their methods, and any parameters (that is, the method signatures).

As mentioned earlier, CORBA, and thus the ORB itself, has various utility services which are available to your agents. They simply call the APIs present in the *ORB interface.* For example, if you need to convert between an agent's name as you know it and an internal "object" reference pointing to that agent, the ORB interface has a utility that converts between a string and an object reference.

As Orfali and Harkey describe it, the way things work is pretty straightforward. When your agent sends a message to another agent (or requests a service from any other registered software entity), the request goes to the ORB server. "Servers can't tell the difference between static and dynamic invocations. The same message semantics apply in both cases. The ORB locates an object [the agent able to provide the service], transmits the parameters, and transfers control to the object implementation through the server IDL stub (also called a skeleton)" (Orfali and Harkey, 1995).

The *object adapter* is the entity that actually processes the incoming messages. In this way it is very similar to the functions envisioned by some of the "facilitator"-like objects in agent-based architectures discussed in Sec. 5. For any object or agent not already up and running, the object adapter will instantiate it, providing the run-time environment and assigning a unique object reference to them. The class from which an object or agent is instantiated is listed in the *implementation repository*. (Remember that in most OO implementations, the class has the methods that will be executed, whereas the instances have all necessary "instance" variable values and a pointer back to its class.

If the invocation of a service is based on a statically known provider, the static skeleton provides the information necessary to get to the right object or method. If the invocation is dynamic, the dynamic skeleton binds the information object or agent message at run time by inspecting the incoming message. The dynamic form is used when you have interpreters and scripting-language-based objects (such as, Smalltalk, or Java-, or Telescript-based agents).

4.4.1 Interoperability between ORBs and agents over the Internet

One important issue that you face is that of interoperability within your agent system, which can be developed using a single infrastructure, say, a CORBA-based ORB, from a vendor whose implementation crosses many platforms. But as you require your agents to acquire and use resources of another developer's agents or, for example, a Cyc-based db implemented on a platform where a different ORB is resident, you will require interoperability between ORBs. As Brando points out:

> Interoperability between ORBs (i.e., between ORB vendors) is becoming a reality. Vendors are beginning to ship ORBs that support the Internet Inter-ORB Protocol (IIOP))....Thus we expect widespread interoperability among ORBs via IIOP very quickly, even with ORBs that use DCE for intra-ORB communication, although interoperation between DCE-based ORBs may be provided via DCE-specific protocols specified by the OMG as well. (Brando, 1996)

If you think you are going to use your agents across the Internet, Orfali says that the CORBA 2.0 Internet interoperable ORB protocol (IIOP) was specified just for that purpose. Figure 4.6 shows how this facility can be used for interagent communications over the Internet.

Essentially, IIOP is just "...a scary name for TCP/IP with some CORBA-defined message exchanges that serve as a common backbone protocol. The messaging part of the backbone supports common data representations for all OMG IDL types, interoperable object references, and common message formats and semantics optimized for ORB exchanges" (Orfali and Harkey, 1995).

Actually, CORBA over the Internet using IIOP is just a first-level, simple implementation. Several industry players are rushing to deliver CORBA-based products, underscoring the move to CORBA as the standard for distributed objects and agents over the net. The Object Management Group is trying to make IIOP the basis for replacement of the Hypertext Transfer Protocol (HTTP)—the protocol used over the web. The Object Management Group is also working on merging the underlying architecture of CORBA and IIOP—the

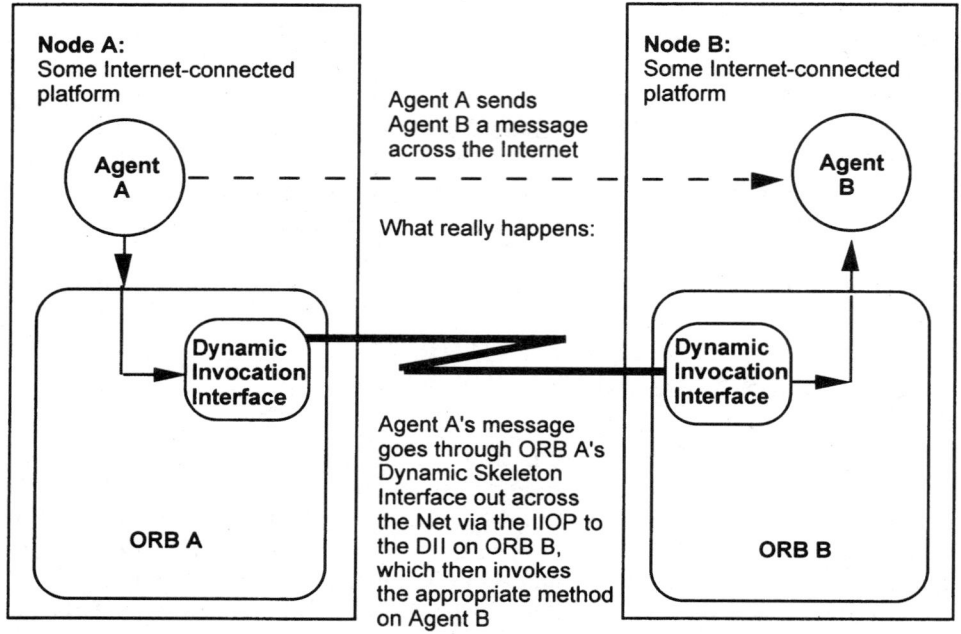

Figure 4.6 Agents over Internet using CORBA 2.0. (*Modified from Orfali and Harkey, 1995.*)

object management architecture (OMA)—with other Internet protocols and standards.

An interlanguage unification architecture being worked on by the Web Consortium aims to do just that. Netscape is also firmly in the CORBA camp with its Java-based IIOP-compliant Open Network Environment toolset. NeXT's (now Apple's) OpenStep WebObjects and IONA's Orbix are being integrated for those of you developing on NeXT's platform. Sun and IBM are also integrating product to enable Java applets to work with CORBA-based corporate applications.

DCE can be used as an IIOP-compliant implementation that provides a more robust environment. You would get services like Keberos security, wider-spanning object directories, distributed time services (for use when your agents have to do more critical scheduling functions, for instance), and an RPC implementation that includes authentication (useful if your agent is dealing with very sensitive information).

Another important aspect of CORBA that is directly related to agents is the task management common facility (TMCF), a part of the CORBA common facilities discussed earlier. That TMCF includes an "agenting infrastructure" that will help you organize and manage roaming, or mobile, agents. Orfali and coworkers summarize as follows:

The CORBA agenting infrastructure will provide capabilities for agents to advertise their services and for subscribers to register for these services. *Broker* agents know how to recruit other agents to create task forces. The broker knows how to delegate or *forward* work to agents it has recruited for the job. In addition, mobile agents will have *home agencies*. An agency provides a script engine, an execution environment, and a registration database. Mobile agents will register themselves with their home agencies and have *visitor privileges* with other agencies. Because agencies are active entities, they may be grouped in different ways. For example, several agencies may be part of a group where one agency plays the role of a *receptionist*. So it is possible to create a virtual network of agencies. The CORBA agenting infrastructure will define open protocols that let agencies communicate with agents, their brokers, and other agencies. (Orfali, Harkey, and Edwards, 1995)

4.5 The Distributed Computing Environment

The DCE from the Open Software Foundation is an infrastructure that you may want to consider in support of your agent application. Brando (1996, p. 53) describes this infrastructure as follows: "DCE provides a set of C programming language interfaces that support the construction and integration of client/server [you can read as peer-to-peer] applications in a heterogeneous distributed computing environment." Vendors of DCE-compliant infrastructure implementations include: AT&T, DEC, Hitachi, HP, SCO, Siemens Nixdorf, Tandem, Bull, Gradient Technologies, IBM, Pyramid, SGI, Stratus, and Transarc.* Figure 4.7 shows the DCE architecture.

DCE's components and their relationship to your agents are as follows:

- *Directory services* support the administration of DCE domain, or cells. Your agents will have a name in a particular application domain, which will reside within one or more DCE cells. Directory services provide lookup of your agent's name, or ID, within one cell or across cells.

- *Distributed time services* synchronize time clocks on all nodes within a cell and across cells. This is helpful when your agents depend on a single time base across a network, such as when scheduling activities or reporting time-stamped events.

- *Security services* support your agents' use of authentication, ensuring that the message an agent receives is the one sent by another agent. These services also provide secure data access between agents.

- *RPC* is a mechanism that, like CORBA, enables agents to send requests for service messages to other agents, independent of location. The location-finding details, as well as formats, representations, and networking details, are considerably hidden from the agent developer. Bernstein (1996) describes the RPC mechanism as follows:

---

*For more detailed information on DCE navigate to: http://www.osf.org/dce/index.html.

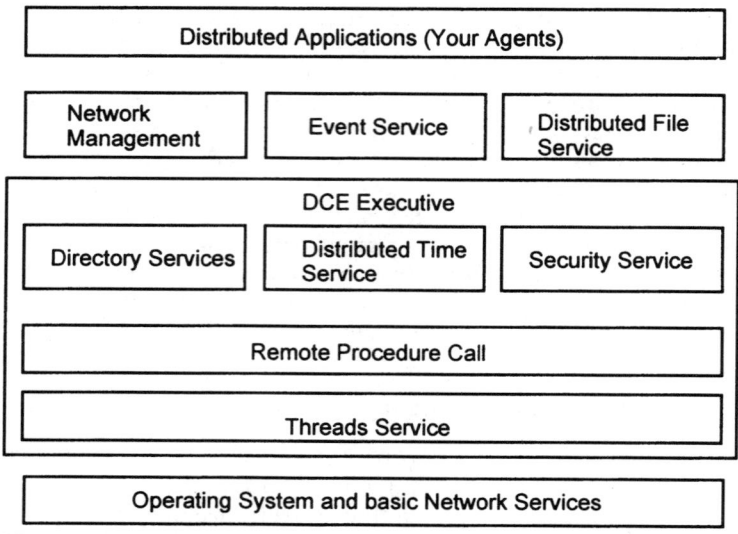

Figure 4.7 Distributed computing environment architecture. (*Brando, 1996.*)

An implementation of DCE RPC includes a compiler that translates an interface definition into a client stub, which marshals a procedure call and its parameters into a packet, and a server stub, which unmarshals the packet into a local server call. The client stub can marshal parameters from a language and machine representation different from the server's stub, thereby enabling interoperation. An RPC implementation also includes a run-time library, which implements the protocol for message exchanges on a variety of network transports, enabling interoperations at that level.

One aspect of RPCs that you should consider is that the mechanism is usually used for large-grained communications between components. Small messages require a lot of overhead using RPC. There are commercial product add-ons to DCE that incorporate a so-called mini-RPC mechanism. There are also architectures that directly support fine-grained object-to-object communications inherently, such as CORBA.

- *A POSIX-based threads package* allows your agents to create and manage independent threads of execution.
- *Distributed file services (DFS)* enable your agents to consider a file as a single logical entity, even if located in physically different nodes or cells across a network. File replications and fault-tolerant recovery support are also available through DFS services.
- *Network management* provides your agents with network status information using industry-standard protocols such as Common Mail Interface

Protocol (CMIP) and Simple Network Management Protocol (SNMP). This is useful for agents targeted toward network troubleshooting.

- *Event services,* although not available from every DCE supplier, allow your agents to generate, forward, filter, and log events.

One thing to realize about DCE is that, as opposed to CORBA, for example, it is not expressly designed for the OO paradigm. Since we are maintaining the "goodness" of OO agent design and implementation in this book, we suggest you carefully weigh DCE's appropriateness to your development effort.

That being the case, your use of DCE would involve a different sort of architectural and developmental approach than those presented in this book. You need to pay more attention to how you partition a procedural program into data and functions into servers, and locate those servers onto the appropriate hosts. Calls to the server's services are via RPCs, embedded, for example, in your C-based "main" procedure.

DCE does provide some OO capabilities and offers capabilities specific to the DCE procedural programming paradigm as follows:

- A DCE-based agent can bind to other agents at run time and make the appropriate RPCs to invoke their services. But you must ensure that these interfaces are defined rigidly at compile time.
- An agent as part of a DCE server can have universal unique identifiers (UUIDs) that point to different agent resources. Therefore a service-requesting agent can specify to a finer granularity exactly what service to invoke.
- Pseudosupport for polymorphism and abstraction via the association of an object-type UUID with a set of RPC handlers. An agent invoking a service can do so with an object type, thus mimicking conventional class-based object-to-object communications.
- DCE supports certain specific data types that are already defined, such as pipes, contexts, and pointers. CORBA supports arbitrary data types.
- DCE's name space is flat, reflecting a nonhierarchical, non-class-based paradigm. Interface inheritance is likewise not supported, as it is in an OO infrastructure.
- DCE has no run-time repository of IDLs.
- The DCE-using developer must anticipate and link any agent's RPC stubs to the server agent. There are no dynamic invocations to arbitrary agents as there are with CORBA.

As opposed to OO development, languages and environments that support CORBA or CORBA-based infrastructures with added capabilities such as OpenDoc, C-based DCE (like OLE) does not support the following important OO idioms:

- Abstraction (agent classes)
- Inheritance (and thus an agent class or message–passing or method-invocation hierarchy)
- Polymorphism
- Dynamic or late binding
- Interpretive-like developer-environment interactions

As to choosing between DCE and a CORBA-based infrastructure, Brando (1996) maintains: "CORBA-based development, with or without additional features and components from CORBA-based infrastructures such as OpenDoc, covers a broad spectrum of application support services and will provide the developer with a richer set of capabilities on which to build…" your agent system.

But in DCE's favor (currently), certain services like security and time are provided in a consistent fashion, whereas, as of this writing, such services will not be available in CORBA until sometime in 1997. If you need these services, you can use a mixture of DCE and CORBA facilities, albeit at higher cost, both mental and financial. Two vendors do provide somewhat OO DCE packages: DEC makes DCE++ and HP has OO DCE. If you feel your agent system will require these services, or others that are not in any vendor's CORBA implementation, then by all means seek out these OO DCE varieties. As Brando (1996) states: "This will position you to make the transition to CORBA more easily in the future."

4.6 Other Infrastructures

The next few subsections address several areas of potential interest to the agent developer. Each of these areas deserves a book or books on its own, and there are many sources of information on these topics already available or being developed. Many of the products and developments are quite recent. We include this catch-all section to ensure that we cover most of the territory involved in agent development. The information in this subsection is intended to serve as pointers to products and developments that you may want to research further, if your interests or agent domain warrants.

4.6.1 Networked objects

Sun Microsystems has an implementation of CORBA called Networked Objects (NEO). Some developers believe NEO is better than other CORBA implementations. As Horiuchi (1996) recently explained: "It [NEO] has more tools to help you code faster, including the ability to graphically link two objects and give you an association." NEO has close ties to Sun's OpenStep (a version of NeXT's

NextStep* environment) development environment. NEO's power inheres in that it provides both the run-time network environment and the development environment for OO applications.

Although it has just been released (as of this writing), if you develop on Sun's Solaris platforms, the NEO ORB can provide communications between Solaris-resident and network-resident agents. NEO's broker supports both C- and C++-based agents, that is, its IDL compiler accepts those languages used as agent specifications.

4.6.2 Portable distributed objects*

NeXT's implementation of a distributed object environment is called portable distributed objects (PDO). PDO is based on the Objective C language. Using NextStep and PDO, you assemble components, providing relationship information as you go. A PDO-based agent would communicate using the standard Objective C messaging facilities, which are similar to what an ORB would do.

In addition, PDO-based agents have the ability to communicate with other CORBA-compliant ORBs, as well as OLE 2–based components or agents. Another interesting point concerning NextStep—it is available as OpenStep on other platforms such as certain UNIX variants, Windows 95, and Windows NT. Even though NextStep is the only major development environment on the NeXT platform, its OpenStep incarnation on other platforms gives it an interesting position as the ORB wars continue. Also, with the recent inclusion of NeXT into Apple Computer the "next" OS coming out of Apple will be based on many NextStep/PDO features.

4.6.3 Publish/Subscribe and AppleEvents

The Publish/Subscribe mechanism is part of the Mac System 7 operating system's interapplication communications (IAC) capabilities. This is what we mean again by that fine line between infrastructures per se and services provided via other means—in this case as an operating system service. This service allows Mac applications and, therefore, agents to share data and send messages to each other. By conforming to the grammar, or message format, of AppleEvents, a Mac-based agent can depend on IAC to deliver the message.

Many Mac applications have registered the messages that their applications can understand, both general messages that all applications should respond to (called required and core AppleEvents), as well as domain-specific messages (called functional-area and custom AppleEvents).

A System 7–based agent could make use of such facilities by, for example, defining and registering the agent's methods as AppleEvents that other appli-

---

*Note: As we go to print, Apple Computer Inc. has incorporated NeXT Software, Inc. into its enterprise software organization.

cations could understand. This would work well in scenarios where you could collaborate with the developers of vertical-market applications to enable their application to become more intelligent via your agent. For agent developers it must be noted that your agents must be System 7 friendly to use IAC.

4.6.4 Operating systems

All agents, in spite of form or function, must execute within, and rely primarily on, an environment which (hopefully) provides the support they need to accomplish their goals. The executable agent itself is only a small part of the picture. Of much greater importance are the development tool set and the agent's execution environment and operating system.

While an exhaustive study of operating systems and execution environments is beyond the scope of this book, the fundamentals of typical host environments must be addressed. A good agent execution environment will offer:

- Multitasking and multiprocessing support
- Resource management (memory, processing cycles)
- Mechanisms for distributed access
- Error handling
- User-interface support based on the major standards (such as Win32, Mac, X-Window)

Each of the major operating systems has strengths and weaknesses in these areas. More information on each of these considerations is given in Section 6. For specifics about a particular operating system, contact its vendor.

Something to keep in mind is the relative preponderance or market share of the various operating systems. As of this writing, the vast majority (about 83%) of Internet servers are UNIX boxes, with Macintoshes at about 9% and PCs running Windows, NT, Netware, OS/2, and so on, making up the remaining approximately 8%.

Currently there is no dominant force in the web server market. Be aware, however, that the use of Windows NT is accelerating, as that operating system beefs up its capabilities, and InternetWare from Novell will also be a big factor in the future of intranets due to NetWare's huge established user base.

With the preponderance of UNIX servers, UNIX-based tools such a Perl and TCL have definite appeal. Because Perl runs interpretively and is a relatively powerful and easy to use scripting language, it is used widely for CGI programming on many web servers. Since C is compiled, it offers better execution performance and is still the most widely used UNIX server language. Of course, Java is threatening to replace Perl, C, and all other languages used on web servers since essentially all types of server will run Java.

Of course, the freeware-based Internet is not where the commercial action will be, as Netscape's strategic move from Internet browsers to intranet servers attests. Due to the power and ubiquity of the new universal client, for example Netscape Navigator, corporations both large and small are adding intranets to their information-management strategies. Indeed, intranets threaten to displace other information-sharing technologies in the corporate data center. Intranets will be built not only on UNIX, but using InternetWare and other operating systems working together. In the future, the major projects will be built and the big money will be made supporting corporate intranets. This is where agents could be cost-effective as buyers, sellers, and liaisons between these walled cybercities.

4.6.5 A potpourri of agent environments

With the acceleration of network computing, the often repeated observation that "the network *is* the computer" is becoming more and more of a reality. In addition, robust computing capabilities are finding their way into more and more nontraditional and so-called embedded markets. As personal digital assistants, set-top TV/computer combinations, and smart appliances proliferate, agents will have whole new worlds to conquer. A small number of noninclusive examples of those worlds follows.

As an agent developer, pay particular attention to those developments that have access to, or can run on, many operating systems by virtue of a platform-independent language-execution engine. Examples include Java, Smalltalk, and Lucent's Inferno/Limbo/Lingo-based environments. Also note the processors being used in these environments and what operating systems and infrastructures and networking capabilities are addressed. Research the tools available to do the kind of agent-development work you envision.

4.6.5.1 Database management systems. Often DBMSs are part of an infrastructure in which an agent exists. Agents will need to deal with protocols for database retrieval, validation, and verification, and work with other applications and agents defined to access the database. Agents will need to deal with OO and relational databases. OO DBMSs have facilities that serve as CORBA object adapters. Some more information on DBMSs and agents is given in Section 7.

4.6.5.2 Collaborative computing and groupware. IBM/Lotus Notes is an environment that combines several capabilities. It offers the advantages of a distributed database based on a document model. It provides robust security, distributed document and data support, custom applications scripting language, centralized, hierarchical management with extensive networking capabilities, and interoperation with other applications. The central concept behind notes is a particular implementation of paradigm of user computing usually referred to as "groupware."

Groupware enables the collaboration by many distributed users on problems and projects. Notes implementation enables the collaboration on documents within a shared workspace. Notes 4.0 has a rudimentary facility for creating and executing agents.

Another avenue to explore in the realm of collaborative or group computing is the use of the Java agent template (JAT). Sections 7 and 8 contain more information on JAT.

4.6.5.3 Expert system environments and agents.
Several expert system development environments exist that have the potential to incorporate agents, or interact with agents. Sections 2 and 3 discussed existing stand-alone expert systems, including Cyc. In addition systems like G2 from Gensym and Elements Environment from Neuron Data Inc. are also worth looking into.

Elements Environment is a combination of an expert system tool (a rule-based generator), a cross-platform (several common platforms are supported) GUI builder, and, most pertinent to agent developers, a set of intelligent software routers that route information to multiple points in a distributed system. These routing agents function as go-betweens shuttling data among modules that make up an expert system application.

G2 is an OO framework for building intelligent real-time systems. It is used in several process and production-control domains. Like Smalltalk and Java, G2 is an interpretive environment, which enables good programmer productivity and cross-platform mobility.

G2 is one environment that explicitly represents connections, as discussed in Section 3. Agents making use of G2 connection objects can connect to other agents and, perhaps most importantly, can reason in terms of the agents and objects they are connected to. For example, cause and effect analysis can be carried out by an agent if it can navigate an event or message from point to point as it passes through a system, observing its effects. This is very effective for propagating alarms, diagnosing faults, and the parallel delegation of tasks by a coordinator or facilitator agent to lower-level agents across a distributed network. G2 also enables soft computing approaches to be encapsulated within its objects. For example, Gensym's NeuroOnline neural networks can be embedded into G2 objects, and are thus accessible by G2-based agents.

Section 5

Agent Architectures

5.1 Introduction

"Software architecture describes the high-level configuration of a system's constituent components and the connections that coordinate the activities among those components" (Abowd, Engelsma, Guadagno, and Okon, 1996). In this book the architecture is the way agents are put together, the form or structure of their relationships and interactions. This is in contrast to the architecture of a complete system, which usually includes the relevant parts of an operating system, any infrastructure components (see Section 4), the user interface, "glue," databases, and so forth.

In this section we present several approaches to putting together your agents and other components—in other words—architectures. Many of the architectures discussed here originate from academic researchers in agent technologies, and we show you some (but by no means all) of the leading-edge thinking in this area. As you read about these different architectural approaches, you should find an architecture, or pieces or ideas from several different architectures, that can be used to help you develop your agent system. We look at the basic structures and summarize the important ideas contained within the architectures, and we find out what issues need to be addressed when putting systems of agents together. In addition, you will see that certain facets of these architectures can be used in other contexts or applied to domains different from the examples presented with the architecture.

Since this presentation is a summary of various architectures, in order to make a more informed decision about their potential usefulness to *your* agent development efforts, please obtain and read the referenced material closely, and contact the respective authors for more details (and maybe some demos and code). Of course, to reiterate what we've said elsewhere, the information presented in this book is but a small percentage of what is available on these

152   Section Five

subjects. Our hope is to stimulate the mental juices and point you in some fruitful directions for further investigation.

5.1.1 Architecture and infrastructure—a fine line

The utility of federations or "societies" of agents depends in large part on the underlying paradigms implicit in the architecture of the federation. Figure 5.1 shows the relationships between major system components. It is very similar to Fig. 4.1, the difference being that the *agents and agent architecture* component is shaded in Fig. 5.1, indicating that this component is the thrust of discussion in this section.

Such architecture, in turn, depends heavily on the available computing and networking resources assumed to be part of the infrastructure. Savvy agent developers like you should rely on as much infrastructural support as is practical and relevant to the specific agent application. Section 4 discussed infrastructures in general, as well as those major mechanisms, components, and

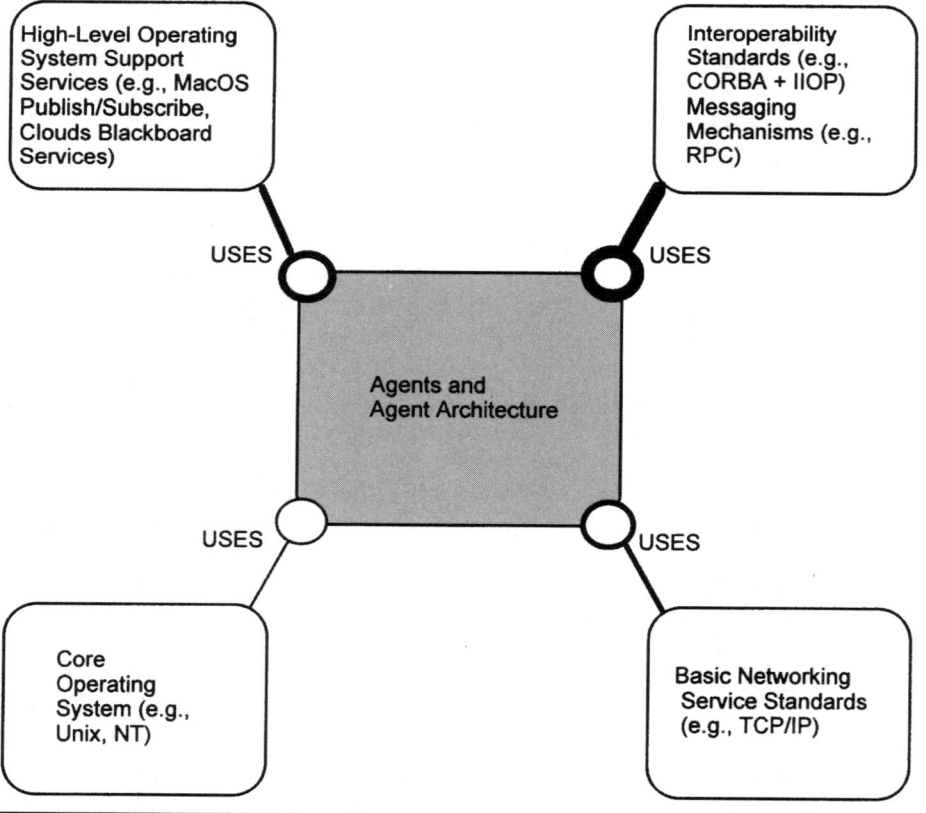

Figure 5.1   Relationship among agent system components.

standards specifically addressing a major requirement of all distributed IA systems—interoperability.

On top of these infrastructural technologies we layer an architecture that specifically and directly supports (and may be constructed of) agents. Of course, the dividing line between what constitutes the collective capabilities of a federation of collaborating agents and what is considered an "architecture" is sometimes very fine. Indeed, with the ability to encapsulate mechanisms inside objects (agents), you could argue that an architecture is nothing more than a suitable combination of the infrastructure mechanisms necessary to an agent's (or an agent federation's) functioning.

But in this book we extend the "nothing more than is necessary" definition of an architecture not only to combinations of existing mechanisms, but to implementations of concepts not inherent in the infrastructure mechanisms. It is your task to study the architectures presented here for ideas you can use in developing your own agent-based systems. Again, though, the architectures presented here are by no means exhaustive, either in their diversity or in the level of detail—investigate further.

5.1.2 What makes a good agent architecture?

After you have read this section you'll still have to design your own agent architecture. Of course, some might argue that what constitutes a good architecture versus a not-so-good one is perhaps subjective. We believe, however, that, computer science being a science, there are concepts that can help you get on the right track from the start. So what criteria should you use to determine whether any of the ideas and elements embodied within the architectures presented in this section are appropriate to your problem? While the details of such a discussion depend on the details of your specific agent application, some high-level rules of thumb can be delineated.

Mowbray (1995) states that the key characteristics of a good OO architecture (OOA) include the following (embellished a bit with our own experience):

- *Simplicity:* If you want to spend your time developing agents, then the architectural ideas and the components you consider should be easy to understand, implement, and maintain.

- *Functionality:* Obviously, any architecture needs to provide the agents with the functionality they require to get their jobs done. Choose architectural ideas and development tools that focus on the specifics of your problem. For example, if your agents need access to large amounts of distributed data, with the ability to reason about these data, then consider the use of a combination expert-system development tool with an underlying ORB or distributed ODBMS.

- *Extensibility:* Since any system, including agent systems is rarely a one-shot deal (that is, build it and forget it, or applications that will never be updated or extended), you'll eventually plan and implement an upgrade or revision to the first release. And you'll have to troubleshoot and maintain any existing agents and architectures. This usually requires an architecture that can be extended, or at least does not preclude any options you can realistically see you'll be needing in the future.

- *Isolation or portability:* To the extent possible, evaluate the use of any architectural ideas (and also infrastructural components) on the basis of whether their implementations will be relatively independent of the underlying operating system's languages and nonstandard or proprietary implementations.

In addition to those high-level, commonsense notions, we can provide some more guidance in analyzing architectures.

5.1.3 Analyzing an agent architecture

To choose an appropriate agent architecture (or infrastructure) you need to develop an understanding of the current and future needs of your agents. How is this most easily done? Most OO experts now agree that scenario-based, also known as use-case or responsibility-driven, analysis and design (see Subsection 3.3.2, which summarizes scenario-based design, and references) provide the best foundation upon which to build your system.

Detailed scenarios in which your agents participate will not only help you describe the agents and their architecture, but will also help you decide which operating system, infrastructural and architectural components, and ideas to use in your system. Scenarios can unveil and help you optimize interagent communications profiles, appropriate agent-distribution and mobility schemes (for example, based on resource expense and location), interdependencies, and user-interface issues.

As an example of what criteria and tasks to consider in such an analysis, the software architecture analysis method (SAAM) provides an outline of the major tasks that you, the developer, should complete. The following is based on SAAM, and has been "agentized" to make it more applicable to your concerns.

- Develop agent-collaboration scenarios representing and describing your agents' behavior in all unique, typical situations. The scenarios can take the form of descriptive narratives together with object-interaction diagrams that show messages between agents and other components in your design.

- Describe the candidate infrastructures and architectures in a consistent fashion. You should capture the essential components and relationships in the architecture, and, by using the scenarios, describe how the infrastructure or architecture facilitates the agent's behaviors. Use the information in this book on infrastructures and architectures as a starting point and gather more information on each that interests you.

- Discover whether any scenarios will require modifications or enhancements to the infrastructure or architecture. You will then be able to gauge the difficulty of adopting any particular infrastructure or architectural idea, based on how much extra work you'll need to do to get everything working to satisfy your design and user requirements.
- Discover if any of the scenarios interact. This means, find out if any of the agents and components involved in scenario A are the same as in scenario B. The infrastructure or architecture with the fewest overlaps would be at the top of the list.
- Rank the infrastructure or architecture according to how well each one supports each scenario. It helps if you first rank the scenarios in importance as well. If your agents spend a great majority of their time in one or two particular scenarios, then weight these more than infrequent scenarios. Then you can determine whether an infrastructure or architecture will be best suited to your needs by multiplying the rated fit (for example, 1 to 5) by the scenario's relative time of use compared to other scenarios (such as those used 75% of the time) or importance relative to other scenarios (for example, the system must have this scenario versus others that are merely wanted but not necessary).

5.2 Spectrum of Architectural Complexity

Agent architectures can range from the very simple to the very complex, according to the requirements of the domain, the users, and the degree of agent sophistication or level of intelligence. Since this book's focus is on IAs (and, by extension, intelligent *systems* of agents will probably have complex architectures), this section focuses on several architectures at the complex end of the architectural spectrum. For example, in this architectural context we do not look at the many single, simple user-interface assistant agents (such as "Bob"), which are starting to accompany shrink-wrapped software.

But don't be dissuaded from reading about these architectures. Many of the ideas can be applied to less complex systems of agents and, as you may start your design and development at the simple end of the spectrum to get your first agent out the door, soon enough you'll want to see if that last agent you made can interoperate with the next agent or add-on feature you're planning.

5.2.1 Simple, single agent equals simple architecture

The simplest agent system could be composed of a simple, single agent performing a focused, specific task, for example, sorting E_mail. Obviously, such a "system" would be structurally simple in that no interagent communications are necessary and the types of things the agent could get involved in are limited.

Although an E_mail agent could have a sophisticated internal structure and encapsulate the latest neural networks, fuzzy logic, or genetic algorithm technologies, these would not necessarily contribute to architectural complexity.

For example, the E_mail agent pretty much stays put on a single node, is a single process, and waits for incoming mail; it is in the idle mode most of the time. Further, its working parameters are likely established via a simple interface to the user (for example, configuration menu as to sorting criteria). Although it must work with an E_mail package, it needn't be net aware, and has no use for multiple threads.

5.2.2 A few simply interacting agents equal a moderately complex architecture

In the middle of the spectrum we may find a system of several agents, still all designed to do the same thing, but for different users or on different nodes. They may be designed to collaborate on a task that may affect multiple users, for example, scheduling a meeting. Their communications scenarios are still fairly predictable, and thus, although being complex enough to have communications protocols, the agents themselves could still be fairly simple.

They would need to work with some scheduling package and be able to resolve conflicts (for example, by voting or via an embedded optimizing rescheduling algorithm). But they would still have little need for broad-based autonomy, would still remain on a single node, and, in addition, their need for a more generalized type of intelligence is limited.

For example, a scheduling agent wouldn't have a need to consult a complex ontology. And you probably need not consider soft computing techniques in this kind of agent. The constraints on scheduling are discretely specifiable. Whereas scheduling problems in other domains, such as production control, involve optimal utilization of many variables representing limited resources (raw materials, equipment and pipes, utilities, time, and so on), scheduling a meeting deals with room availability and people's time—a much easier problem.

5.2.3 Many complex agents plus complex interaction equals complex architecture

We have not yet speculated as to what a "most" complex IA system would be capable of, or would consist of. That is left for Section 9. But here we could say that a complex, yet eminently buildable, practical IA system would presume a sophisticated architecture. At this complex end of the architectural spectrum, you would probably find many different types of agents, each imbued with some autonomy, having the ability (either individually or collectively) to exist, cooperate, and perform within multiple platforms, and operating systems and, at the extreme, in different contexts or domains via learning.

5.3 Rieken's M Architecture—A Complex Architecture of Integrated, Diversified Agents

The complexity of many interacting agents is reminiscent of Minsky's "society of mind," which involves coordinating many differently goaled agents which are collaborating and contributing multiple reasoning processes to accomplish a complex task. Riecken (1994) says, it "takes many 'integrated agents' to create an 'assistant.'" A system of agents so created could serve in many capacities because of the contributions of many different approaches and technologies made by the diverse agents.

As Riecken (1994) explains: "This approach has resulted in the realization of M, a software assistant 'who' attempts to recognize, classify, index, store, retrieve, explain, and present information relating to human-computer interaction in a desktop multimedia conferencing environment."

5.3.1 Integrating diversified agents—the issues

A number of issues arise from taking an integration-of-diversification approach:

- How are the agents managed and coordinated?
- How do the agents communicate with each other, with their environment, and with their "master"?
- How is knowledge partitioned between the agents?
- How do agents share each other's resources? Which resources are kept private, if any?
- How do you specify the relationships among agents?
- Is intelligence, whether or not perceived by the user, built in explicitly, or does the developer depend on it emerging somehow from the synergy of the interactions of the agents?

These are issues you should be prepared to consider in designing your own architecture. Features of Riecken's architecture will help you understand these issues, and will point out some approaches you will find helpful in your deliberations.

5.3.2 Basic structure of integration

First, let's discover what kind of components make up M. The "society of mind" concepts and other technologies used to construct M (and agent systems like M) were intended to provide "aspects of spatial, structural, functional, temporal, causal, explanation-based, and case-based reasoning capabilities" (Riecken, 1994).

When developing any kind of architecture, the first task, and one that OOA lends itself to nicely, is task decomposition. As illustrated in Fig. 5.2, different

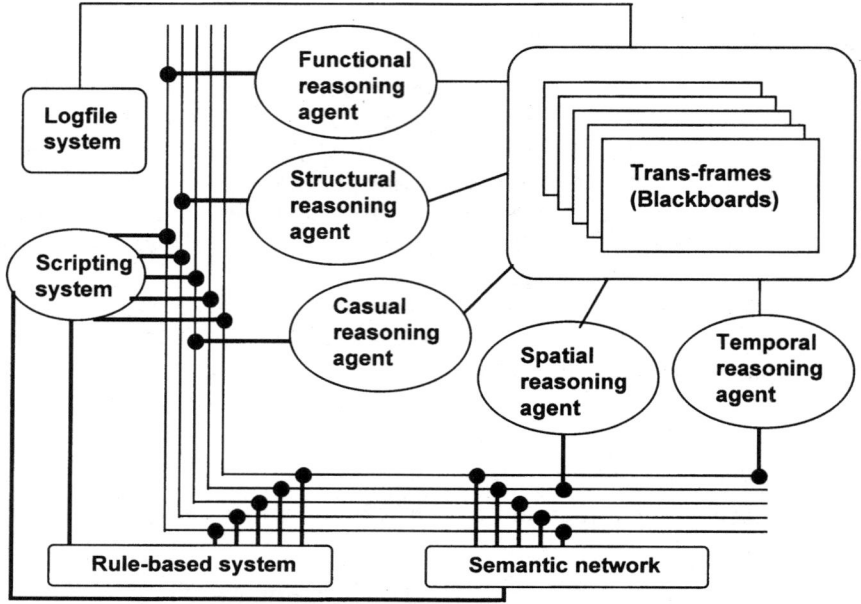

Figure 5.2 M software architecture. (*From Riecken, 1994.*)

types of tasks are assigned to different types or classes of agents and other functional components. M is also a good example of an agent-based application using several AI techniques. In addition, as we discuss M, notice and compare the approaches taken by the integrated Cyc-based ontology and its various reasoning and update mechanisms, and similar components and mechanisms in M.

Figure 5.2 shows the following components used to construct M:

- *Blackboard:* A globally accessible structure within a system used to store data related to a task and the entities collaborating on the task; implements transframes (see fourth bullet). An example of agents using a blackboard mechanism is given in a modified, distributed-blackboard approach taken in an application for mobile software agents in a telecommunications network. Essentially the approach borrowed a method used by ants, namely, make a change (a chemical change to the path in the ants' case) in your environment that other agents can detect as they pass (Appleby and Steward, 1993). This is similar to a blackboard-based operating system (like Clouds) or a blackboard-based application-communications architecture.

- *Semantic networks:* Specifically, a spreading activation semantic network. Semantic networks encapsulate pieces of knowledge in nodes connected by relationships. When a node is activated (knowledge is accessed by another agent or component), other nodes may also be accessed (activated) via the relationships, thus "spreading" the activation throughout the network.

- *Rule-based systems:* A knowledge base or ontology consisting of facts (about some domain) and rules that can be fired when certain conditions become true or false. The firing of rules can be thought of as a production. When some condition is satisfied, the rule produces the consequence. Typical rules include constraints (such as, legality constraints, completion constraints) and explicit descriptions about some domain.
- *Minsky's K-lines or polynemes, transframes, and pronomes:* K-lines represent mental experiences. Certain mental agents are active during K-line production such that they are reactivated to remember that experience. A *polyneme* is an agent that activates other agents involved in some task. A *transframe* represents various aspects of an action, and a *pronome* is an agent related to some aspect of a complex representation.
- *Script:* A structure representing a sequence of events that have worked in solving problems in the past. An agent follows a script, or a script can historize activities within an agent system. Scripts are weighted according to their potential to improve upon the theories that classify and explain actions, objects, and relationships. If a script's usefulness reaches some threshold, you could update an agent's ontology or a shared ontology appropriately.

Riecken (1994) posits that, using these technologies, "an assistant that can classify and explain actions applied to objects within a highly dynamic world should be functionally effective if it can simultaneously generate and test multiple domain theories in relation to a given goal." This sort of reasoning capability is analogous to the commonsense mechanisms Lenat is trying to accomplish within Cyc. Aspects of reasoning in one domain (or microtheory in Cyc parlance) can be *translated* into other domains (or subdomains within a single domain). This enables agents to participate in larger, more complex scenarios, increasing flexibility in the pursuit of goals and also promoting reuse.

Some other interesting aspects of M's architecture and its development that can be applied to your agent-development efforts include the following:

- The most important directive: Never take control away from the user.
- The use of sets to classify objects within a domain based on properties and the kinds of actions that the objects can perform or that can be performed on them. This makes agents more efficient by enabling actions on a group.
- The architecture supports simultaneous theories of a domain which can be "generated, ranked and modified." These theories are posted on the blackboard, which represents a live, dynamic, shared ontology.
- Information about an action in the system is represented with an input record that includes the action type, the actor, the object acted on, and a from-to qualifier.

- A logfile system is used to historize knowledge or theories from the blackboards once a task or problem is finished. This history is available to future agents attacking new problems. You could use the logged information to do batchlike updates of Cyc-like ontologies, and incorporate the new ontologies into blackboard (shared ontology) updates.

5.3.3 Application of M—the virtual meeting room

In this example the M architecture classifies and manages objects in a collaborative, multimedia, electronic work environment—a virtual meeting room (VRM). Meeting participants manipulate a virtual world's objects, including electronic documents, electronic ink, images, markers, white boards, copy machines, and staplers. M's task (one personalized M per participant) is to recognize the objects and their interrelationships and their states, following the changes imparted by the participants.

Object orientation also comes into play within M. There are five general or root classes, which are *composed* (remember the importance and expressive power of composition from Section 3) of other, more primitive classes representing (the actions and definition of) specific types of objects in the VMR. The composite classes are:

- Primitive drawable objects, such as paper, bulletin boards, post-its
- Composite drawable objects like a notebook composed of pages
- Drawing tools (or input devices), such as keyboards, pens, paint brushes, scanners, cameras
- Primitive editing tools that can cut, join, copy, transform
- Composite editing tools, such as a copy machine that can cut and copy and move

The different reasoning agents (structural, spatial, temporal, functional, and causal) in M share knowledge of the VMR, as represented by the instantiations of the classes. This enables the agents to perform useful actions for users.

For example, M could determine that two documents were associated by applying spatial reasoning to the fact that they are near to each other and to a circle a user drew enclosing them. The association could also be inferred by applying structural and functional reasoning to the circle "enclosing" the documents, and finally by applying causal reasoning to the semantics of a participant's action of drawing the circle around the documents.

Using the concepts of M-like architecture to build an agent-based application that improves the performance of participants working in a VMR was the aim of Riecken's work. However, you can easily see that you could apply an M-like architecture in different domains, in which agent-assisted collaboration would increase individual and group productivity.

5.4 Genesereth's Architecture—Architectural Concepts Emphasizing Interoperability

Instead of hard-coding the interoperation between a number of programs, agent-based software engineering provides certain benefits. Software engineers don't have to concern themselves with the details of other programs or hardware platforms; the operating system or infrastructure ensures interoperability. The resulting individual agents and the overall application are more robust, and they are more easily maintained and migrated to new releases or domains.

"In agent-based software engineering, the idea is for programmers to write their programs as individual software agents. Communication is assured through the use of a standard agent communication language. Agent interaction is assisted through the services of a coordinated set of operating system programs called facilitators" (Genesereth, 1995).

Of course, this is the major benefit of coding to interoperability standards, such as E_mail's SMTP, network management's SNMP, the graphics interchange format (GIF), and the computer graphics interchange (CGI),* and, more broadly, interapplication mechanisms such as OLE, Apple's Publish/Subscribe, and more recently OpenDoc and CORBA as well.

Genesereth professes that agents can become the basis for the component-based software paradigm change we have all been waiting for. In Genesereth's view, agents and facilitators work together toward a goal in what is referred to as a federation. Figure 5.3 shows the architecture. The facilitators receive a message from an agent or facilitator and route it to another facilitator or a suitable agent.

Various existing infrastructures or frameworks, such as those covered in Section 4 (for example, CORBA), do this under the guise of a broker, which keeps the objects or agents and their resources (such as their method signatures) cataloged and sends interobject messages (requests for services) to the appropriate object.

The agent-based architecture detailed by Genesereth goes a step further in that its broker, the facilitator, also "translates messages...decomposes problems into subproblems, and they can schedule work on those subproblems" (Genesereth, 1995).†

5.4.1 The importance of a metaprotocol

Genesereth proposes that there are two distinct phases in the operation of an agent—initialization and normal operation. We suggest that the number of

---

*In this context, CGI refers to the computer graphics interchange. Elsewhere in this book another CGI acronym appears, and it is short for common gateway interface, used to get, for example, Java applets to talk to a host's resources.

†This architecture has some aspects in common with the agent-oriented flexible information network (AFIN) architecture discussed later in this section, that is, the facilitator is somewhat analogous to the organizational agent in AFIN.

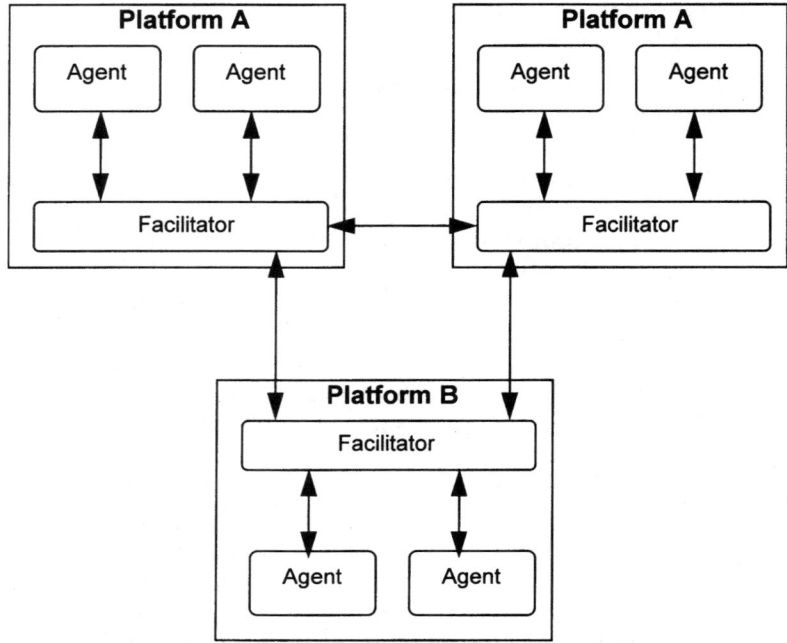

Figure 5.3 Architecture of agent interoperability. (*From Genesereth, 1995.*)

phases depends on how you break up the problems an agent will attempt to deal with.

For example, in Smalltalk it is customary to categorize similarly and formally the different types of things an object or agent can do by established categories of methods called protocols. Smalltalk (and other languages or class libraries, as well as infrastructures such as CORBA and OLE/ActiveX) embodies a number of existing protocol idioms, including the initialization protocol. What Genesereth defines as "normal operations" can be viewed as a higher-level or metalevel description of a range of activities more usefully broken down into several protocols.

For example, another phase of operation would usually be required as well—the abnormal condition or exception-handling phase (or protocol). These protocols are composed of one or more actual methods, procedures, or functions as appropriate. Genesereth suggests that the internals of the normal operations phase of an agent consist of what is in effect a getNextEvent loop which, depending on the event type, would vector the agent to the appropriate method, including an exception-handling method.

A metaprotocol is a delineation of the categories of protocols, and an abstract way of representing the kinds of syntax and semantics that would be particular to each metaprotocol. For example, a protocol consisting of a number of methods that establish and maintain a communications link between two agents would typically have a name and possibly criteria that determine whether a method belongs there or elsewhere under a different protocol. Other metaprotocols could be specified for other categories of agent behaviors and interactions including resource and knowledge queries (for example, ASK...), task delegation (for example, TELL...), agent initialization, shutdown, and so on.

A metaprotocol would be a way of expressing those different categories of protocols such that agents could more easily, and unambiguously, determine what kinds of behavior or resources are available from other agents.* If agents in a small group, or within a very large society, were to share a metaprotocol just as they might share ontologies or particular states of their being, the prospects for true autonomy and more elaborate and possibly emergent interactions would be far greater. This is because agents could more easily find out what other agents are capable of at more abstract levels, without entering into detailed communications.

5.4.2 Language issues

"Agent-based software engineering was invented to facilitate the creation of software able to interoperate...[in a heterogeneous environment]" (Genesereth and Ketchpel, 1994). As in OO programming (OOP), agent-based programming (ABP) makes use of agents as prebuilt components that are plucked from a library and put together in a pattern appropriate to solve the problem at hand. The agents communicate with each other via some communications language.

Although the simplest agents could be simple subroutines, Genesereth considers real agents to be more complex entities with persistence, existing as independent control threads or processes on single or multiple machines. Since agents *are,* or are like, objects, their language would have much in common with mechanisms used for interobject communications. Such communications are based on a message-passing paradigm. Internal details of how an agent deals with a received message are private and therefore preserve the interface independence from implementation.

Recall from Section 3 that the notion of polymorphism means that a message's meaning can depend on the recipient or the context. In contrast, with ABP, not only the syntax, but the semantics of the language's parts are fixed and therefore agent-independent. In other words, the agents must understand and conform to an explicit message format that includes parts (such as message verbs like ASK) defined outside and independent of the agents' context.

Since a standard is being imposed, what is to prevent different, inconsistent standards for different applications or domains? The answer, according to Genesereth, is: "...mandating a universal communication language, one in

---

*This is similar to the approach that the salutation protocol in Subsection 5.6.3 uses, only on a wider and more abstract level.

which inconsistencies and arbitrary notational variations are eliminated" (Genesereth and Ketchpel, 1994).

Genesereth suggests that there are two approaches to creating such a standard language—procedural and declarative. Examples of procedural languages appropriate to ABP are Smalltalk, Telescript, TCL, Java, and the interapplication commands available on the Mac operating system called Apple Events. The advantages to declarative languages (also referred to as scripting languages) include the ability to send or communicate by sending single commands or whole programs to a recipient. The commands or programs can be executed immediately or delayed, and can be compiled (on the fly as well) or interpreted (as with VM-based languages), thus executing efficiently. Disadvantages include some need of foreknowlege of the messages' intended recipient, knowledge the agent developer may not possess.

In addition, typical agent-based systems would involve multiple agents doing their respective tasks on one or many platforms more or less simultaneously. With a procedural language it is most difficult to synchronize and merge results without resorting to complex semaphore-like mechanisms. Declarative languages, on the other hand, if sufficiently expressive, enable the communication of all sorts of information, including states, definitions, assumptions, rules, data, and commands or programs (just like procedural languages). A particular declarative language, built with the intent of facilitating agent programming, is called, appropriately enough, Agent Communication Language (ACL) and is discussed in detail later.

Genesereth, although giving scripting languages their due, maintains that there will be declarative aspects intertwined in the ultimate agent language. In our own work, which involves prototyping communications between objects in advanced real-time systems for process and production control, we've found a command-based language with parameter passing, used within a message-passing hierarchy, to be optimal when compared to using variable read/write mechanisms to initiate processing.

The reasons herald back to the way OOP mechanisms work. The command interface fosters reuse and information hiding—you needn't expose internal variables to external entities, and agents with the same command signature can execute entirely different behavior (polymorphism). Performance, robustness, and elegant design are also facilitated through a message-passing hierarchy; command vectors needn't be explicit and NO-OPS are inherently passed upward to superclasses with implementations. Knowledge sharing is also more explicit in that mini-ontologies, together with passed parameters whose values *are* complete, executable scripts, can be passed.

5.4.3 Communications issues

There are essentially two different approaches to coordinating interoperable agents: (1) direct communications, where agents coordinate themselves, and (2) assisted coordination, where there is some external, non-agent system or object providing coordination (or scheduling). The advantage of the direct communications approach is that you don't depend on an external entity to properly tell your agents what to do and when to do it. There are two ways of providing agent interoperation via direct communications.

- *Contract net:* An agent requests that other agents submit their proposals with appropriate bids to do the work. The agent then reviews the bids and awards a contract to the winning agent (or group of agents if the request required a work breakdown and multiagent collaboration).
- *Specification sharing:* Agents supply each other with their interests or needs, and their capabilities. The agents use this knowledge to coordinate themselves. You can see why, using this approach, a command-based declarative language would fit well within this approach to intercommunications. Indeed, one of the advantages of this approach is that it cuts down on the number of commands that must be sent, which is apt to grow very large as agents proliferate (for example, the Internet).

In fact, it is due to some of the disadvantages of direct communications that you find one that is most efficient. For example, implementing direct communications is very complex. Each agent must have the complete protocol required for all negotiations. If some of this protocol could be implemented elsewhere—in one place—some of this complexity could be eliminated. This leads to a federated architecture.

Examples of common, extant pseudofederated systems are operating system services such as directory assistance. Examples from the OO world include object brokers such as CORBA, the OLE mechanism, and DSOM. Other brokerlike activities are provided by Apple's Publish/Subscribe and Microsoft's Dynamic Data Exchange (DDE); both find out who is to receive a message, but also send the messages and handle problems and return answers to the senders.

What separates these systems, which are already in use, and agent-based federated systems is the degree of sophistication and richness. For example, using ACL, a facilitator can be made aware of many more details about the sender and intended receiver, thus narrowing the routing possibilities. In addition, facilitators can translate messages if they have knowledge of syntax and semantics. Ontologies can get translated this way.

Facilitators also make use of certain AI techniques to facilitate more efficiently. Techniques such as automated reasoning, search control, and automated message routing can be used within a facilitator to remove that burden from lower-level agents.

5.4.4 ACL: the agent communication language

When agents interoperate, they do so using some type of common or standard language, as mentioned earlier. In the case of Genesereth's agent-based interoperability architecture, "...the agents use a far more expressive language than usually seen in existing systems of this type. It provides for the communication of constraints, disjunctions, rules, quantified expressions and so forth" (Genesereth, 1995). This rich expressive capability can be especially useful when incorporating other AI technologies discussed in Section 3, such as fuzzy logic, expert systems, neural networks, and genetic algorithms.

Another important advantage of this architecture, according to Genesereth, "stems from the use of this language to encode general knowledge about application areas (such as definitions of concepts) and programs (for example, their specifications) and the application of automated reasoning and automated programming technology in the implementation of facilitators..." (Genesereth, 1995). Since facilitators in this architecture can be responsible for so much more than merely passing messages, ACL enables the fulfillment of these advanced responsibilities.

To give you a flavor of ACL, let's first look at its structure. A message sent from an agent to a facilitator takes the form of a series of expressions. The first expression indicates what kind of message is being sent. This would include declarative (state an axiom or assert a truth), interrogative (ask a question), and imperative (tell or direct) type of messages. The last expression is the name of the message, and all expressions between first and last are the arguments of the message.

Genesereth identifies the following predominant message types:

- *Assert* and *retract* communicate a declaration or negation thereof.
- *Ask* and *unask* query for truth or cancel a query.
- *Achieve* requests that the accompanying argument or sentence be made true as a goal.
- *Forget* requests that the accompanying argument or sentence be no longer a goal.
- *Reply* transmits answers to previous messages.

In addition, developers can add their own message types, thereby providing a great deal of expressive extensibility.

ACL's arguments are based on KIF, the semantics of which, in turn, are based on first-order predicate calculus. Without going into specifics, suffice it to say at this point that these semantics support variables, operators, constants, rules, and definitions. The combination of these elements allows you to build knowledge about objects in your problem domain.

One way agents can use a language such as ACL is to communicate knowledge and actions about a particular application domain. Genesereth suggests using "an open-ended dictionary...each word [within the dictionary] would have a KIF annotation for use by the facilitator in mediating disagreements of terminology" (Genesereth, 1995) and, presumably, enable more accurate collaboration among agents.

Now let's see how agents actually use ACL and how they function in an environment that includes facilitators. In Genesereth's proposed architecture, a software agent *is* a process capable of communicating with other processes (that is, agents) using ACL.

An agent first initializes by announcing, via an ACL specification message, its existence and telling the facilitator to register its capabilities (by specifying its behavior in response to certain message types) and interests. Then the agent enters its getNextEvent loop, waiting for a facilitator to pass it an event for which it has registered an interest. If an agent needs further information to carry out its processing, it asks the facilitator to secure it (from other agents or components). Depending on the implementation, the requesting agent can stop its own processing (suspend its thread) while waiting (synchronous), or it can continue processing if possible (for example, by suspending its requesting thread and forking a new thread).

5.4.4.1 ACL and agent behavior.
One of the criteria Genesereth believes must be met for an agent to be called an agent, is to be able to communicate in an agent communication language such as ACL. This not only means that the agent must read and write ACL messages, but it must also conform to behaviors explicated in those messages, together with other behavioristic principles imposed via the architecture or domain. These extraneously imposed principles, like Asimov's robotics laws, might include:

- *Veracity:* An agent must always tell the truth. Of course, truth is relative to the agent's knowledge base. But the principle still holds: Unless the agent is a participant in a sleuthing game or, possibly, a bidding situation where the object is to mislead other bidding agents not to bid on the thing you want, an agent should not be coded to lie or otherwise cause damage to bits or atoms.
- *Autonomy:* An agent may not stop another agent from performing a service unless the other agent has indicated its willingness to be stopped.
- *Commitment:* If an agent says it can do a task, and then is asked to do it, it must oblige the request.
- *Domain precedence:* If an agent is enabled to cross domains, the constraints of the current domain take precedence over previous domains or future or scheduled domains.

So the question becomes, how does a developer ensure that an agent conforms to such behavioral constraints? Genesereth claims that by the correct use of an ACL with the ability to express such constraints, such assurances can be granted (Genesereth and Ketchpel, 1994).

5.4.5 An interoperability facilitator

Now let's discuss in more detail the functions of a facilitator. The facilitator, as mentioned before, not only serves as a message-passing medium, but can translate messages from one format or language to another. Genesereth states that this process has two aspects: vocabulary translation and logical translation.

An example of vocabulary translation would be conversion from a graphical object's specification in, say, GIF to an equivalent PICT definition, or some proprietary format, such as the ParcPlace's VisualWorks graphic specification to a data exchange format (DXF), an AutoCAD-based standard format. Another example might be translating one agent's method signature in Smalltalk into another agent's analogous or corresponding method signature in C++.

An example of logical translation would be converting a message with arguments composed of instructions in some specific language, with the notion that the facilitator would know to pass that message to an agent that can understand that language, that is, is "coded" in that language. This enables agents written in different languages to communicate with one another.

This logical translation is similar in concept to the development packages available that map IDLs and object definition languages (ODLs) to one another. (In fact, in Section 7 we examine an example made with a tool called Orbix, which composes such a mapping enabling a VisualBasic-based agent to talk to a C++-based agent executing on a UNIX platform via OLE and CORBA brokers.)

A developer needs to ensure that when making a facilitator's translation functions, that information and intent are not lost. Some languages are less expressive than others, so it would be necessary to explicate the "least-common denominator" of understanding. The interest specification (an agent's declaration of what is important to it) of agents registering with the facilitator enable it to determine appropriate mappings. As Genesereth (1995) explains: "A Prolog agent might restrict its interest to Horn clauses; a relational database might restrict its interests to ground atomic formulas."

Another function you might give a facilitator is that of buffering messages and combining their semantics to construct a message meeting the specification of some other agent's behavior. Another task for facilitators is scheduling. Of course, beyond simple FIFO or other simple scheduling methods, there are many software packages available that do scheduling. Especially when considering priority-based schedule optimization and rescheduling algorithms based on dynamic system goals, the astute developer would take advantage of existing scheduling software that can be incorporated or called from within a facilitator, instead of coding these functions from scratch.

When we discussed doing an OO-based analysis of architectures in Subsection 5.3.1, we said that scenarios and use cases could help you optimize interagent communications. When planning communications in a facilitator-based architecture, you can decide whether a facilitator is needed at all times. Although there are attractions to maintaining architectural purity—which is usually defined as rigid adherence to the original intents and specifications of the architecture—there is sometimes a cost for purity, a decrease in efficiency.

For example, if you see, in the design of a system, that many messages of the same type and format will be conducted between two agents on the same node, you'll realize that a lot of facilitator overhead will be generated. Instead, you can decide to let a facilitator set up the "pipe" and session between the agents and let them communicate directly using a full-blown RPC-like or mini-RPC (lightweight) mechanism. You may want the facilitator to be involved when exceptions occur and during termination of the session.

5.5 Leveraging Existing Intelligence

The issue we discuss in this subsection is how to take advantage of existing programs and other code snippets that you might want to include in your agent-based applications. This is in contrast to designing and implementing new agents from scratch, that you can verify work correctly and in accordance with the principles described in various architectures.

There are several different approaches to incorporating existing, so-called legacy code into your agents or agent architecture.

- *Encapsulation:* Wrap your existing code in a structure that makes it look and act like any other agent. This wrapper can look at internal data and can make calls to any of its methods. In addition it can intercept external requests for services (messages) from other agents and determine whether or not the legacy code can perform the service and send an appropriate reply. This wrapper approach is especially appropriate if the legacy code does not have interprocess communications capabilities to begin with. But it requires that the legacy code be available.

- *Transducer:* Create an object that mediates between your agents and the legacy code. This is akin to the facilitator functions presented earlier. The advantage of this approach stems from its not requiring a detailed knowledge of the legacy code other than an interface specification. This is also similar to the standard idiom, or an architectural pattern known as a proxy. Most ORBs use local proxies to communicate with remote objects or agents. Genesereth claims that this approach works well for situations where your agent needs to get access to files or to people (Genesereth and Ketchpel, 1994). For example, use a transducer to read and modify or translate a particular file structure to another structure in order to enable an agent to read it properly. Another example is building a GUI transducer, enabling an agent to communicate with a human over specialized interface software or hardware.

- *Rewrite:* Obviously the most radical approach, rewriting the legacy code, or appropriate portions thereof, has the advantage of bringing that functionality up to date with respect to implementation. Another advantage is that you may be able to improve its efficiency or capabilities, or you may want to explore more optimum designs for the code's functions. This approach is possibly best for situations when you want an agent to interoperate with "time-monolithic" programs, that is, programs that are not preemptive and whose operations run to completion without sharing results externally. For example, Genesereth and Ketchpel (1994) state: "Many automated design programs work to completion before communicating with other programs. For example, the output of a logic synthesis program is passed as input to a printed circuit board layout and routing program, whose output is in turn passed to an assembly planning program. This process is repeated down the line. Recent work in concurrent engineering suggests that there is much advantage to be gained by writing programs that communicate partial results in the course of their activities and accept partial results and feedback from other programs." This situation is ideal for agent intervention. If certain portions of an engineering work flow were made more modular, they could be run as semi- or totally independent threads of execution and could not only work on tasks in parallel with other modules of the total engineering "program," but could then be more amenable to the other two ways of making software interoperate with your agents. In fact, if you are doing a complete rewrite, why not write the software as OO and therefore have made a great leap forward toward making it even-more "agent-friendly," not to mention more reusable.

5.6 Rosenschein's Approach: An Architecture for Agent Negotiations

If most or all of the devices with "intelligence" built into them (in the way of microprocessors, microcontrollers, and so on) are to communicate effectively, an agreed-to protocol is necessary. This protocol must be rich enough to enable communications at the many levels hardware and software devices can interact.

Not only must devices coresident on one network communicate, but they must communicate across networks of different types. For example, say you want to use your work phone to turn off your home coffee pot (which you inadvertently left on). You presumably use plain old telephone service (POTS) to dial up your home control system and dial in the number for the coffee pot and another number representing the off command. The control system sends a signal over the power line to the coffee pot, which has appropriate circuitry to sense the command and turn off the pot.

Updating the scenario a bit, maybe you have a representation of your home local operating network (LON)-based control on your PC at work. Via an ethernet connection you remotely log into your home computer, which is connected to LON. You again access or change the LON boolean network variable representing the coffee pot's on/off control. Of course, if you can do those things, you'll eventually want an agent to attend to these "housekeeping" details. So it will have to know how to navigate through the different networks and negotiate with resources resident on the network devices.

Within an agent's relatively homogeneous environment, the agent had better know how to get around, ask questions related to its goals, and otherwise command or negotiate for the resources necessary to achieve that goal. In the future, agents you build or configure and send off to locations other than your home or local intranet (where presumably you have a handle on what environment exists and what doesn't) will be "strangers in a strange land," so to speak. With trends continuing as they are, we don't think it's practical, or possible, for a truly mobile agent to perform in a heterogeneous world in the next few years. The infrastructure, whether implementations of CORBA-based ORBs, a common operating system, or Java, Telescript, or Smalltalk VMs must be there for successful agent mobility (or, in the case of the ORBs, interagent communications). Although we suppose an agent could work within heterogeneous systems if it were prepared to do so, that is, if an agent were given a specific mission, and it were prepared to deal with specific environments, it could go from one intranet to another, the idea of wandering in a heterogeneous world is not practical now. For example, a script written for one flavor of UNIX might not run on another, such as C shell, Korn shell, or Bourne shell scripts. In any case, for now and the near future (less than 5 years) the agent must rely mostly on the infrastructures and a resource- and knowledge-sharing mechanism to get things done.

5.6.1 Negotiation protocols—the issues

The principal aim of a negotiating protocol is to enable differently goaled self-motivated agents to collaborate on problem solving and tasks. An issue here is that it is not necessarily the case that all agents contribute positively to the scenario. This may be especially true if agents from several different developers are negotiating for resources or bidding to complete the subtasks. Some agents have ulterior motives, so to speak. They may consume too many resources, or they may slow down completion of the overall task. For example, if all agents have the highest priority to acquire a printing resource, they all might deadlock waiting for an opening.

Trade-offs to consider include the following:

- Low or high resource costs, versus the complexity to resolve resource conflicts
- Use of a central designer or developer, so that the agents have shared goals or exist in environments that impose constraints conducive to a specific goal, versus multiagents with no central designer, so no shared sense of purpose or goal.
- Domain specificity versus generality
- Master-slave relationships versus autonomy
- Speed of agreement in negotiations versus achievement of an optimum state
- Nonexploitation of resources versus "you snooze—you lose"

To ameliorate situations that are too competitive, one idea to consider is to constrain behavior such that only beneficial and desirable outcomes occur. This is done via the creation of a "social" environment conducive to agent interactions. Since agents are seeking their own goals, the question is how to make them cooperate and put aside (temporarily) their direct pursuit of a goal to strike a deal with or help another agent.

One way to accomplish this is to develop domain-based rules, that is, rules and constraints specific to, for example, computer-aided design (CAD) work flow or scheduling. Another way is to develop your agents to follow standards based on stability (no chaotic agents) and simplicity.

5.6.2 Negotiation protocols from game theory

The bottom line is, when designing agents that may enter a system, world, or domain with other agents, that you can't assume either cooperation or competition. What to do? Use rules of interaction (in multiagent systems) derived from game theory.

Different protocols have different effects. For example, suppose that you are about to place a telephone call. After you dial, a number of different carriers bid to carry the call. The bids are taken by a computer in your telephone, with the lowest cost winning. This encourages programmers to invest their bidding agents with strategic capabilities, or at least to ensure that the agent performs toward strategically determined goals. According to game theory, this may not be the most efficient use of resources or outcome.

Another possible prearranged outcome might be that the winning bidder actually gets paid the second-lowest bid amount. This scenario is called Vickery's mechanism. It eliminates purposeful under- or overbidding to gain (possibly surreptitious) advantage. Detailed consideration of these ideas and much more is found in Rosenschein and Zlotkin (1994).

5.6.3 Quest for an agent salutation protocol

There is a need for an architecture that allows intelligent machines and the agents within them to find out what other devices, applications, and services are available to them, and how to take advantage of those that are needed. This is known as the discovery-and-utilization problem. The salutation architecture (SA) by the Salutation Consortium provides a basis for the negotiation of form and the format of interoperation among devices, applications, and services, to describe capabilities and availability, to discover the capabilities of others, to search for capabilities, and to request and establish interoperable sessions with others to use those capabilities.

Lest you think this effort might not go anywhere, consider that the Salutation Consortium members include Advanced Peripherals Technologies Inc., Canon Inc., Eastman Kodak Co., Fuji Xerox Co. Ltd., Fujitsu Ltd., Hewlett-Packard Co., Hitachi Ltd., IBM Corp., Integrated Systems Inc.,

Konica Corp., Lexmark International Inc., Matsushita Electric Industrial Co. Ltd., Microware Systems Corp., Minolta Co. Ltd., Mita Industrial Co. Ltd., Mitsubishi Electric Corp., Murata Machinery Ltd., Novell Inc., Oki Data Corp., Ricoh Co. Ltd., Sanyo Electric Co. Ltd., Sharp Corp., Toshiba Corp., and Xerox Corp.

The SA is to be processor, operating system, and communications protocol independent as well as being scalable (usable within very small, embedded systems such as personal digital assistants (PDAs) and toasters, to mainframe computers and automobiles). The SA has:

- *Clients* (things that want service)
- *Servers* (things that can provide service)
- A *salutation manager* (*SLM*), which functions as a transport-independent service broker for application interfaces
- *Transport managers* for transport-dependent communications
- *Function units* (FUs), described in a functional unit description record (FUDR), that serve to define abstract capabilities, type, and attributes (such as print, double-sided, four sorters, A4 paper).

As Fig. 5.4 illustrates, a client interrogates a server's FUs. The information exchanged is called a personality. There are three types of personality:

1. *Native:* Defined outside of SA; SLM not involved
2. *Emulated:* Defined outside but used within SA framework (wrapped)
3. *Salutation:* All aspects within SA

A service description record (SDR) (composed of an equipment description record and a set of FUDRs) is where a client describes services in which it is interested and passes the information to the SLM. The SLM responds with an SDR containing the FUDRs registered by SLM from service-providing servers. Here is the typical flow or exchange of message:

1. The client SLM initiates a capability exchange by sending a Query Capability packet with the SDR to the server or servers.
2. The server SLM compares the client's SDR with its own SDRs (FUDRs).
3. If they match, the server SLM sends a Reply Capability packet to the client's SLM, containing the SDR with all the matching FUDRs.
4. The client chooses the server with the appropriate SDR and sends an Open Service packet to establish a service session with an FU. It contains the FU's handle, the personality to be used, and the service-session handle [client-server handle (CSH)]. Subsequent Salutation packets include the CSH to identify this particular session, which could be one among many.

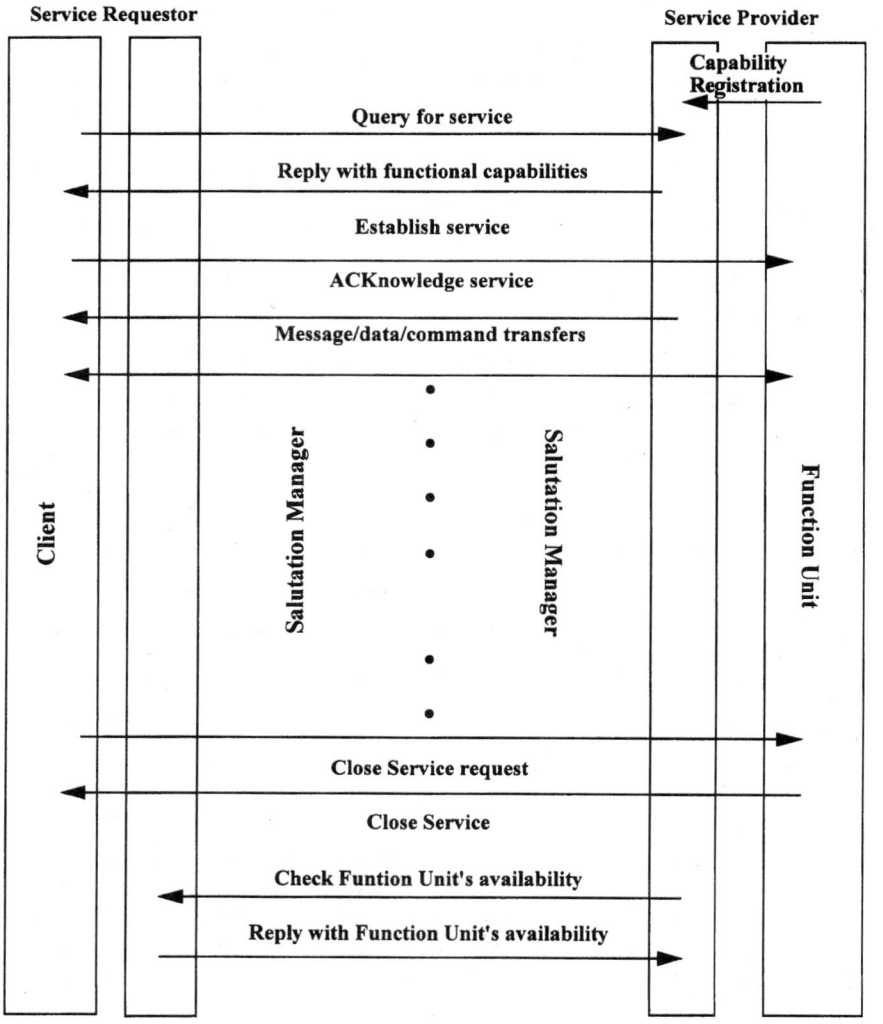

Figure 5.4 Salutation protocol. (*Based on Solution Consortium.*)

5. The server returns an ACK(knowledge) Service packet with acceptance or rejection and the CSH.
6. The server and the client then consummate the service activity with Transfer Data packets containing commands, responses, and data messages, until the service is completed. How does SLM mediate between SA packets and the messages to the application services within its purview? The answer is that the SA API must be used by those applications.

7. SLM checks the availability of other applications with FU Available message, reporting which FUs are alive, or FU Unavailable message for those not ready to service a request.
8. The client sends the server a Close Service message.

To use this kind of protocol effectively, accurate device or system abstractions are necessary. Devices and systems are first grouped within a personality (such as salutation document systems), then within the FUs (such as document-delivery services; for example, FAX Data, Send FU, Print FU, DOC Storage FU). Client-server messaging provides for the exchange of all relevant information necessary to the execution of the task. For example, within the Print FU exchange, a print description language (PDL) or image format can be specified. Some usage examples include the following:

- When Java is linked to these FUs, the PDL interpreter could be downloaded as well, or the Acrobat reader could be downloaded for a document viewing FU.
- IBM has an SLM for Windows and OS/2 that works with copiers and FAX machines.
- System abstractions exist for personal-information systems that encompass address book and schedule FUs, PDAs, tele/fax equipment, and personal information management (PIM) applications.
- In the works are VOX, data-format conversions, and speech-recognition abstractions.

All this over local-area networks (LANs) and wide-area networks (WANs), POTS, and local PC buses (Firewire/1394, etc.)! One aspect that bears more investigation (we'll leave it to you) is how this protocol would fit in with local "device" networks such as LON, CEBus, or FieldBus. There also remains some question as to overlap and competition with Microsoft's proprietary "At Work" approach and the more open approach of embedding web servers into devices. But our sense is that this is a development worth keeping an eye on, especially if you plan to design, for example, agents for embedded systems, or mobile work-flow agents that might traverse several devices. One example of this might be an agent that monitors incoming faxes and E_mail, finds something important, sends or carries the information to a printer or copier for distribution to others on a list, then "hand carries" output from those devices to the requester's devices.

The following ideas about salutation protocols are important for agent-related work:

- Production and widespread acceptance of an agent resource metaprotocol, as discussed earlier
- Architecture for interagent communications across all manner of systems
- Ability to request agent services (capabilities of agents included in list of FUDRs)

5.6.4 Agent negotiations—considering the domain

When developing protocols for collaboration and negotiation among agents, consider the relationships between those protocols and the domains in which the agents will exist or operate.

Rosenschein and Zlotkin, for example, propose a hierarchical domain categorization scheme. The categories are represented, at each level in the hierarchy, as classes, each higher level in the hierarchy being more general (or abstract) than the subclasses below it.

As an appropriate aside, consider the following design note. In Section 3 we talked about OO agent and domain representation and development. The example domains-as-classes approach taken by Rosenschein and Zlotkin facilitates the implementation of domain models (for agent testing and simulation, as well as domain implementations) using OO principles.

For example, remember the concept of containment and composition? An OO agent could be developed as part of a system that represents the agent as "contained" within a domain container object. This is the simplest and most explicit notion of the concept of containment, as presented in Section 3 on object orientation. The box contains the objects but is not made of them.

The domain, as a subclass of the root container object, would partake of all the attributes and behaviors accorded a container. For example, a root container object would be able to answer to the query listAllContainees. A root domain object could then be produced as a subclass of each root container object that had additional domain-specific attributes and behavior. The root domain would then be subclassed and further specialized as appropriate for the specific domain.

Of course, using the notion of a containment hierarchy, a domain containing certain types of agents appropriate to that domain could itself be contained within a higher-level domain (for example, E_mail agents contained within an E_mail domain object, perhaps the E_mail application-as-container object). Then this E_mail domain application object itself is contained within an object representing an operating system, which, in turn is contained within an object representing a network, and so forth.

In this view if a representation of a community, federation, or "society" of agents is needed, it could be represented as a composite object. In other words, the "federation" itself is an object that is composed of individual agents and possibly other federations in a composition hierarchy as well. This composite object would then be contained within the domain-as-container object.

Alternatively, as there are many different types of composition (as mentioned in Section 3), you could model the federation of agents as a composite object as before, and also model the domain as being composed of agents and federations of agents, schedulers, maybe a facilitator (as in the interoperability architecture discussed previously), and other objects.

Now back to Rosenschein and Zlotkin's domain-categorization scheme, which is roughly as shown in Fig. 5.5. Note that the representations in the figure are abstract. They represent abstract *types* of domains. Specific subclasses of each abstract domain would be built for each level in the domain class hierarchy.

Agent Architectures

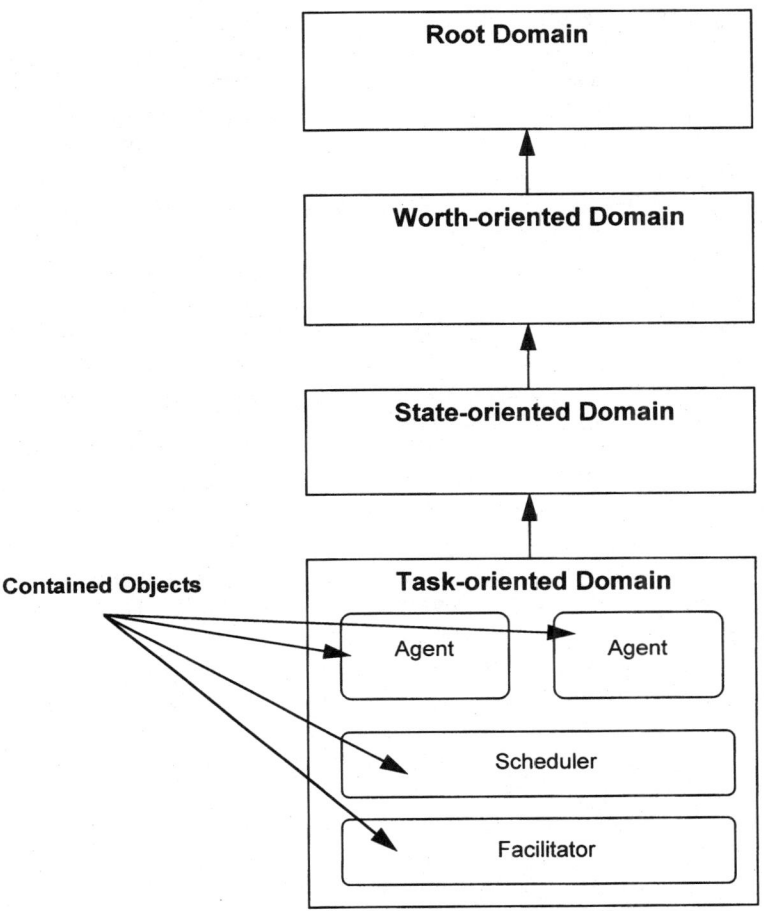

Figure 5.5 Domain class hierarchy.

The first, lowest, or most specific level contains classes which represent *task-oriented* domains. Agents "occupy" this domain as they carry out tasks that do not conflict with each other, and further, the tasks within this domain can be distributed among the agents "living" there. Therefore the primary purpose of protocols at this level is to optimally (and possibly dynamically) distribute (and possibly schedule) tasks.

The next level in the domain class hierarchy is the *state-oriented* domain. This domain class represents goals. As such the behavioral protocols involve joint planning and negotiation, scheduling, and mutually beneficial collaboration. This is similar to some of the behavior of the "facilitator" object discussed earlier. The term "state-oriented" indicates that a goal is represented in this domain object as a specific state to be reached.

Regarding domain containment or composition hierarchies, you can project lower-level states to higher-level or containing objects, thereby arriving, via appropriate logical operations, at "superstates," states representing a composite of the underlying states. In this way you can ensure, through proper representation and implementation, that the goals of a domain do not conflict with the goals of the individual agents within the domain. This superstate approach has the further benefit in that superstates can combine and simplify lower-level matters, as you develop a code for higher-level agents that monitor lower-level agents.

The foregoing ties into the idea of what kinds communications take place between agents within and between various levels of an agent hierarchy. In short, the hierarchy also represents, or reflects, the lowest to highest levels of cooperation and responsibilities, as well as event or alarm distribution policies. The low-level agents are managed and must cooperate with each other on their level. They pass summary knowledge upward (or higher-level entities transform low-level detailed knowledge into summary knowledge). Top-level agents know or can find out everything, while lowest-level agents have very focused scopes.

When considering implementations of this state-oriented domain object, notice that, because of all the interagent communications, certain side effects may occur. You, the agent developer, must learn and teach your agents to recognize and plan for or reduce undesirable effects.

The highest level in this domain classification scheme is the *worth-oriented* domain. This domain also represents goals, but a different aspect of them than the state-oriented domain. The worth-oriented domain specifies the relative acceptability of states, that is, a scale or spectrum of desirability of a state representing a goal. Agents get information from this domain that would lead them to evaluate their performance toward a goal, and enable them to determine which parameters or attributes of a goal state need improvement.

Rosenschein and Zlotkin (1994) further analyzed specific task-oriented domains and how domain analysis helps agent designers, as follows:

- They looked at two cooperating agents in a "postal" (*Note:* Our term change from "postmen") domain. The agents collaborate by comparing their sequence of stops and giving each other mail destined for stops on each other's route.

- For a database domain, a database on a network has information that two agents want. The database owner charges for access, so the information-retrieval agents need to optimize their requests by eliminating redundant queries between them.

- For a fax domain, where two agents share their intention (that is, goal—let's not get too anthropomorphic) to send data to the same location. Since the communications carrier charges a flat rate for the linkup between fax machines and no limit on the data, the two agents can bridge their data and send them with one fax.

5.6.5 Agent negotiation protocols—utility maximization

From the example domain analyses, you can see that to build agents that can cooperate in this fashion, they would need to share and understand their respective task lists (which together constitute each agent's goal) and find any similarities. Cooperation implies a negotiation protocol that both agents understand.

"The negotiation protocol involves a definition of what utility is among the agents, and a definition of the so-called *conflict deal*. The conflict deal is the...default deal the agents get if they fail to reach agreement" (Rosenschein and Zlotkin, 1994). In addition, the agents would need a negotiation strategy. A deal is the resulting union of the agent's task lists. The conflict or default deal is the original sets of tasks, and the utility for an agent is how much less the agent needs to do as a result of the deal (cost in number of tasks it does not have to perform).

Agent developers, working within a known domain and using a known communication and negotiation protocol, would seek to maximize their agent's utility and, in a system, maximize the product of all the agents' utilities. The literature is full of theories about how to maximize utility. We'll briefly mention two techniques here.

If the agents all know about each other's tasks, an algorithm can compute the point of maximum utility, and all agents can reconfigure themselves to accommodate that point. Alternatively, in an information-poor society of agents, they can use what is called a monotonic concession protocol, where each agent starts out offering deals in its own best interest and then backing off one "unit" at a time or offering one concession at a time. Refer to Rosenschein and Zlotkin (1994) for more details on task-oriented domains and subtle, game-theoretic, and incentive-based deal-making concepts. In addition, we recommend their suggestions for further reading given at the end of their paper.

5.7 Kautz's Architectural Approach—Bottom-Up Prototyping and Interation

What people now call agents ranges "from adaptive user interfaces to systems that use planning algorithms to generate shell scripts" (Kautz, Selman, and Coen, 1994). Consider that current users have very little patience interacting with so-called agents. This makes it difficult for you, the agent developer, to design agents that perform specific, useful tasks without impeding or annoying the user. One helpful approach involves using mostly GUIs instead of text-based interfaces. Another design criterion should be building in reliability and appropriate error handling. (Also refer to Subsection 6.2.3, where we present more information on the work by Kautz and coworkers.)

But perhaps more important, according to Kautz, is to start building your agents from the bottom up, without first committing to a particular architecture. In other words, per good OO development practice, begin your agent project as a prototype and keep it small and unambitious at first. You can then test your agents in an environment that is less robust and has less resource costs and less of a "cleverness" quotient than trying to do it all at once. This applies both to the initial design stages and during development.

Remember, the typical OO design-development paradigm is based on a iterative, spiral development process. You keep iterating design, development, and user test and release in an effort to improve the system continuously. As you see the real-world interactions of your agents, and the user's reactions to them, you'll get a better idea of what the best architecture is to support a robust agent society. Kautz and coworkers believe that this empirical approach will guide agent developers in more precisely defining the initial requirements toward the end of developing agents that are truly useful and fit unobtrusively into the user's environment.

One example of this approach to empirical or bottom-up design is found in Kautz's recent work. As his group initially designed an agent-based "visitorbot," an agent that scheduled people for visits into their laboratory, they came to some conclusions that led them to change their approach after receiving feedback. Initially they designed and deployed a monolithic "visitorbot" to perform all tasks necessary to scheduling visitors, including visitor announcement via E_mail to the appropriate people, collecting E_mail responses to the announcement, doing the constraint-based scheduling, informing participants of the schedule particulars (such as room), and rescheduling if necessary. The following are some lessons learned from that initial deployment:

- "E_Mail communication between the visitorbot and humans was cumbersome and error-prone" (Kautz, Selman, and Coen, 1994)
- Information was captured through the use of text-input forms, in which trivial errors wreaked havoc. Lesson: employ a graphical interface (for example, selectable buttons) to specify meeting particulars.
- "There is a need for redundancy in error-handling....the visitorbot could become confused by bounced E_mail,...or 'vacation' programs" (Kautz, Selman, and Coen, 1994).
- In addition, unexpected termination of visitorbot-initiated subprocesses led to unforeseen errors and problems. Kautz et al. suggest that agent developers apply certain real-time systems design and software reliability techniques to detect and alleviate such problems.
- Instead of designing a lot of sophisticated scheduling software into the visitorbot, Kautz et al. found that the scheduling problem could be translated into another form and handled by a third-party package. This leverages others' work and lets agent designers concentrate on agent-related activities, quite apart from solving other difficult and already-solved problems.
- Through the use of encapsulation techniques the user never need be made aware that your agent consists of the clever collaboration of a few existing soft-

ware solutions packaged to serve a particular domain with a specific user interface that you have provided as a "veneer" over all the internal complexity.

- "Early discussions with potential users made it clear that privacy, security, and attention to users' work habits are central issues in the successful deployment of software agents" (Kautz, Selman, and Coen, 1994).

- Some related concerns involve the use of replacement applications that users may not feel comfortable with or downright hate. Also, with the advent of filtering capabilities within agents, ensure that your users can configure whether or not automatic filtering is applied to information such as E_mail.

These lessons led Kautz's group to develop an agent personalized to each user. This user agent would then mediate communications with the visitorbot, thus enabling the scheduling activities, while retaining control of private information in the user's hands. Further, this "separation of concerns" is a classical way of reducing complexity and thereby has certain other advantages. For instance, user agents can be tailored to the computing platform, thus enabling various levels of sophistication and user interface.

Most importantly, this agent prototyping effort led to an aspect of architecture that emerged from the lessons learned, as follows. It is important to separate the task-specific agent, namely, the visitorbot, which performed the *what* of the main scheduling function of the society, from the individual and idiosyncratic user agents, which performed most of the *how* of information generation and communications (both agent-agent and human-agent).

This emergent or default architecture led to later refinements in Kautz's implementation, including a communication's protocol, again arising from the bottom-up approach. A simple tagging of E_mail indicates what the receiver should do with the message. Because the architecture separated user-interface concerns from the task domain, as domains are added, the new task agent gains immediate access to a reliable GUI on many heterogeneous platforms. In addition, new modalities of agent-human interaction can be added as appropriate, with no modification of the task agent (for example, develop a user agent extension that allows it to communicate with its user via cellular pager).

5.8 Kuo-Cho Lee's ITX Architecture for Agent Control

Cooperative software applications that handle distributed and heterogeneous computing and networking platforms require a new paradigm—that of a distributive cooperative task. "In this paradigm, an agent supports a user, represents the user to the system, and handles complex interactions with other cooperating agents and system resources" (Lee, Mansfield, and Sheth, 1993). One critical issue is how to meet the application's objectives (or goals) when faced with unpredictable user input and various types of failures.

In addition to all you've read so far, an agent architecture should deal with issues such as the following:

- Failure modes
- Policies regarding propagation of failures, error message, and alarms
- Making sure that there is no single point of failure
- Prioritized goals
- Consistent manipulation of states and superstates
- Dependencies and change propagation

As you read about these different architectures, realize that the optimal architecture for a particular situation will probably borrow aspects and ideas from most existing architectures. One particular architecture, referred to as interactive transactions (ITX) (Lee, Mansfield, and Sheth, 1993), is focused on modeling and controlling the interactions among agents by integrating two important concepts from the real-time world: adaptation via feedback-based control and transaction processing.

As Fig. 5.6 shows, ITX uses shared objects stored in heterogeneous databases to support agents across environments. When two agents need to cooperate

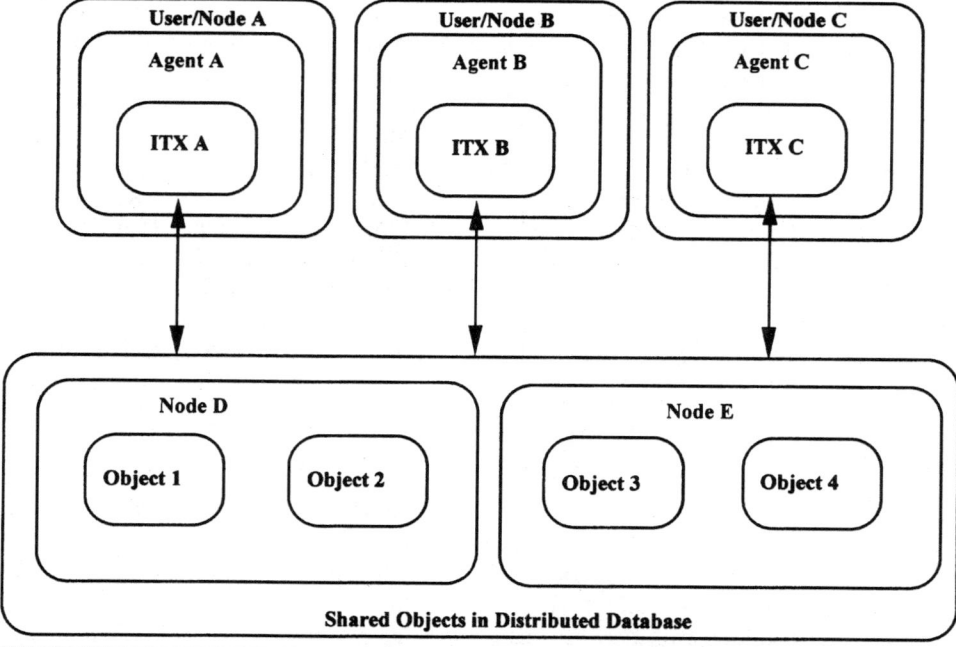

Figure 5.6 Interactive transaction agent architecture. (*From Lee, Mansfield, and Sheth, 1993.*)

on a task, ITX models and coordinates their efforts as an interactive transaction on one or more shared objects. The shared objects represent resource and control information. In this fashion, agents exchange knowledge and collaborate on tasks without actually communicating directly.

This is somewhat analogous to a blackboard approach, where agents coordinate themselves by posting and reading information to and from a shared set of persistent objects. The operational aspects of this ITX agent-interaction architecture are as follows. Each agent has one or more goals or objectives to reach, expressed in terms of an overall system goal that the agents are collaborating to achieve. These objectives are specified as states. The overall system goals are specified as the collective state of the shared objects. States of the shared objects are monitored and differences noted with respect to a known, correct stable (goal) state, called a *fixed point* (similar to a checkpoint in press control). In case of a system failure or other exception, the fixed point can be used to transition back to a stable state.

Each agent iteratively performs atomic transactions on one or more shared objects to achieve its own local goals and, therefore, the systemwide goals. The agent's logic determines which transaction or method to execute. When there are no changes to the system objective after two successive transactions by a particular agent, *that* agent is said to have achieved its local goals. The agent still monitors the appropriate shared objects or system state for changes that violate its fixed-point criterion, and therefore its own goals. An agent can reach a fixed point with respect to its own context, but not in terms of the overall system goal. In that case, different transactions are tried to come into compliance with the overall goal. In this manner, all agents converge on the systemwide goal.

Transactions can be as simple as writing (or reading) a single state variable, to executing a complex series of methods that end up changing a lot of information in the shared objects. Other features of the ITX system framework include:

- The ability to add and delete agents dynamically as needed, without interfering with any ongoing agent activities. As Lee and coworkers state: "This feature is useful for real-time conferencing systems....In addition...we can dynamically add administrative agents to observe the cooperative work for billing or debugging purposes without affecting the ongoing application" (Lee, Mansfield, and Sheth, 1993). We believe this feature would come in handy for any real-time system.

- "...Accommodation of unpredictable user interventions" (Lee, Mansfield, and Sheth, 1993). Dealing with the user is one of the most difficult activities any developer faces. If you allow too much freedom, either the code gets too complex or you end up not accounting for some combination of user actions. The other option is to constrain users at the risk of annoying them, or worse, rendering your system of agents useless in their eyes. Therefore any architecture that has the hooks to help you overcome the unpredictability of end users is bound to be a big plus.

- Robustness in the event of individual agent or system failures. The ITX architecture, because of its "hysteresis" response to a variation from a fixed-point state, allows either rapid recovery or graceful degradation. A system process detects any failures of equipment. For example, it starts the iterative transactions designed to attempt restabilization by the user or application agents and any system agents. If any agent's transaction fails, "...the cooperative work can continue, because all the information required for a cooperative task is reliably stored in shared objects that are updated using atomic transactions" (Lee, Mansfield, and Sheth, 1993).

- "The ITX system also provides a convenient way to enforce constraints on system resources" (Lee, Mansfield, and Sheth, 1993). This is a potentially very important feature. One of the things your agents will have to constantly deal with in a distributed environment, is the negotiation for and securing of resources needed to run. ITX enables this via a resource management agent with its own methods that check shared system resource objects. This agent could execute customized constraint-enforcement methods when its local goal of a relative or a systemwide fixed point of optimum resource utilization is perturbed.

- Provision of reliable communications. ITX use of a pseudoblackboard technique is handled via shared mailbox objects with message queues. As Lee et al. state: "The transaction facility permits atomic write operations to multiple mailboxes. It can also guarantee that messages delivered via transactions are serialized among each other" (Lee, Mansfield, and Sheth, 1993). This results in consistent, in-order mailboxes. As Lee et al. (1993) emphasize: "Using the ITX system, the...[agent] designer can concentrate on solving the problem itself rather than on the mechanisms for reliable communications." We heartily agree.

For those developers interested in ITX, it is currently prototyped by Lee and his group in a distributed workstation environment. The agents are coded in C++, and they use an OO database to manage the shared objects.

5.9 Edmonds' Collaborating Agents—A Federation Architecture

A useful collaborative system of agents should, unless otherwise specified, account for the fact that agents from many developers will occupy that system. In other words, it's a heterogeneous world out there. One of the aspects of heterogeneous worlds is that things—expected and unexpected—emerge.

For example, a society of agents, intended to support collaborative design processes, should be able to support emergence by definition, because new designs emerge from that process. Designs (whether circuit, home, bridge, or other systems) usually result in things or objects that were not in the design domain to begin with. Agents must therefore be able to help with the construction and destruction of novel objects. One approach to designing such agents, presented by Edmonds et al. (1994), imbue agents with pattern-recognition capabilities. Note that pattern recognition is the forte of neural networks, discussed in Section 3 as one of the important technologies enabling agents' intelligence.

In any domain you may be able to design all agents to collaborate on a task in a homogeneous way. This is a viable task if you can assume or design all agents to have (or have access to) a common base of understanding of the domain. But since you may not have control over all aspects of all the agents' designs, you need to know some concepts related to the sources of heterogeneity among agents. You can then decide whether and how to deal with it, or make your own design, avoiding the issues about to be discussed next.

Some of the important ways in which agents can differ include the following:

1. *Knowledge representation techniques.* Requires the use of translators.
2. *Techniques for processing knowledge* (domain or about other agents). Finding out how other agents are doing their tasks may be especially difficult if their designers made such "methods" private or deem them proprietary (possibly encoding such techniques).
3. *Meaning of a piece of knowledge.* The semantics of a particular representation may be different across different domains, or within the same domain, and knowledge can have time- or scenario-derived meanings which, if not correlated by an agent, might be a source of confusion. For example, simple logical inference says nothing about time (when something is true).

When developing agents to help with collaborative or group design tasks, for example, the semantic understanding differences among agents are especially important, because as new things or objects emerge, so do new meanings.* According to Edmonds et al., a "society of agents" and the underlying infrastructure to support groups of interacting people must:

- Provide context-sensitive dialogue, that is, dialogue that is responsive to the current state of interaction
- Provide conference-management functions, such as enabling users to join and leave the group
- Enable group interaction with a remote geographic information system, that is, enable users to invoke an application's functionality and view one another's results
- Provide awareness of other members of the group
- Enable individual preferences
- Provide advice on group administration

---

*This is where ideas from Genesereth's agent communications language and facilitator-based architecture can be very useful.

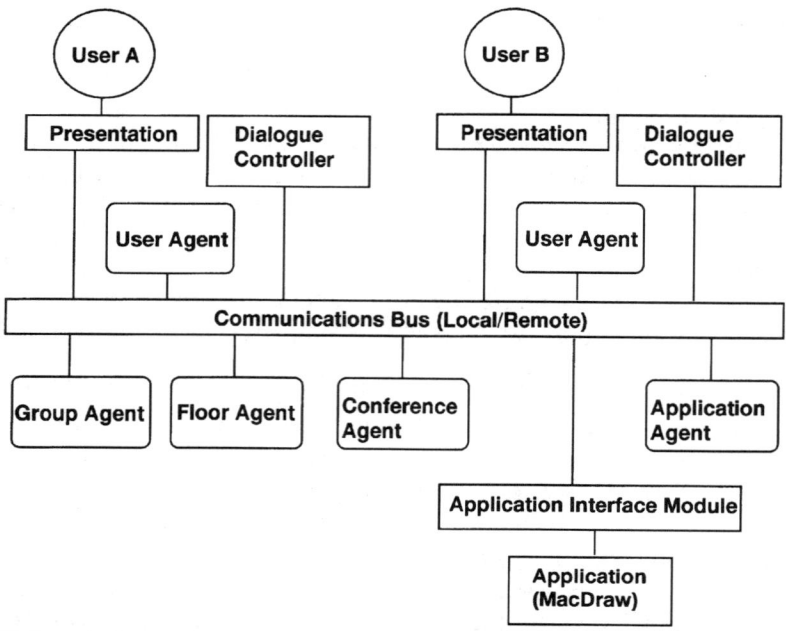

Figure 5.7 Conference support system. (*From Edmonds, Candy, Jones, and Soufi, 1994.*)

The "federation" of agents that Edmonds et al. put together is illustrated in Fig. 5.7. The collaborative conference support system architecture consists of six classes of agents:

1. A *conference agent* helps the person who set up the conference initialize the environment. This includes identifying the other participants and their locations, and specifying what applications are going to be used. The conference agent then creates each user's presentation layer and dialogue controller as well as an appropriate interface for the specified applications. It then starts the other agents and the applications for all the users. During the conference users can change the configuration via the conference agent.

2. A *floor agent* enforces a floor policy such that only one user interacts with the system or application at a time. It interacts with the user agent to accomplish this.

3. An *applications agent* provides services to users from possibly remote systems and applications. It interacts with the dialogue controller and the application interface module (AIM) to ensure that each user's view is updated with changes and delivers application messages to the AIM.

4. A *group agent* represents a group of users. The group has "global" attributes that no single user has. For example, the group agent is configured to allow a single user to end the conference.

5. A *user agent* is the interface between the user and the presentation layer, other group agents, or dialogue controllers. It controls the interaction of a user with the system. It handles syntactic heterogeneity, that is, it interprets different people making changes to shared documents via possibly different means.

6. Although not directly shown in the figure, an extension of the user agent, but maybe contained within the useragent, called the *emergence agent* enables the system to accommodate semantic heterogeneity—how the meaning of work in progress can change depending on what is done to it.

5.10 Agent Architectures and Emergence

In Section 2 we mentioned emergence, the synergy of the whole producing behavior and attributes that are greater than the sum of the individual parts. Certain AI technologies, coupled with appropriate architectures, or the way in which agents are put together or relate to each other in a system, could give rise to a form of emergent intelligence. Some of the issues related to emergence are as follows:

- How to develop robust agents capable of making local decisions?
- How to ensure that local decisions are globally optimal?
- How to program for emergent behavior?

One promising part of the whole answer involves the next type of architecture we'll look at, called subsumption architecture (SA). The subsumption architecture is based on the primacy of the sense and act paradigm (versus a sense, plan, and act paradigm), which governs agent behavior. This is essentially a reactive mode of existence. Yet using a society-of-agents model, surprisingly complex and sophisticated behaviors can arise from a subsumption architecture. The important aspects of a subsumption architecture include:

- Complex, intelligent behavior "emerges" from simple elements (presumably pursuing local goals) interacting within a complex environment.
- Agents intercommunicate.
- Change of state may occur upon timer expiration or message receipt.
- Agents do not wait for responses to messages they send, so no hanging or dependence on other agents.
- Agents are organized into levels: simplest behaviors at lowest level; more complex behaviors that "subsume" (take priority or cause a lower-level

agent's state change) as you go up the hierarchy (like Minsky's society of mind agent—a common thread).

- No planning is involved, so the system is capable of real-time responses.

In the federation architecture discussed earlier, collaboration among agents is partially supported by the emergence agent, which tracks changes in meaning as the system unfolds in time. Edmonds' example use of the architecture details a design process that leads to shapes, shapes "not directly implied by the existing components of [existing] images" (Edmonds, Candy, Jones, and Soufi, 1994). Emergent shapes occur when a change in description or structure occurs. Therefore an agent needs to recognize these changes and facilitate interactions thereof.

To see what is involved in a computational approach to emergent agents, consider that they "must handle the deconstruction and reconstruction of objects.... [they must be able to] track, predict or, at least rapidly, follow a member of the design team as they perceive emergent forms in the design representation" (Edmonds, Candy, Jones, and Soufi, 1994).

The approach by Edmonds and coworkers to making an emergence agent with these capabilities lies in modeling the action and perception aspects of the design domain. For example, a drawing modified (the actions) using MacDraw is converted to a pixel representation and then presented to an image analyzer to see what can be seen (the perceptual process), which is more than originally drawn. This study, although not detailed, provides a direction for others to follow in constructing emergence agents.

5.11 Sugawara's Architectural Concepts—Dealing with Change

Dealing with change is a constant requirement in today's fast-moving, networked computing environments. Developers are unleashing all kinds of new functionality, applications, and agents. New tool sets and OO development promise even faster release cycles. And change hits from another front as well: unexpected hardware and software "occurrences"—you know the type; someone runs into the power pole and the node doesn't have an uninterruptable power supply (UPS). Some network "newbee" hoses up the works with a circular posting to <everyone>. A robust agent architecture needs to take into account these unexpected annoyances and, if not weather them intact, at least degrade gracefully and recover once the system is functioning again.

A partial answer to some unexpected occurrences, Sugawara et al. (1995) agent-oriented flexible information network (AFIN) is based on a multiagent paradigm. This paradigm has the ability to evolve and produce satisfactory behaviors when new requirements from the environment occur and new objects are introduced into the networked system. This architectural approach posits three properties essential to realizing an AFIN: intelligence, homeostasis, and evolution.

- *Intelligence:* FINs need to understand and interpret user requirements, even if in nonstandard form, or if incomplete or ambiguous. If FINs cannot fully understand requirements, then at least they should not degrade the network and other components into dis- or nonfunctionality. They should degrade gracefully or ask for more information to complete the task or operation. FIN-based agents (objects or software components) should have access to global accumulated knowledge and be able to increase that knowledge from experience.

- *Homeostasis:* FINs and their agents should detect changes in the system and its interconnections (such as by comparing a snapshot of the system with a comparator agent). If the changes are negative, then agents work to minimize negative effects on the overall environment or users. If a positive change occurs, then environment-modification agents are called.

- *Evolution:* Information network (IN) changes controlled by an evolution mechanism (EM). FIN-based agents implementing the EM would reorganize the system's configuration autonomously (for example, presumably updating topologies, graphs, or routing tables, adding available new methods and function or property descriptors to tables, addressing changes, and so on). Then a new snapshot is taken and stored.

A FIN with these characteristics can be modeled or implemented by considering the network and its environment as a set of autonomous agents that intercommunicate within and across agent hierarchies. Modular agents can be replaced or added without untoward effects (a tenant of OOA/D/P). All agents understand the coordination, negotiation, and communications protocols within the system.

A primitive agent could be specified with the following modules or protocols (groupings of similar methods, as in Smalltalk)—communicator, domain specifier, cooperator, and task processor, as shown in Fig. 5.8. These components have the following features:

- A *communicator* protocol has methods enabling interagent and agent-to-environment communications.

- *Domain specifier* protocol methods provide the agent with knowledge about that particular agent's roles in the system and its expertise within those roles. A method within this protocol accepts or rejects incoming messages (problems to solve).* Polymorphism, overloading, and message-float or discrete message-send mechanisms "upward" along agent class and composition hierarchies. For example, with inheritance, agent superclass variables and methods are available to the subclass agents as well. These mechanisms can potentially greatly expand an agent's capabilities.

---

*This run-time determination is in contrast to compile-time binding, where every module, object, or agent knows what kinds of things it is going to do ahead of time.

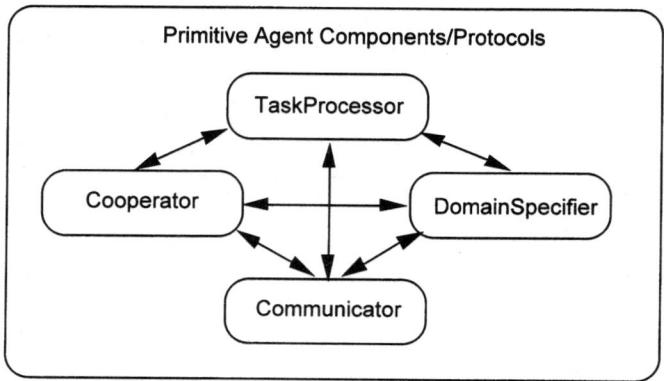

Figure 5.8 FIN agent components. (*From Sugawara et al., 1995.*)

- *Cooperator* protocol methods communicate with peer agents to plan the cooperation and recovery of problem situations, as well as conflict negotiations (such as voting) and resource sharing.
- *Task processor* methods are the rules or other logic and algorithms (such as fuzzy rules, neural network, I/O, conventional, programmatic, and so on). This includes any behavior, implemented within an agent, that solves the problem and responds accordingly.

Another concept in this architecture is the notion of hierarchies of agents, implemented with agent groups.* Organizations can be created on the fly, grouping available agents and assigning tasks according to a preprocessing step that breaks a problem down into smaller, more primitive problems. This capability can be more easily enabled by using a discrete representation of connectors and connections, as pointed out in Section 3.

As shown in Fig. 5.9, an organizational agent would have the same types of protocols as any other primitive agent, plus it would be responsible, under its task processing protocol, for dynamically assembling or disassembling primitive agents appropriate for the problem or message given. It would be the agent that, for example, would issue the ConnectTo or DisconnectFrom messages that would be sent to a connector object and executed there. This agent would also be responsible for maintaining a group "superstate," as a composition of states of its "contained" primitive agents. Superstates are helpful in determining progress toward higher-level goals.

In Fig. 5.9 note that a group agent can have, as one of its contained agents, an organizational agent that takes care of the coordinating protocols or states

---

*This notion of groups of agents is similar to concepts presented earlier: the "society of mind" relationships and neuronal groups discussed in Section 2.

Agent Architectures 191

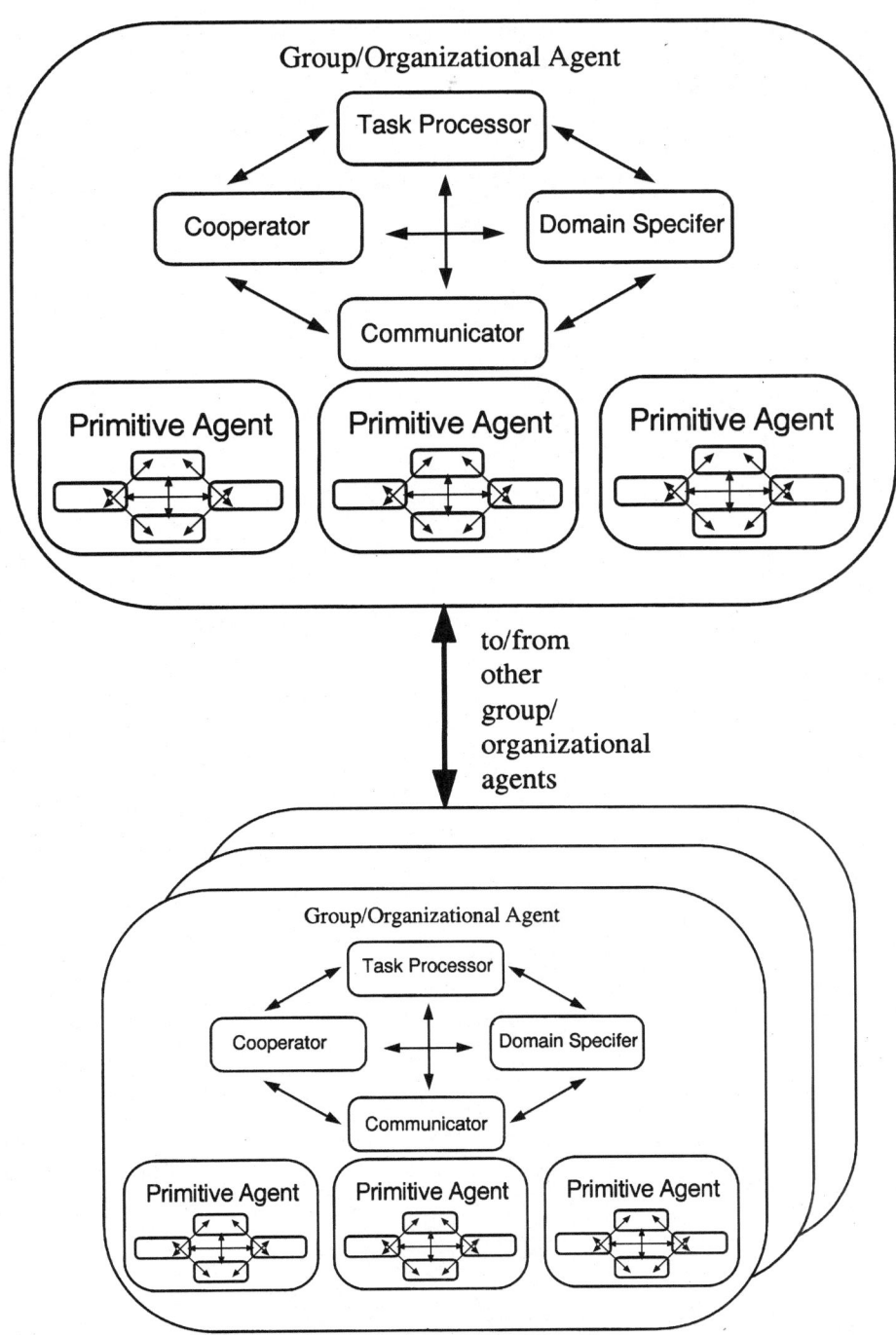

Figure 5.9 Agent organizations.

of the group's primitive agents, *or* those protocols can be part of the group agent itself. It all depends on the desired level of modularity.

Once a designer has specified the types, or classes, of agents, using good OOA/D techniques, one can see how implementing these agents within a system that enables grouping (composition or containment) as an explicit representation of hierarchies and other relationships would give rise to a robust and rich system.

Section

6

Agent-Design Considerations

This section presents a collection of relatively independent topics that should also be considered when designing your agents and agent system. Together with what you already have read in the previous sections, this information is helpful in analyzing what is important in your agent application and what is not. There is no particular sequence to these topics; some of them are expansions, summaries, or follow-ons to information presented earlier.

Some topics are included here because there was really nowhere else they fit—yet, as we researched this book, we came across nuggets of information that we thought were important for you to know. Where appropriate, we also point back to relevant sections where the topic is covered in more detail or in a particular context.

6.1 Introduction

As we already discussed, IAs have some degree of autonomy and the ability to make decisions on behalf of themselves and their creators an users. To be truly effective, agents must be able to sense or perceive relevant aspects of their environment, then rationally control or act on at least one of those environmental aspects to affect their creator or users. Many of these activities are goal-based.

For example, an agent at an on-line auction can analyze the merchandise up for bid, the other agents at the auction, and its owner's needs to come up with a bidding strategy that could win at least some of the merchandise its owner wanted. The agent could then pay for the merchandise and arrange to have it sent to its owner.

Designing IAs with such abilities and attributes is no easy task. As we've seen in previous sections, no less than the bulk of computer science, and specifically AI and other cognitive sciences, have been devoted to making more intelligent machines. Now we ask that you shove all that into a few software entities and push them out the door to do your bidding.

Hopefully, as you progress in designing agents, you will narrow your agents' environment or domain of application, investigate and use appropriate AI techniques, and decide on a supporting infrastructure and a way of organizing your agents into a suitable architecture. But even as we've reviewed some important ideas and technologies in those areas that will help you in that seemingly unattainable endeavor, there remain a number of topics that do not fit within the categories so far covered. Let's get started.

6.2 Designing Agents—The Big Picture

One of the first things you would do in creating any software package is a requirements analysis. You analyze the domain, talk to existing and potential users, and generate lists, narratives, scenarios, user profiles, and so on. The following gives some ideas for requirements analysis possibly unique to the intended creation of agents.

6.2.1 Agent requirements analysis—general aspects

The phrase "intelligent agent" implies autonomy, and the capability to set and accomplish goals to ensure survival and prosperity. A human agent is empowered, commissioned, and educated to turn information and other, physical resources into useful products and services. To be successful, human agents must monitor and understand changing conditions and adapt to them. For human agents, such as those dealing in talent, insurance, real estate, or secret documents, decisions made and actions taken based on an understanding of what is going on around them are the difference between success and failure. To better understand what it takes to design successful software agents, let's look at the general attributes of successful human agents. Intelligent human agents generally:

- Are licensed, commissioned, or have other bona fide qualifications.
- Are empowered to access and analyze information and take action in their fields.
- Can act independently or in concert with other entities.
- Are trained in their fields and have the knowledge necessary to accomplish needed tasks.
- Have access to needed information.
- Have the necessary resources, that is, capital, communication, transportation, office space, and so on.

- Can create and work toward meaningful goals.
- Can meaningfully communicate and route results to appropriate collaborators or clients.
- Can take action to change their environment.
- Can set new goals to reflect a changing environment.
- Can measure performance against goals and learn or change behavior to reduce the error or distance between them.

Somewhat analogously, an intelligent software agent generally:

- May be launched by a user or another agent. Once launched, Newton's second law of motion applies. The agent executes or waits until termination conditions within a system are satisfied, or until an explicit termination message is sent, the nodes where it's executing crash, or its task is completed.
- Obtains rights and permissions to use resources (CPU, memory, electronic money) and access data.
- Contains algorithms, conditional statements, fuzzy logic, neural networks and the like, for making decisions about its environment and executing its tasks.
- Can run as an independent, asynchronous task.
- Can work toward the users' goals, and may create subgoals.
- Communicates results, accepts input from users or other systems.
- May take action to change its environment.
- Can measure performance against goals.
- May set new goals to reflect a changing environment or to reduce error between a goal state and the agent's current state.

6.2.2 Agent requirements analysis—specific aspects

To determine the specific requirements for a particular type of agent, a good practice is to reduce the purpose of the agent to an operational narrative about what the agent is supposed to do. Start with a simple paragraph, then expand each sentence with details. In developing the narrative, it is often helpful to ask and answer some of the following types of questions:

- What data will the agent deal with? More specifically, analyze the data to be monitored. Consider meaning, volume, timing, and type.
  - *Meaning:* What is the context of the data, what are its semantics?
  - *Volume:* How much data will your agent have to deal with?

- *Timing:* How often do the data need to be refreshed?
- *Type:* What formats must be supported?

- How will the agent execute its tasks? Autonomous IAs are, by our definition, independent asynchronous tasks. Depending upon how tasking is supported in the agent execution environment (processes, threads, lightweight threads, and so on), tasking must be planned to take advantage of the tasking priorities and types of processes provided (threads versus tasks, for example). In the case of mobile agents, the actual launching of the task is handled by agent brokers (such as a CORBA-like ORB) at the respective host sites, but you must be aware of the tasking options available on the hosts.

- What are the trigger conditions? Agents can be explicitly launched by a user or, more commonly, set in motion as a consequence of system events such as the arrival of a specific E-mail message or the updating of a value in a database. As discussed in Section 5, agents are also commonly launched and monitored and controlled (to varying degrees) by other agents. Or the coordination of societies of agents is a self-organized activity. Agents usually monitor their environment and then take some action when specific data values are encountered.

- What action is to be taken when a trigger condition occurs, and how does that action get the agent closer to its goals?

- What is the goal hierarchy of the society of agents? For example, if the user's goal is to make money, a financial advisory agent may accomplish this goal by setting and accomplishing subgoals, by delegating to transaction agents. Each transaction agent may have a goal to monitor, buy, and sell individual securities at a profit.

- What are the authorization and security concerns? More on this later.

- What resources are needed? How are they obtained, and how does the agent negotiate for them or otherwise obtain them if necessary?

- With what other agents or systems does this agent interact?

- What protocols are used for communication among agents and between agents and other parts of a software system?

6.2.3 General design considerations

Kautz, Selman, and Coen (1994) identify a list of major design considerations, of which each must be addressed differently for agents than for interactive software packages. These include the following:

- A user-friendly interface. This includes an agent builder, an agent launcher, and means of communicating either results or errors to the user.
- Redundant error handling. When an error occurs in most interactive software packages, a message box is immediately displayed, allowing the user

to take appropriate action. Since virtually all autonomous agents would execute asynchronously, other means of error handling need to be employed. One error-reporting method, for example, is to write errors into a log file, which may then be mailed to an error address, along with the agent itself.

- Use of simple existing utilities. Agents, especially mobile agents, should have a small memory footprint, so redundant processing definitions are an especially bad idea. The agent must rely on an execution infrastructure and essentially make API calls to perform the basics. Otherwise the server or host upon which the agent executes may not have space to run the agent. Taking into account that agents may simply run forever as they monitor a data stream or streams, they need to be unobtrusively embedded into an environment. In crafting a successful agent, it is of the utmost importance to know the functions and utilities that are supported by the hosts, including operating systems, infrastructures or middleware, and, if applicable, a VM, upon which the agent executes.

- Reduction of complex problems to many simple interactions between agents. We do not know yet how to build a generalized agent, and specialized agents are extremely limited. The real power of agents lies in their ability to cooperate in order to make the results greater than the sum of the parts. We do this in other contexts by combining the functionality of one utility with that of another. For instance, we can chain a data gathering function with several data processing applications and output formatters by using data pipes and I/O redirection. Work-flow agents use this strategy to model user-defined processes such as waiting for E-mail from a specific person, printing the message, then notifying others by E-mail and sending a file via fax.

Probably the hardest part of designing an agent from scratch is to define tasks that are beneficial to end users, yet can be accomplished using available technology. One key, therefore, is to identify and develop low-level functions which can be combined to perform useful tasks. Examples of such functions are APIs for E-mail and interprocess communication.

The ideal agent seamlessly blends into the way a company or an individual already conducts business, and responds to domain-related goals. Today agents are capable of performing combinations of low-level clerical tasks on a user's behalf, but users, for the most part, are not yet able to safely assume that these tasks will be done correctly.

As agent technology improves, end users will expect agents to enhance productivity and reduce complexity. As with most new types of software, user expectations will greatly exceed anything developers and products can deliver. Eventually, as Isaac Asimov predicted of robots, agents and human users will develop realistic expectations and begin to work together productively.

There are, of course, vast differences in the perspectives of a software developer and an end user. While the end user is concerned only with ease of use and correct, timely results, the developer cares about run time and development infrastructures, among other things. To be successful, an agent system must work for both groups. When integrating IAs into an existing system or application, the most important factor for both the user and the developer is, of course, how well the agents and the agent run-time environment fit into the current mix of systems and applications.

Ideally, both the agent and the systems with which it interacts execute within an open, consistent framework. This makes it possible for agents to interact with each other regardless of how they were developed, and for agents to monitor and control a variety of applications and devices.

You will likely design your agents based on several criteria, not the least of which being the real-world environments and tools you intend to use to create an agent-based application. The major components of an agent system include an agent builder or suitable development environment (Section 7), an agent infrastructure (Section 4), and, of course, agents, together composing an architecture (Section 5) within an execution environment. Sometimes you will find many of these components provided within one overall environment. For example, the Telescript-based environment from General Magic provides an agent-oriented architecture, an agent specification and communications language, and a VM-based execution engine.

The spectrum of development approaches to creating advanced, effective IA systems is perhaps bounded by the two extremes mentioned in Section 5, where we discussed architectures. One approach is to create agents that have a great deal of embedded sophistication and complexity. This may also involve designing your agents within an equally sophisticated architecture, which tries to anticipate all the relevant situations that might confront the agents, and typically hard-coding the responses to them. The other extreme is to develop relatively simple agents within a simple architecture. This approach emphasizes robust and sophisticated communications mechanisms, combined with the ability to learn, the ability to reach out beyond themselves for knowledge and reasoning capabilities, and the use of appropriate, hierarchical goals within the agents and the overall agent system.

The components of an agent are explicated by analyzing what an agent is going to do—its behavior. First, just like its human counterpart, every agent must have the ability to observe conditions and respond to them. And as pointed out in AIAMA, the ability to respond rationally is of the utmost importance. To do this, an agent must have an appropriate model of the world in which it executes, and the ability to reason about that world. World modeling and reasoning can take several forms. Typical AI-derived approaches include using common if-then production rules, constraints, frames, various types of knowledge networks (such as petri networks, navigable graphs), assertions, and state-space representations using variables.

An approach we favor is the object-oriented method of analyzing, designing, and implementing both the agents' environment(s) and the agents themselves as collaborating objects. All the other approaches can be embedded or encapsulated within suitable objects and their methods. Even if your agent environment is not OO, you can model the relevant environmental features as an object, thus preserving the architectural purity of your overall agent model. The environment objects then serve as repositories for the interfaces between your agents and their world.

As illustrated in Section 3, the various agents collaborating on the manageMyStockPortfolio task talk to objects and functions in the world that are not necessarily agents. For example, the stock broker's system, the servers dispensing company financial information, and databases holding general economic data might be neither agent- nor object-oriented. But you can nevertheless model those entities as objects and encapsulate the interfaces between them and your agents within appropriate protocols and methods that are part of specific agents.

How well the agent models and responds to the user's world and real needs is an important indicator of an agent's intelligence. An IA may, for example, observe that you periodically get mail asking for a specific report. It may also observe how you generate the report. It can then ask you if you want it to generate and send a report to the requesting party whenever you receive a mail message containing certain key words from a specific set of people. Depending on your response, it then acts on your behalf.

Besides rules and a model of the world, an agent must have one or more goals. Otherwise the agent may simply proceed on its merry way, doing its master little good, and perhaps not a little harm. Using goals, an agent can check its results against the goal to make sure it is doing what it's supposed to, or at least improving. An agent may also simulate a series of actions to determine how close to its goals the actions take it. A user may have a goal to make a network more efficient, for example. Imprinting this lofty goal upon an agent requires translating the user's needs into precise, formal, sometimes quantifiable expressions. Some commercial applications have a rudimentary form of user-driven configuration of preferences, much like the filter preferences in an E-mail package.

Someday it may be common to find software which can "automagically" translate more sophisticated user goals into agent-understandable goals. Perhaps with the advent of commercial applications of robust natural-language processing (NLP) techniques, users will be able to express their desires more naturally to an agent, and lo and behold, an embedded NLP capability will interpret and transform that expression in terms the agent can deal with. If incorporating NLP processing into your agents is not part of your design, users must provide and explicitly represent simple goals, such as "find and execute all test scripts against version 4.2 of software image xxx." Developers must, in turn, translate the explicit goal into a method call or variable value with which the agent can work.

The root of an agent system's goal hierarchy is the user's main goal. Below that, there may be supporting subgoals. For example, the goal could be to plan a successful ski vacation in Utah. The agent could look up "ski vacation" in its dictionary and determine that a subgoal would be to get the best deal on two round-trip tickets from Kansas City to Salt Lake City. Other subgoals would be to get hotel reservations, check weather predictions and snow conditions, reserve a rental car with a ski rack and snow tires, and buy lift tickets. Depending on the user's priorities, the agent could monitor conditions until a great ski day was found, then finalize all arrangements.

6.3 Agents, Platforms, and Environments: Where Agents Fit In

Since agents are going to play a big part in future computing, properly fitting agent technology into an application or an operating system can provide big advantages. Successful agents must, of course, fit into the existing operating environment and applications. This can largely be accomplished by a comprehensive agent execution layer (such as Telescript, Smalltalk, or Java-based virtual engines), which is integrated with an agent builder or launcher along with agent support class libraries, scripting languages, and an API. Once a good agent development and execution environment is in place, agent designers are faced with their greatest challenge: designing agents so that they work with and benefit users.

6.3.1 Execution environments

Most of the work involved in getting agents to behave in ways beneficial to users is done in the run-time system, or execution environment. For the most part, agents cannot be trusted to simply execute on a standard operating system such as UNIX, Windows, or NetWare. There must be an agent support layer or agent execution environment that provides mobility, verification, necessary APIs to provide processing functionality, communication, security, and resources.

6.3.1.1 Requirements for agent execution environments. In conventional compiled systems, an agent would need to worry about memory allocation and the early binding of variables. To be most effective, especially in a distributed system, an agent must run in a dynamic execution environment. The system should be capable of dynamically loading an agent and launching it without forcing the agent to worry about memory allocation. A strictly interpreted system could suffer intolerable performance penalties, so a dynamic, late-binding VM environment, such as that used by Smalltalk, Telescript, or Java, offers significant advantages. The major requirements for an *ideal* agent execution environment are as follows:

- A standard, portable class library (including GUI classes).
- An execution environment must allow only safe methods to be executed on safe objects. For example, mobile agents must not be allowed to access sys-

tem resources such as files and directories for which they do not have specifically granted privileges.

- Automatic memory allocation and deallocation (garbage collection).
- A code loader and checker that loads the code into the correct address space, binds agent variables to the appropriate environment objects, and checks the code for unsafe constructs.
- Support for multithreaded execution. Agents must be able to execute autonomously, and many agents must be able to execute in the same environment. Support for asynchronous threads and rendezvous primitives like semaphores.
- An infrastructure that facilitates agent-to-agent, agent-to-environment and agent-to-user communication. This can be implemented with blackboard-like, semaphore-protected data structures, for example.
- The environment must provide support for persistent objects, since agents need to store information, possibly outside themselves, for later use.
- Protection system resources, and provision for agents to negotiate for and obtain (possibly through purchase) resources and services as needed. This may include digital cash or credit.
- Move or distribute agents safely as per-agent needs and requests.

To boldly go where no agent has gone before, agents need to navigate themselves to, and be at home in, a wide variety of environments, including the following:

- Combination TV/PC/Game-Player/net-access devices, the so-called set-top boxes, and network computers (NCs)
- Local operating networks (LONs) such as LONWorks, which are fast becoming part of many distributed systems based on "embedded" architectures
- PDAs, such as the MagicCap-based ones
- Java-chip-based thin client machines
- Internet and WWW wide-area networks (WANs), including all the supporting machinery, such as bridges and routers
- Mobile wireless communication networks
- Environments running (perhaps Java-supporting) real-time operating systems, such as pSOS and OS-9

These environments require advanced development tools. In Section 7 we provide details on some of the more promising environments.

Java applets, for example, must have a Java-enabled application, such as a browser (for example, Netscape Navigator or Communicator) or HotJava, on which to execute. Java-based agent-building tool kits like Aglets from IBM further enhance the execution environment by providing agent watchers and agent building via configuration. A browserless environment from Marimba Inc., called Castanet, promises a slightly different approach to content distribution (content meaning applications, movies, software, audio, and so on). The slight paradigm shift in the structure of distributed software launched by Castanet is one composed of a tuner (the receiver and interpreter of the content), a transmitter (the content provider), and a channel (the content itself). The execution environment (the tuner) is contained in your machine and receives content updates as sent by the transmitter. In this case the tuner needs to ensure that it is well behaved on your machine.

Other systems, such as Telescript and distributed Smalltalk (DST) (as integrated within ParcPlace/Digitalk's Visual Wave), furnish their own complete execution environments. We'll discuss DST in more detail in Section 7. Still other dedicated expert system environments, like Gensym's G2, and the various tool kits for building soft computing applications (ANNs, fuzzy logic, genetic algorithms) provide complete encapsulation within a proprietary run-time engine, or allow access to the functionality via a well-defined API. For each of these types of tool kits we advise you to look closely at the environmental facilities they provide.

6.3.1.2 Multithreaded, multitasking operating systems. Multithreaded, multitasking operating systems are essential for successful agents. Many available agent development tools focus on building search agents. The agents' ability to search the Internet or a data stream, such as a news wire service, is well known. What often is not considered is that data searching capabilities are only the beginning, and are only part of a truly IA.

Linked together by a good agent scripting or communications language (such as ACL or Telescript), cooperating agents can analyze the information found and take action based on it. As already discussed, we know that in most cases an agent executes as an independent task. Therefore it is of utmost importance that each agent execution platform support multipriority multitasking. As cooperating agents, one set of agents can respond to triggers, launching or delegating to other agents instructions to perform related tasks. For example, when your paycheck appears in your account (trigger), one agent can provide a list of all your creditors. Another can perform a mail merge, while another writes checks or pays electronically where possible.

6.3.1.3 Mobile agents and environments—initial concerns. Beyond true multitasking, an execution environment suitable for agents must provide support for program mobility, security, and resource allocation. As autonomous entities with narrowly focused goals, agents cannot be left on the honor system to act responsibly within a network. Mobile agents have great potential to cause problems within a system as they travel from node to node. A mobile agent can run amok through either maliciousness or erroneous programming, and can become what amounts to a dangerous virus.

Mobile agents are practical only in secure systems, in which the places to which agents can move are restricted according to the authorization an agent has. Once an agent is accepted by a host, system resources must be allocated and protected. An agent must be forced by means of a restrictive language and by using only host or system APIs, or via an agent builder, not to hog or vandalize memory, CPU time, and other resources.

If mobile agents are to be used, both the infrastructure upon which your agents are based (such as CORBA) and the language or the architecture of the agent execution environment must provide a means for agents to move from one host to another—agents cannot be trusted to do this on their own. As of this writing Telescript provides the best capabilities for agent migration, allowing an executing agent to move and retain all of its current execution state. Other approaches simply E-mail or copy a static representation of the agent (that is, the source code, class, or a serialized object in the case of Java Remote Method Invocation (JRMI)) to a location where it can be launched or relaunched.

Most of the agent-moving work must be done in the host or network. A typical host might perform the following:

- Network and mobility support is responsible for moving files to and from addresses. These files contain agents, which must be executed at each host.
- Verify that the agent has all the right codes embedded in the header.
- Discern the capabilities of an agent, and determine whether an agent can execute or whether it is safe to execute. This could mean parsing the agent's source code to find any references to forbidden global variables or functions that aren't understood. The agent must know what functions the host supports, and call only those functions, or the host must be able to load a personality that is indicated in the agent header.
- Perform information, resource, and data access functions, such as getting current time, customer record, and so on.
- If an agent runs amok, stop it and send it back to its user to be analyzed, then change permissions such that the user's agents are no longer welcome.

The following several subsections contain more detailed information on agent mobility.

6.3.2 Distributed computing paradigms

As we mentioned in Section 1, over the last two decades, computing power has become increasingly decentralized—from a single-batch access mainframe to dumb terminals to smarter terminals to very large networks of 300 + MHz desktop workstations. Systems supporting business applications, however, still remain relatively centralized, largely because of data storage considerations. The two most prominent distributed computing paradigms are client/server systems and point-to-point or peer-to-peer architectures.

The traditional view of personal computer and client-server software is highly interactive. Users expect near-instant response times from popular packages such as word processors and spreadsheets, and they expect results to be displayed and updated constantly. Agents present a different interaction model.

Agents may be interactively defined, but are launched by either a user, another agent or a system event and execute as independent and asynchronous processes. This creates a challenge when it comes to returning status or results to the user or to other agents. Some techniques for returning values from executing agents include E-mail, semaphore-protected shared queues, and RPCs.

"In accomplishing its goals, an agent must have not only consistency and security but efficient, safe access to a rich set of applications and functions, which ideally will already be widely used in the enterprise" (Atkinson et al., 1995). As we professed earlier, a consistent agent execution environment, such as Java, Smalltalk, or Telescript, running on a VM tailored to a variety of platforms, is crucial to effective agent systems. Especially in highly distributed systems, following either client/server or peer-to-peer paradigms, this type of environment must provide security such that agents:

- Cannot change or delete system components unless specifically authorized
- Do not monopolize the CPU
- Do not use up an inordinate amount of network bandwidth or memory
- Do not conflict with the memory needs of user applications

Attention to security makes it possible for agents from a variety of developers to interoperate safely, and to include agent capabilities within a variety of products.

6.3.2.1 The client/server paradigm. This paradigm designates specific nodes in a network as servers, where relatively large amounts of computing power, as well as data and software resources, reside. Attached to these servers via a network are many client machines, which are essentially single-user workstations. If this sounds somewhat similar to the old idea of a mainframe computer hung about with thousands of dumb terminals, it is.

The computing world, especially for business applications, is rapidly embracing the client/server computing paradigm. (Even so, the client-server paradigm will soon be eclipsed by the far more agent-enabling peer-to-peer paradigm, where agent applets or servelets are oblivious to location or the notion of a home base—they live in the net, as it were.) Within the client-server architecture there are two places where computation can be done—either in the server machine or in the client machine. There is a growing controversy over where the bulk of such processing should be done. The debate is whether to put more power on the desktop (a trend going back to the beginning of the personal computer revolution) or create monster servers. Most corporations lean heavily toward a fat server and a lean client approach for several reasons. Among these are the fact that a fat server with many thin clients is very similar to the old mainframe systems many data processing managers know and love, and the reality is that in most companies, it is politically, financially, and administratively easier to beef up the server than multitudes of clients.

Some users, however, almost invariably use fat clients—software developers, graphical designers, power users, and so on. It's not cost-effective for these users to be inconvenienced during server downtime, or wait for busy servers to process information. Besides, the Internet is moving toward a fat-client network scenario, since the WWW is essentially read-only, and the best way to cruise the web is to download chunks of it and read them at your convenience. Data buffering and filtering are often best done on a fat client.

The IA answer to the fat or thin question is a smarter network. Components such as agents can execute on any node within the network having idle processing capacity, via an infrastructure that supports distributed agent execution (such as Telescript), not on specific clients or servers. A report by Bloor Research Group says that we are heading toward a more centralized, yet networked computing model that is giving rise to thinner but "growable" clients. This trend is being facilitated by technologies like Java, where software applets and servelets (agents) travel to where they are needed, execute, and dissolve (Keynes, 1996).

A truly effective agent execution environment could mask a large part of infrastructural details underneath, making it potentially irrelevant to the agent developer. Eventually many of the applications now executing in client-server or mainframe environments will run in an increasingly peer-to-peer network such as the Internet, WWW, or intranets. Agent execution environments that provide services approaching peer-to-peer support can help smooth the transition. Environments supporting Java applets and servelets, and Telescript agents can move from machine to machine, independent of whether they are considered the client or the server portion of the application. Indeed, an agent could be both at the same time, taking advantage of the power and resources of any machines with the necessary resources. This approach is becoming more evident in complete portable environments that support the peer-to-peer paradigm, as we'll see next.

6.3.2.2 Agent mobility as extension of the client/server paradigm. With a projected 100 million regular users by 1998, the WWW, not to mention private intranetworks or the Internet as a whole, represents not only the greatest opportunity, but the greatest challenge distributed computing technology has yet encountered. While a mushrooming number of networked computers worldwide lure companies to invest billions on the web, the diversity in operating systems, hardware, data representations, performance, and security considerations stands as a major obstacle in the path of distributed systems development. Stationary search and filter agents are already integral parts of the Internet, since there it would be nearly impossible to find information otherwise.

With virtually all significant resources under the control of server nodes, many of the same inefficiencies and bottlenecks plaguing mainframes emerge in both client-server and peer-to-peer systems. Users depending on a specific server or set of servers, as do most users, can experience a chronically slow response time and highly inconvenient server downtime.

One of the culprits of server-access woes is the way resources are activated, and the way responses get sent back to the requester. Server resource access across networks usually involves some form of RPC in the case of DCE, for example, or remote object method invocation if the OO paradigm is employed, as in the case of, for example, CORBA (refer to Section 4). There are some severe limitations involved in this method of resource sharing:

- The latency and the possibility of communication failure that is part and parcel to the RPC approach can be a serious drawback and should not be hidden away from the programmer.
- With RPCs all interaction between client and server must go through the network, sometimes taking the long way around.
- Typically, an agent, or a program or set of programs acting as server agent (sometimes unbeknownst to the user), according to the client-server protocol paradigm, cannot *initiate* a message.
- The server interface is static and therefore must account for all possible requests.
- The one-response-per-request server functionality limits effective inter-agent communications.

One alternate strategy to invoking processing on remote nodes is to pipe data across a network from server to client and vice versa. Sometimes there is simply no substitute for storing and processing chunks of data on the client machine. While this may be the best solution where relatively small blocks of data are involved, in most cases this is even more impractical than the RPC approach since several large simultaneous data transfers can use up all the bandwidth in a network, leaving other users unable to use the network.

Of course, an infrastructure or middleware approach that is becoming most common is the idea of using request brokers. One agent makes a request of another agent on some distant machine. A local interface to a CORBA-compatible ORB, for example, marshals the request (together with, perhaps, other requests) and sends them to the remote machine, where another CORBA-compatible ORB breaks down the communications and directs the individual messages to the intended receiving agents. You'll see an example of this approach in more detail in Section 7, when we discuss distributed Smalltalk.

Another promising alternative to the current distributed-resource models is the idea of software entities whose structure and function are encapsulated within scripts—messages with data and executable content. This is similar to the practice of sending executable SQL scripts to a database server and getting results back. This is the whole idea behind Java applets and servelets (using JRMI), and is at the core of the Telescript language. This approach is being further enabled at the infrastructure level by, for example, CORBA's TMCF for scripting. Instead of transferring huge blocks of data or attempting to invoke remote processing by sending only executable content, this solution transfers small message packets from platform to platform.

These message packets are really objects containing methods, data, and, in the case of Telescript, the current execution state. When one of these (lightweight) objects or agents arrives at its destination platform, it executes (usually) within a self-contained execution environment that facilitates and moderates any access to local resources. When the agent reaches a goal and finishes processing on the remote machine, it can return, with analyzed and condensed information, to the location and owner (agent or human) that launched it on its way. In this paradigm all computation is carried out on the network nodes where the needed resources (computational and informational) reside. When a remote resource is needed, the computational entity moves to the remote node and continues seamlessly (at least in theory).

Sending executable agents down the pipe also tends to be more asynchronous in nature. Once the agent goes on its way, the sending entity can forget about it until it comes back with results or an error report. Therefore unlike RPC, for example, the communications channel, sockets, or ports can be closed down until needed again. Remote agents also execute unattended. Your local resources are not consumed keeping track of what the agent is doing.

This approach also naturally implies that resources reside somewhat amorphously in the network as a whole, and are accessed via go-betweens, or agent brokers, as opposed to residing and executing on any specific piece of hardware. While there is great potential in this approach, there are a number of problems inherent in it, including security, resource protection, communication, and compatibility across heterogeneous platforms. No solution is perfect, but in the best of cases, agents would execute on the least burdened server, storing most (or all) of the data and other resources they need. Agents can duplicate efforts and bog down networks if processing cycles on a server are at a premium, or a large result block must be sent back to the user.

As part of the solution to these problems, the agent must be in an infrastructure or operating-system-independent format and execute on an agent execution layer such as VM, or interpreter. Several existing languages and execution environments could work well in the mobile agent paradigm, including Telescript from General Magic, Java from Sun Microsystems, and VisualWave or distributed Smalltalk from ParcPlace/Digitalk. These are addressed in the next subsection and in more detail in Section 7.

6.3.2.3 When the Internet/intranet is the computer—agent collaboration and the impending peer-to-peer paradigm. Agent collaboration becomes increasingly important as resources to solve problems are distributed all over the Internet, the web, and within intranetworks. Providing all agents within an agent-based application—a society of agents—with the same communications and knowledge-sharing capabilities necessitates the use of standards at various levels of the application's architecture. Some of these standards relate directly to the burgeoning use of the Internet and, particularly, the WWW.

The web is based predominantly on two important standards that are not immediately or directly amenable to transparent, large-scale agent collaboration and knowledge sharing (that is, fast transport of many lightweight messages). The first standard is the HyperText Markup Language (HTML) [and follow-ons relating to other, more complex or dynamic structures like Virtual Reality Markup Language (VRML)]. The other is the HyperText Transport Protocol (HTTP). HTML is a markup language governing the look, the format of the web pages you see with your net browser. (Of course, with the advent of Java, Java-enabled browsers can now view and manipulate applets running within an HTML-based web page.) HTTP governs the way information gets packaged and sent to and from across the network, using lower-level protocols, like TCP/IP.

Developers creating agents that need to communicate with other agents on the web, using multiple messages, sometimes have difficulty in practice when using the standard HTTP single-request and reply mechanisms. Indeed, part of the problem inheres within the way you now must interact with the web—through browsers. You see browsers function as a client mechanism and, therefore, even with Java-enabled browsers, a Java applet–based agent can only communicate with the server that created it. If the downloaded Java applet-based agent needed to talk to some other server (or "peer" agent) on another network node, forget it (without some fancy and potentially brittle programming).

The only recourse as of this writing is to use the server push HTTP function, effectively keeping the connection between the agents open. In future releases the standard may specify that HTTP includes a client pull and server push functionality. Server push means that the server can send information to the browser. The browser displays the data, but leaves the connection to the server open. When the server or server-based agent detects new or relevant information, it can send these new data to the browser. Client pull means that the server sends information to a browser, including a directive (in the HTTP response or the document header) that says "reload this data in 5 seconds" or "go load this other URL in 10 seconds." After the specified amount of time has elapsed, the client agent does exactly that, either reloading the current data or getting new data.

As the documentation to Netscape Navigator puts it, "in server push, an HTTP connection is held open for an indefinite period of time (until the server knows it is done sending data to the client and sends a terminator, or until the client interrupts the connection). In client pull, HTTP connections are never held open; rather, the client is told when to open a new connection, and what data to fetch when it does so."* These new mechanisms open the Internet to a *semblance* of peer-to-peer communications capability, without the specific need

---

*Refer to the HTTP documentation at HTTP//www.netscape.com/assist/net_sites/pushpull.html for more information.

of other more difficult or costly to maintain programming techniques, nor the necessity of interobject communications support from infrastructures like CORBA/IIOP.

However, until a fully peer-to-peer protocol is developed that functions within the bandwidth offered by the Internet, agent applications requiring relatively instantaneous communications between agents will either have to wait, or be restricted to intranetworks or LANS. Although this restriction seems onerous at first, we feel, for many types of agent applications, that it is not. Many agent applications can be built based on predominantly asynchronous communications, and agent autonomy should let you launch a mobile agent, let it do its thing where it must, then report (or travel) back to other collaborating agents, or to a coordinating agent. For example, a travel agent could send out many slave agents to seek out the best travel arrangements in several different service or product categories (air, hotel, auto rental, local amusement tickets, and so on). Each of these slave agents would go to one or more service or product servers, collect data, and come back to the master travel agent which can then filter, collate, and prepare one optimal or several alternative itineraries.

Continuing with our discussion on existing net/web inadequacies, another problem with web-based standards is that HTML, while an excellent markup language, does not in itself consist of the kind of information that agents can use. HTML describes the format of documents; no task-based semantics are recoverable by an agent. Of course, the agent can parse the HTML page for content that includes a message, a script, or attributes that can tell the agent what to do next. But this is getting close to the mobile agent paradigm, as discussed earlier. Making an agent parse HTML-based pages to extract task-based information (as opposed to content, as in a search agent) is an onerous burden to place on agent developers. The developer of the source of the data, and the developer of the agent that uses these data, must agree ahead of time on certain aspects of the structure and content. This limits reuse and is not the way to do autonomous-agent design.

Some solutions are in the offing. One we mentioned, and that is covered in more detail in Sections 7 and 8, is the Java-based JAT, developed by Robert Frost's group, now at CrossRoute Software Inc. Agents built with JAT are Java applets (or servelets) that can communicate with a standard ACL, like KQML. The JAT 3.0 version (by the time you read this) should have the capability to send messages not just to its mother server, but to agents all over the Internet via an *agent router* (or, more accurately, an agent message router) which keeps open the connection between collaborating agents. This allows multiple browsers (or browser agents) to communicate with multiple servers (or server agents), thus coming very much closer to the peer-to-peer paradigm over the net.

Another solution to the problem of useful agent collaboration over the net relates to the limitations of HTML. There are, as of this writing, proposed extensions and transformations to HTML-based documents. First, as we hinted at earlier, an agent [or any program that can access an HTML document via the

Common Gateway Interface (CGI) or other mechanisms] can parse the content of the page and extract or transform the content into a document for use by other programs. Another approach being worked on at Stanford is to extend HTML to include tags that include the meaning or semantics (metadata) of what the content represents.* An agent could then understand the structure of a document and extract meaning from the structure. A somewhat parallel approach would be to add tags to HTML that specify some ontology or knowledge base that could be shared among agents. The ontology or the inferencing capability of the ontology could then be used to extract meaning from the HTML document and inference on that content to come up with answers to queries, for example.

6.3.3 Considerations for a common agent platform

When we say an agent runs on a network today, we usually mean one of the following:

- It runs on a specific server.
- It runs on a client, using specific server resources.
- It runs on a specific server which references other servers' resources.
- It runs on a client-based browser and server (that is, Java applets).

Each of these paradigms greatly limits an agent's power and flexibility. What agents need to run effectively is an agent-tailored execution environment, or common agent execution platform. Through this uniform agent support layer, the network becomes an execution of each specific computing platform. In other words, the network itself becomes the execution environment.

Agents will be written in a variety of programming languages. This creates a need for a common agent platform, a language-independent means for agent interaction and mobile agent migration. A server's implementation of the platform might offer a full range of services; a personal computer's implementation might offer a subset. An application protocol would interconnect the implementations of different systems so that their agents can interact and their mobile agents can migrate between systems. Such a platform, or agent execution environment, can also ensure that an agent won't do anything destructive to the system or hog system resources. The platform could provide services like these:

- *Agent naming:* Assign names to agents so as to distinguish one from another.
- *Launch authentication:* Authenticate the implementers or originators of agent programs.
- *Agent authentication:* Authenticate the authorities of agents.

---

*The World Wide Web Consortium is investigating several metada approaches or standards for describing data. When implemented, these standards will make life much easier for you and your agents!

- *Agent migration:* Transport agents between computers. The mobile agent's destination might be specified indirectly as the computer that hosts specified stationary agents.

- *Agent permissions:* Assign names to permissions, enforce an agent's assigned permissions, and let one agent grant permissions to other agents or renegotiate its own. The name of a program might be used to detect that it was already present at its destination.

- *Agent relationships:* Assign names to relationships, let agents begin and end relationships, and enforce rules of conduct for relationships (all preferably via explicit representation—for example, using connectors).

- *Agent creation:* Let agents create other agents locally and remotely. A new agent might have the authority of the existing agent and either the same permissions or a subset of them.

- *Agent termination:* Terminate agents in an orderly fashion; for example, by notifying other agents with which they have relationships. An agent may be terminated in several ways, including receiving a termination message from an outside force such as a user, a system, or another agent; or it may check its current status (elapsed time, resources remaining, mission completed, mission not possible, and so on) and decide to terminate its own execution after notifying all interested parties.

- *Agent waiting:* Store (cache or write to disk) agents that must wait for long periods of time for events to occur (for example, for relationships to be established or networks to come back on line).

- *Agent checkpointing:* Checkpoint agents to disk so that they survive crashes of their host systems. A transactions service might coordinate the checkpointing of related agents.

- *Agent interaction:* Let related agents interact. The means of interaction (events, pipes, RPC, mediated method invocation) might depend on whether the agents occupy the same or different systems.

- *Agent management:* Let network management and debugging agents monitor other agents. A debugging agent, for example, might be able to suspend and resume agents under development.

There are several contenders for such a common agent platform, but in a restricted sense. The virtual engines supporting Java, Telescript, Inferno, and DST all provide an environment where applications and agents written in a specific language can execute, once removed from the operating system and network infrastructures.

The companies supporting these and other virtual engine technologies are collaborating to varying degrees to try and get their agents, codified in one language, to run, or at least be recognized by the virtual engine dedicated to a different (usually competing) language. For example, Inferno, the fixed VM from

Lucent Technologies, Inc., may offer execution of Java agents as well as the native Limbo language entities. And Sun Microsystems and General Magic are collaborating on integrating their respective offerings such that Java applets and servelets can interoperate with Telescript agents. Indeed, it may be that extensions to CORBA-based middleware can translate between agents and agent environments couched in different languages. After all, the CORBA IDL is intended as a middle ground for interobject communications. Its usage can be extended to a true lingua franca between agents written in different languages.

Another alternative to agent communications is to make use of a common ACL, based on some standard, such as KQML, and a shared knowledge base or ontology, perhaps also based on a standard such as KIF. KQML-based agents are sometimes referred to as *typed-message agents.* In Section 5 we discussed architectural approaches that illustrate the use of this approach. For now let us say that KQML-based agents collaborate to achieve some goal, the message semantics are typed and are independent of the application or domain in which they are collaborating, and, because of the heavy use of collaboration, the peer-to-peer protocol is essential to the efficient performance of their tasks.

6.4 Agents and Humans

Agents are created to serve human needs, and truly useful agents can be modeled on human agents. Without proper direction, resources, permissions, and so on, it is impossible for either a human user or an agent to achieve the goals they strive for. If either a human or a cyber agent labors under false assumptions, or doesn't have sufficient support, failure is certain. One of the keys to a successful relationship, as stated earlier, is for humans to develop a realistic understanding of what agents can and cannot do.

With cooperation between human users and agents, success is possible, given that the scope of an agent's mission is usually narrowly and completely defined. (This is not to say, however, that the agent is incapable of learning how to better carry out its mission.) While an agent can model some of the behavior of a good butler, for instance, it will be some time before even clever AI-based agents have the same kind of human-based understanding that comes from *being human,* and a human user of agents should not expect miracles in this regard. That being said, let's look at some of the anthropomorphic considerations associated with good IA design.

6.4.1 Agents and humans working together

When dealing with user requests, agents can exhibit some degree of intelligence via the following characteristics:

- *Goal-driven:* The user provides a high-level goal, which the agent figures out how and when to meet, using the tools at hand. An IA can create achievable subgoals and enlist the help of other agents to achieve them.

- *Open-minded:* The agent can tolerate a reasonable degree of ambiguity—alternate spellings, synonyms, multiple locations for the same person, and so on. With the addition of AI capabilities provided by fuzzy logic, agents can handle a range of seemingly amorphous data.
- *Practical:* In addition, there are the costs to be paid in time, disk space, money, and the user's time in providing information commensurate with the potential rewards.
- *Well-connected:* The most successful agents can access an execution environment, providing uniform interface to a wide variety of services.

Consider the human equation when designing an agent for a typical business application. In most cases a business consists of material, processes, people, and, most importantly, information. The processes and information a business uses in departments such as finance, human resources, and purchasing are some of its most important assets. Often tasks such as processing a loan, purchasing supplies, or hiring a new employee are codified somewhere in a dusty (paper-based—remember paper) volume, leaving human workers with the responsibility of making the processes work, sometimes by hit-or-miss or overlapping methods. In addition, when human workers leave the company, they take their hard-won knowledge with them.

Agents can provide the greatest leverage in such a context, when they are used to automate business processes by accessing information across an enterprise. If humans can work together with agents, the agent learning from the human and performing menial tasks such as monitoring, searching, and filtering, while the human workers supply experience and judgment, a company can profit greatly. Of course, this type of human-agent integration won't happen overnight. It will happen only as users identify places where agents can fit in, and these automated assistants prove themselves useful. As agents proliferate, information can be more easily shared and duplication of efforts reduced across an enterprise.

Only if agents are widely accepted by human workers can they genuinely be considered a help, rather than an extra layer of complexity in automating and streamlining processes across an enterprise. Agents with the best chances of gaining acceptance have the following attributes:

- They are unobtrusive. Agents must not get in a user's way, especially while he or she is performing a complex task. Even if the agent offers helpful advice, most human workers who know their jobs won't appreciate it until they have a reason to trust the agent. Agents must be ready to help, but not try to force themselves on users, or dominate the way a user works.
- They are easily customized. Even the smartest agent isn't going to be able to fill a niche perfectly until a human instructs it in the specifics of the job. For example, an information-retrieval agent must be told when it would be most useful to perform a query, approximate target server name, and log in

information, what fields to check, and so on. The ideal (end-user) agent builder or launcher would provide a GUI for defining agent rules, parameters, and constants.

- They are useful on progressively complex levels. Ideally, an agent builder would provide not only a GUI, but a scripting language and an extensive API as well. This would allow users to instruct an agent to perform more complex tasks as it gains users' trust, and as agent and human learn from each other.
- They promote information sharing. As agents cooperate in a distributed environment, they can find out what other agents are doing and what information is available that their users may not know about.
- They fill an empty niche. Human acceptance of agents, and therefore increased agent adoption, is much more probable if an agent can perform tasks that are not the kind of thing humans normally excel at or enjoy. Agents are suited for many human-complementary tasks, including the following:
  - Database queries that require shuffling large amounts of data across a network are prime candidates for agent intervention. An agent can arrive at a database server and make a series of queries, then analyze the resulting collection and return only the information the user really wants.
  - Agents work well for monitoring and reporting on many things at once. Agents can monitor activities in a distributed system, such as checking for trouble in a LAN and providing reports for use by network management software or monitoring nodes in a network. This also works for constantly checking a variety of data sources at once. By acting as go-betweens, agents can also make navigation and resource location seamless within a distributed system.
  - Agents are a good idea if you want software to learn from experience within a restricted domain. The genetic algorithm and neural network approaches discussed in Section 3 are good ways to make an agent capable of evaluating and learning.
  - Agents can find information and share it with each other. With many specialized agents collaborating toward some higher, human-specified, complex goal, chances are greater that an agent may find a new utility or data source that can improve the agent system's chance for success.
  - Monitoring and analyzing data streams in real time. Embedded system agents have long been performing this type of activity, since it's virtually impossible for humans to do on their own. These data streams can be anything from flight-parameter values to radio frequency (RF) signals from a distant galaxy.
  - Continuously executing background tasks, such as checking for errors in a database or illegal conditions (rule violations) in a system.
  - Repetitive, periodic tasks triggered based on timing or system conditions.

- Agents are well-suited for performing business procedures for which the work flow is well defined.
- Agents work well at managing system resources such as memory and databases. They can restrict access, and alert system management to problems and potential problems.
- In systems requiring negotiation for resources, for example, auctions, agents can provide the ability for all interested parties to be equally represented.
- Agents, such as those used for decision support, can be just the thing for ferreting out and analyzing information from complex data relationships.
- Agents can assist in maintaining security in distributed systems. Users desiring to connect to resources must deal with an intermediary agent, or proxy, which finds and retrieves what the user needs without allowing a user to access system resources directly.

6.4.2 Anthropomorphic considerations

Currently even the most intelligent software agents still know relatively little about complex human needs and desires, and as a result can do nothing to help fulfill them without being given fairly explicit instructions. As agents are imbued with goals that mirror the subtleties of human-universe interaction, they come closer to the ability to *anticipate* their masters' needs and act *proactively* to fulfill them. This requires close agent-human interaction, not to mention some clever programming. Proactivity means things like creating one's own goals, replanning, deductive and inductive reasoning, and learning.

McKie (1995) compares software agents to their human counterparts, explaining that like human intelligence operatives, software agents can be trained and briefed to perform specific missions, deployed, and instructed to wait and monitor conditions until a specific event or set of events occurs. After activation from the wait, an agent (whether software or human) carries out the mission and reports the results.

We don't yet have the technology to develop a true general agent which can automatically react to a user's every need, whether specified or implied (a combination butler, research librarian, accountant, travel agent, and so on). However, we are close to being able to reasonably model the most important attributes and operational characteristics of a human assistant dedicated to a specific domain or context, for example, a travel agent.

The first step in designing such an agent is to define, using techniques borrowed from OO analysis and design, such as narratives and scenarios, a specific task or set of tasks that an agent could perform. One way to do this is to look at the tasks the human counterpart performs and attempt to iteratively build an agent to do the simple, repetitive (perhaps onerous) parts of it. The steps might include:

1. Look at a job a human service provider performs.
2. Select a task or set of closely related tasks from the human job.

3. Examine the possibility of an IA performing the same tasks.
4. If an appropriate task or set of tasks is found, design an agent to perform the most basic part of the task.
5. Prototype and test the agent.
6. Continue to iteratively improve the agent until it sufficiently approximates the performance of the human assistant with respect to the specific task.

Over time, agents doing simple tasks can be made to cooperate to perform more complex missions. Consider what a human assistant would need to do to schedule a meeting between busy executives working in different (remote) locations. The first step would be to negotiate a meeting time suitable to all parties. To do this, the assistant would first need to get a list of all available meeting rooms with the needed facilities (white board, video, or teleconferencing) in each potential location, and provide possible meeting times to all potential participants or their representative assistants. Your assistant would probably refer to a list, and send each executive an E-mail message. Each executive would then need to return a list of ranked dates and times. Next your assistant would need to match available dates and times with room availability. Finally your assistant would send out the official meeting notices and collect the responses.

To accomplish this task, a software agent might follow the same steps as its human counterpart. The agent might be invoked with a list of possible meeting times and potential participants to contact via E-mail. It may then need to search each remote node for available meeting rooms with the correct facilities. A more efficient approach would be for your agent to contact each of your colleagues' agents with a list of available times and figure out the intersection between your list and theirs, and then return the resulting list after all had been visited.

Once the agent returned with a list of times and days that would work for all potential attendees, it could then contact each node (assume each building-control system knew all of its own rooms and facilities) and ask it for rooms available with the needed size and facilities. A software agent could also consult with its fellow agents to determine which meeting location would be optimal, for example, given a goal of shortest distance from each prospective attendee. To achieve maximum efficiency and user-friendliness, for most of the meeting planning scenario, agents should not interact directly with users, but with users' agents.

6.4.3 Considering the agent's audience—the user

What a user and an agent need most is communications between them. A user must be able to figure out in advance what an agent does, and keep tabs on the agent's progress in accomplishing a given task. A relatively simple input form may be the safest agent interface—what a user can order an agent to do is severely limited, but is more likely to be successful. The most advanced agents respond to something closer to English, or natural-language input, allowing users to use keywords to describe data and processing. This is probably too complex for a novice user, and too dangerous for an experienced one.

Of course, some agents cry out for a graphical interface, for instance, pointing to a map destination in either cyberspace or the real world or programming work-flow phases into agents. Unlike many applications, there is often no immediate feedback loop associated with an agent interface. After an agent is launched, it may go about its business, continuously collecting and analyzing data, tuning a system, or enforcing system rules without providing much in the way of output to a user. On the other hand, agents may provide periodic E-mail reports just to let a user know they're on the job.

Whether or not agent and human keep in touch, there will be misunderstandings. Humans want to give general instructions, or no instructions at all, and have the agent do what needs doing. Agents, on the other hand, still require relatively explicit instructions.

As with any technology new to the desktop, users may believe hype generated about the coming agent revolution, and expect far too much of recently delivered agent software. "The heart of the agent developer's problem is that most people, quite naturally, expect supposedly intelligent software agents to perform the same kind of tasks that, for instance, human travel agents or insurance agents do. Such knowledgeable servants achieve your goals without forcing you to learn the details of their work. In fact, the software agents being built promise nothing of the kind" (Browning, 1996). Browning goes on to compare the agent technology being delivered today to nothing more than instinct-driven insects and animals trained to do simple tasks. He then goes on to add: "Encouragingly, the kinds of instincts that make a good bird dog, or even a crumb-crunching cockroach, may yet make a good information retriever" (Browning, 1996).

Of course the ideal agent would be something like a butler in a P. G. Wodehouse novel, offering wisdom on a variety of topics, and intuitively going about making his employer's life easier in a hundred ways. This would require an agent or team of agents to understand what humans do and why they do it. (Often, humans don't know themselves). Today's technology is nowhere near this. What we can do is produce some limited-domain agents, which, when combined, can enhance a user's productivity. For example, desktop agents, such as Edify, can provide a unified interface for filtering phone calls, faxes, and E-mail, coordinating appointments, and information searching.

6.4.4 Agent-user interaction—trust, competence, and learning

If agents are to help us change the way we work with computers, the fundamentals of the way we currently interact may have to change. For example, the current direct-manipulation metaphor, first popularized by Apple Computer's

Macintosh, "requires the user to initiate all tasks explicitly and to monitor all events" (Maes, 1994). Use of autonomous agents may lead user-interface designers to consider indirect management instead of direct manipulation. The method of computer-human interaction (C-HI) enables collaboration between the agents making up part or all of the system, and the human user.

Some issues arise when designing agents within the new paradigm of indirect management. As Maes (1994) states: "The first problem is that of competence: how does an agent acquire the knowledge it needs to decide when to help the user, what to help the user with, and how to help the user?" Second is the issue of trust: "...how can we guarantee the user feels comfortable delegating tasks to an agent?".

One approach to the first issue is to have the user program semiautonomous agents, or configure them and their rules and attributes. This end-user type of task is usually domain-specific, for example, configuring filtering, storing, and forwarding rules for incoming E-mail. This end-user-as-programmer approach relies on some least-common denominator of user competence, placing an onus on the agent designer to get it just right.

You would not want to assume that the user is a so-called power user, one adept at most, if not all, tasks and thus not interested in a coaching agent's interruptions, harassment, or boring, repetitive suggestions for automation or observations. Neither would you want to assume that the user is a complete idiot or a naive or new user who might welcome any and all help offered.

Either approach would require you, the designer, to assume the user has the ability to "recognize the opportunity for employing the agent [appropriately], take the initiative to create or configure the agent, endow the agent with explicit knowledge,...and maintain the agent's rules over time (Maes, 1994). Of course, trust is less of an issue in this approach, because the user has most of the control over the agent's creation and behavior.

Another approach is the knowledge-based approach. "It consists of endowing an interface agent with extensive domain-specific background knowledge about the application and the user" (Maes, 1994). This idea ties in with Cyc-enabled or Cyc-ic agents, agents that can gain access to a knowledge base of commonsense information about a domain, and can rely on the reasoning capabilities of that knowledge base.

Using its knowledge of the domain and its master, the user, a Cyc-ic agent would know the when, what, and how of contributing to a user's goal. For example, you could make an agent that knows virtually everything about Microsoft's Windows 95 (in its user-visible, operational aspects). Such an agent could then assist a user in most C-HI tasks running on that operating system. One way this could be implemented is to extend the Cyc ontology to include a Windows 95 knowledge base.

However, even in this knowledge-based approach, the two issues of competence and trust still remain. In addition, the agent ends up being, by design, very domain-specific and unable to contribute to other user tasks and goals. Of course, the developer's inner control mechanisms and representational schemes would still be reusable, if appropriately designed, within other agents. For example, you could produce a host of operating system help agents by reusing the core user and Cyc interfaces, and provide the appropriate operating system–specific Cyc ontology.

Yet another approach to building user-friendly agents makes use of machine-learning techniques. Some of the technologies described in Section 3—neural networks, fuzzy logic, genetic algorithms, among others—can be put to good use here. The idea is to enable the agent to start with some fundamental knowledge and a rich and powerful set of adaptive capabilities, then let it loose to learn from the user. This technique works especially well in areas where the agent is trying to learn-thence-automate a user's repetitive behaviors.

The learning approach satisfies the competence issue because the agent's capabilities spring from the relative level of sophistication of the user. A naive user begets simpler and quicker actions and suggestions from the agent, whereas a power user soon convinces a learning agent that help is needed in only the most demanding or complex tasks. The learning approach also satisfies the trust issue in that the interaction between the agent and the user builds gradually. In addition, if the agent can give explanations for its suggestions and actions, more trust and comfort is built. This approach works only if the learning domain is sufficiently limited—remember, we're not building a general-purpose android.

In addition, the learning approach has several other advantages over other approaches. "First, it requires less work from the application developer. Second, the agent can adapt more easily to the user over time and become customized to individual and organizational preferences and habits. Finally, the approach helps in transferring information, habits, and know-how among the different users of a community" (Maes, 1994). Learning agents become competent (and thus gain the trust of their users) by:

- Monitoring the user and finding patterns appropriate for automation. For example, if I always send an E-mail message with the buzzword "Asynchronous Transfer Protocol" in it to three associates, the agent can readily pick this pattern up and automate the search and forwarding.

- Accepting and adapting to indirect and direct user feedback. An example of direct feedback is a user explicitly communicating to an agent (such as via rule or configuration) to never do a certain action again, for example, emptying the trash when shutting the computer down. Indirect feedback to an agent would be ignoring an agent's suggestion for an action and taking an alternative action.

- Learning by explicit example. If designed to do so, an agent could present the user with a training protocol whereby scenarios and cases are presented with options for the user to choose. The user's choices are then recorded and carried out in similar situations.

- Learning from other agents. This implies that some interagent communications protocols and mechanisms are in place, as discussed in Section 5. For example, under the M architecture, a simple blackboard technique is used for agents to post information about themselves. An agent can average responses from other agents or can accept the advice from a particularly good agent—one that has offered advice that, through experience, worked in the past.

6.4.5 Helping humans work: designing work-flow agents

Simple agents can handle tasks such as filtering information from data streams, performing searches, and notifying users of events, such as arriving E-mail from a specified sender. None of these simple activities would seem to have much impact on a user's world, but it turns out that when simple agents are combined using work-flow technology, a number of relatively mundane multistep tasks can be accomplished.

Linking applications and utilities through scripting languages has long been a mainstay of simple agent technologies. Simple agent builders, especially those designed to facilitate and enforce work-flow rules, enable a user to visually create script files. These scripts automate some business process, such as registering a patient in a hospital, processing an insurance claim, or approving a loan. The ideal agent builder would include rule-based building capabilities, features to handle information streams, and the ability to graphically represent agents and the entities with which they associate, such as fax machines, modems, and directories.

A good work-flow agent builder must also support connections between agents and the work products. Each entity represented graphically may have a number of potential connections or connectors. Rules for making connections must be enforced; for instance, connecting a printer directly to a fax machine might be disallowed, or piping the output of one type of agent into the input of another (Atkinson et al., 1996).

Agent creation can be done graphically, provided the agent's purpose and domain are sufficiently restricted or the tool is sufficiently versatile, for example, a work-flow agent. Creating such an agent using an agent builder such as GroupWise Workflow, Edify, or Verity can be as simple as clicking the mouse a few times, provided the agent capabilities you need are already included in the builder.

6.5 Incorporating Agent Capabilities in Shrink-Wrap Software

When the personal computer revolution began, microcomputer software had few value-added features. If it worked and did a passable job of performing the most basic functions, it was a success. Over the years, as machines have increased in performance by orders of magnitude, the software industry has generally responded by adding stacks of features to software packages (creeping featuritis) in an attempt to outbid the competition. This practice has lead to feature burnout, with customers often using only 10 to 20 percent of the features in a package (Berst, 1994).

Because of this, customers increasingly decline to upgrade their software, reasoning that it is foolish to pay for more features when they don't use the ones they have. Many vendors are learning that the best way to impress customers is not by dumping new features on them, but by making the software smart enough to figure out and do what the user wants without forcing him or her to wade through multilevel menus and complicated dialog boxes.

While most major software vendors have been doing serious research into making software smarter, it wasn't until the concept of an IA emerged that significant progress was made. Now many commercial software packages include IAs, or assistants, with the core offering.

Intelligent assistants can walk a user through difficult tasks or monitor how he or she uses the system and offer shortcuts and suggestions. Software agents can find out what you want to do and perform the task for you. Simple agents can be found in products such as GroupWise and BeyondMail. These agents watch for conditions such as receiving E-mail from a specific person, then perform some predefined action.

Agents also crop up as assistants, coaches, or wizards to help users make features work together to perform specific tasks, such as formatting documents, querying a database, or creating a project schedule. While not especially intelligent, these agents can nonetheless act on behalf of a user and make software features useful, which would otherwise have gathered dust while the user went about things the usual pick and shovel way.

6.6 Topics on Intelligence

"Intelligence requires knowing those differences that make a difference and recognizing patterns that connect different domains." MICHAEL KNAPIK

6.6.1 Simple agents versus intelligent agents

Agent intellect can range from the simple-minded to the more intelligent variety. A simple agent works within strictly defined boundaries (such as: "Turn air conditioner off when temperature = 75 degrees Fahrenheit."). An IA has more flexibility and can work with ambiguity, such as "Maintain temperature within defined comfort zone." Such an agent may have to incorporate soft computing techniques, such as fuzzy logic, to properly and more naturally represent such a directive.

Simple agents—for example, today's information filters—operate by sitting in memory and connecting with an information source at regular intervals to determine whether some condition has been met (such as: "Stock price dropped to $50 per share" or "New information available on selected research topic"). When the event occurs, the agent provides its master with the information, or takes a specific action such as: "sell immediately."

Other examples of simple agents include network traffic managers, print monitors for open printing, fax redial, and others. These examples are grouped under simple agents because users (in this case, LAN administrators, word processors, and people sending faxes) instantiate simple agents to notify them when an action has been completed. Simple agents can contain knowledge of how to accomplish their task, including using rules and functions, but these agents cannot modify their behavior based on changing environments and user whims.

Agents with more intelligence (possibly composed of many simple agents) do not simply perform as explicitly directed, but instead gather facts from various information sources, compare them, then perform one out of many possible actions. These agents need access to many data streams, and must also coordinate efforts with other agents. Instead of being controlled from a single computer or software entity, these agents often reside on the network itself and operate from different servers, depending on user needs.

Are current agents intelligent? Agents or, really, task-automation facilities may be an integral part of some current applications and system software, taking over some of the most mundane desktop and LAN-management chores. These automatons have access to a knowledge base that assists them in their tasks, and they are able to add to that knowledge base. But to say they learn from their mistakes, or have any degree of autonomy—both attributes of IAs— is stretching it. In addition, the reasoning capabilities of most of these assistants is rudimentary. If you wanted your agents to exhibit robust commonsense reasoning capabilities, one option would be to enable your agent to gain access to a Cyc knowledge base (discussed in Section 3).

6.6.2 Designing in intelligence and autonomy

6.6.2.1 Goals and plans. While relatively intelligent agents may become pervasive in the coming years, at this point the technology is still in the early stages of development. As Etzioni (1994) explains, "Advances in computer, information, and telecommunications technology have made software agents both necessary and possible. However, the role of AI and AI researchers in these developments has yet to be determined." We maintain that AI techniques *must* become ubiquitous within agents if they are to fulfill their promise.

At its heart, an IA is an autonomous, independent, asynchronous computational task. Independence and autonomy require many of the AI-based capabilities that we discussed in Section 2, working in unison to attain the primary goal of the agent. As it executes, almost every IA, or society of IAs, centers upon monitoring a stream of information or the state of a system. In addition to monitoring, data filtering and rechanneling are often required. Filtering means that only information meeting user-thence-agent-established criteria is selected out of the stream. Then the resulting reduced stream is channeled or routed to a user or another agent, where it can be analyzed. For example, several monitoring agents can feed one analysis agent. The analyzer determines the state of the system, which can then be compared against a goal state. If the agent has not accomplished its goals, it continues to monitor the data stream and the agent can determine what actions, depending on the state, should be taken to fulfill its goals.

It is in determining and carrying out the subgoals and actions to accomplish them that intelligent behavior can be simulated. Intelligence consists, in large measure, of doing more beneficial things for the user than the user can command directly. A truly general IA would understand the user and a broad range of the user's needs so well as to anticipate and fulfill them before being asked.

Agents vary as to the degree of intelligent behavior displayed, and the amount of autonomy each has. Some agents are trusted to replicate themselves, move around a distributed environment, invoking functions on a variety of clients and servers. As with any human job description, specific performance goals can be set and measured. As goals are met, new goals contributing to the success of an enterprise need to be set. An IA can monitor a data stream, then act independently upon what it finds based on goals, and generate new goals when necessary to accomplish the user's larger objectives.

Often the degree of autonomy is a function of how secure the expected execution environments are. If an agent is fire-walled such that it can cause problems only on a single platform, it can be afforded substantial freedom. Autonomy also stems from intelligence. If an agent is intelligent enough not only to make good decisions, but to modify its goals and work toward them in the process of achieving a user's ultimate goal, it may be trusted to do its job essentially without user intervention (cruise control), provided its execution environment prevents it from doing any harm.

Goals must be constrained—no user can allow an agent to search forever or use endless resources trying to tune a system to the optimum performance level. One of the worst cases would be an agent attempting to optimize disk space by deleting needed files. An agent must be constrained to use only appropriate means and resources. Thus a successful agent may be only as autonomous as constraints allow.

Another manifestation of intelligence is cooperation and sharing among agents. Another is learning from the past. For example, a database query system learned from experience what a user was interested in and automatically searched for synonyms and related topics, or a control system learns what the best flow rates are for greatest efficiency.

What an IA does best is to use defined procedures for achieving well-defined goals based on a shared world model. The hard part, the part that takes advanced intelligence, is defining and representing the appropriate goals. A goal must be something achievable and verifiable, yet something that makes a positive impact in the user's world. Simple goals are trivial: "Find all persons in the personnel database with the last name of Thomas." Such a goal is not only easily achieved and representable, but its achievement is easily verified as well.

Of course, the most advanced agents can create their own detailed goals needed to achieve a general goal supplied by the user or a coordinating agent. Such a goal might be: "Fly this jet airliner more efficiently." An airplane-piloting agent (an intelligent autopilot) might then set for itself the subgoals of optimizing all flight parameters such as drag, angle of attack, altitude, thrust, fuel flow, fan speed, turbine speed, and exhaust gas temperature, all while maximizing the comfort of passengers and crew. In general, you need to create goals that are achievable (the agent can understand them) yet can make a positive impact in a user's world.

As in Asimov's laws of robotics, certain axioms with respect to goals must be applied by the agent designer and developer to ensure successful, nonthreatening agents, or software robots.

- The agent should refuse to make destructive changes to the world.
- When its mission is completed, the agent should restore its world to its original state.
- The agent should limit its use of valuable resources.
- The agent should attempt to block human actions that have unintended consequences.

Achieving many of these goals depends not only on the agents themselves, but on the execution environment where the agents run.

6.6.2.2 Etzioni's Softbot example of goals and plans.
Researchers Etzioni and Weld have developed an IA, called the Internet Softbot, at the University of Washington in Seattle, Washington (Etzioni and Weld, 1994). Ideally, the Internet Softbot can translate a user's high-level goal into more specific goals, which in turn is accomplished via program functions or system-utility calls, such as file transfer protocol (ftp), gopher, or netfind. A "safe" instruction or configuration form can be used as an interface to the agent, serving as a guide to prevent the user from ordering the agent to do something deleterious.

Most complex agents are built using computer languages such as Java or Smalltalk. These require an agent developer to code a solution essentially on the computer's terms. A more advanced form of agent building would be to use a goal-oriented interface. Such an interface would accept a goal represented as a well-formed logical expression and translated into executable statements. The Softbot planner acts as a very high-level compiler, taking a logical expression describing the user's goal as input. After searching a library of actions, the planner breaks the user's goal into specific calls to available services. Imagine the potential—in the future mobile agents may consist only of a goal statement, which is understood and carried out by an advanced agent infrastructure.

An intelligent search process is at the core of the Softbot planner. It locates action sequences called plans, which are then combined to fulfill the users' goal. As subgoals are found to be necessary, they can be generated at build time or as the agent executes. "If the Softbot cannot satisfy its goal directly, it will automatically generate a sub-goal to indirectly satisfy the goal. For example, if looking up the phone number fails, an agent can find alternative numbers, such as that of a co-worker, supervisor, secretary, etc." (Etzioni, 1994).*

6.6.3 An agent's environment and planning models

"The map is not the territory, and the name is not the thing named." ALFRED KORZYBSKI

Perhaps one of the most difficult parts of creating IAs is to adequately elicit and represent those aspects of the IA's environment, including the domain of the agent's expertise, critical to each agent's task. It would be difficult to design a task scenario for an IA if you knew little about its environment. So the idea of planning, and the notion of accurately representing appropriate parts of the IA's environment, go hand in hand.

Note that although we speak of environment and domain as though they are separate things, this may or may not be the case. For example, if an IA is being created to seek out and fix network problems, the domain and the agent's environment are the same thing (or very nearly so). But a financial advisor agent's domain is clearly the financial world, whereas its environment would include things not remotely concerned with finance (like the platform on which the agent runs).

We mentioned earlier that, if you are creating agents within the OO paradigm, an OO model of the domain and the agent's environment is advised to preserve purity. But this is not just empty philosophical pabulum. Architectural and implementation purity offer two principal advantages:

1. *Understandability:* The advantages to following the OO paradigm when crafting a model of an IA's environment is that even though, as Korzybski stated, the map is not the territory, if you can model each discrete thing in an agent's environment as a discrete software object, the prospects for understanding what you've crafted are vastly improved. You'll at least have a clear one-to-one mapping of the environment to the model of the environment.

2. *Consistency:* If you design and implement your agents, and model (or design and implement if you have that ability) the environment and the domain using the same paradigm, the interface and interaction scenarios among agents and between agents and their environment are both inter-

---

*For more information on Softbots and goal-oriented agent programming navigate to the following: www.cs.washington.edu/research/projects/softbots/www/index.html.

nally and externally consistent. When it comes time to test your system, consistency pays off in reduced unit-test and integration-test efforts. Test script patterns are more universally applicable.

But that being said, just how do we go about deciding what objects in the environment and in the agent's domain need representing, and just what does this representation entail? What is meant by planning?

Perhaps the most straightforward way to approach modeling is to prepare scenario diagrams (sometimes referred to as dynamic models or temporal interactions) of your agents and the things that they interact with. Each thing gets its own iconic representation in such a diagram. The actual interactions are represented with arrowed lines connecting the agents and the things they either monitor or control. As you develop the scenario diagram (also called an object-interaction diagram), you can add details such as specifying the actual message content and indicating what each receiving entity does upon receipt of a certain message. These interaction diagrams can be used to model agent-agent interactions, and agent-environment interactions.

Each environmental object would have variables whose values track and represent the actual environmental attributes of interest. If the environment is capable of being controlled by the agent, a method for each discrete controlling action is created within the environmental-control protocol within the agent. These methods would interface to the appropriate external software entities, such as operating system and network managers, to gain access to and reserve resources, file systems to update data, and so on. Obviously, the agent's user-interface, if relying on native user-interface APIs such as WIN32, would have one or more user interface objects within it to control and update status windows, graphics, and so forth.

In AIAMA, Russell and Norvig help us out in our environmental modeling task, as well as our agent-planning tasks by analyzing and categorizing the different types of environments (Russell and Norvig, 1994). An environment affects the design of an IA based on where its environment or domain falls within the following spectra of properties:

- *Accessible to inaccessible:* Define the ends of the spectrum that determines the degree to which an agent has access to its environment. If all the relevant attributes of an IA's environment are accessible, the IA does not have to maintain an internal representation of the environment.

- *Deterministic to nondeterministic:* Define the ends of the spectrum that determines the degree to which an agent thinks it controls the current state of its environment. If an agent cannot access all of its environment, then it may appear to be nondeterministic, and thus the state thereof can be uncertain.

- *Episodic to nonepisodic:* Define the ends of the spectrum that determine the degree to which an agent's perceptions and subsequent actions are cleanly divided into noncausally linked episodes. In other words, if the IA does not have to look or think ahead (beyond its current percept-action) to help evaluate or reason what it should do next, then the environment is episodic. Episodic environments are easier to model and design agents for.

- *Static to dynamic:* Define the ends of the spectrum that determine the degree to which an environment changes during the time the IA is reasoning about what to do next. Static environments are easier to design IAs for because the agent can ignore its environment while a reasoning method is executing.

- *Discrete to continuous:* Define the ends of the spectrum that determine the degree to which an agent's interactions with its environment are limited and clearly delineated. If an IA has a fixed, enumerable set of percept-action pairs, then your design task is greatly simplified.

In AIAMA, Russell and Norvig couched the foregoing environmental properties as two-valued, either/or in nature. They then go on to explain how the determination of what an environment is (in terms of those properties) could depend on how you look at it. However, environmental complexity may prevent an absolute characterization. That is why we, in our foregoing summary, chose to define each of these properties as a spectrum.

Yet another aspect of environmental modeling and agent planning we mentioned in Section 2 is that of performance measures. Since any interesting, useful IA affects its environment (of course, a really IA may someday wax philosophic to the exclusion of doing something useful), you should include in any IA some way of representing and updating a measure of the IA's success. This usually means devising (at least one of) an agent's goals in (perhaps partial) terms of some value that can be increased or decreased. For example, devising an MEU measure for a vacation-planning agent might include a variable representing total cost and complex variables representing possible destinations and their rated desirability. The MEU measure might then be based on a composite value produced by an expression that factors the destinations with the maximum desirability with the lowest-cost package.

Of course, any time an agent deals with an environment, it can't possibly know everything happening out there. Computer hardware failures, network failures, data-transmission errors, and just plain old programming errors (bugs), all contribute to inaccurate or imprecise knowledge about the state of any system. For that reason, an agent is bound to become uncertain, at some point, about the state of its environment, which includes the state of other agents with which it is collaborating. In Section 2 we discussed some aspects of dealing with uncertainty.

One of these aspects is planning or replanning when unexpected, unplanned-for events occur. In addition, when agents interact with their environment, as Bateson has observed: "From the point of view of any agent who imposes a quantitative change, any change of pattern which may occur will be unpredictable or divergent" (Bateson, 1979). In other words, an agent attempting to control or affect its environment can set up an unplanned-for cascade of events, or a pattern of aftereffects that could cause the whole system to diverge into an unplanned for pattern or chaos.

Another aspect of dealing with uncertainty is planning when the agent just can't get all the information it needs to continue being an IA, namely, to reason and determine the correct course of action. All of these aspects concern the necessity of an IA to be able to plan under uncertainty.

Again, planning under uncertainty is important to developing agents because you don't always know for certain in what state your agent may find itself, and, further, you can't plan for every possible scenario in which your agent may be inextricably entwined. As we said before, all agents sense their environment to some degree, and execute methods or tasks, or control operations to rearrange that environment or themselves per some goal or set point.

One useful way to approach this problem of uncertainty is for the user to "provide the agent developer with a model, called the planning model, which describes the rules of the interactions between the agent and its environment" (Hafez, 1989). This plan can take the form of a simple table of control and sensing operations for each task or goal, which, "...if executed by the automaton (i.e., agent), is guaranteed to accomplish the task" (Hafez, 1989). Other important parts of a more comprehensive planning model might include:

- A specification of constraints on the types of actions an IA can take
- A specification of constraints on the ranges of values of variables
- A fail-safe state which, if achieved by the IA, is guaranteed not to do harm
- A graceful degradation procedure that minimizes effects on other system components
- An emergency shutdown handler or method
- A checkpointed (saved good) state the IA can return to after a system upset

6.7 Agent Components

In Section 3 we discussed the OO way of designing and developing your IAs. We presented a high-level example of a composition hierarchy of agents that, working together, could perform tasks within the financial and investment domains. Those types of components were themselves high-level IAs or other major pieces of software dedicated to a particular task, such as the user-interface components that would be part of any agent that needs to communicate with a human.

What we present here is a (no doubt incomplete) list of those low-level components that you might find in a generic IA or, according to the example in Section 3, a RootAgent. Some of these components were gleaned from our readings, especially King, 1995, and Wayne, 1995. Others we included based on our experience. Of course, delineating all the variables and methods necessary for such agents is beyond the scope of this book. Most of the variables and methods would be domain-specific anyway. However, some of the components that a typical or root agent IA might include follow. In the OO development paradigm, these components could be represented as variables, methods, or other objects and agents, as appropriate.

- *Agent name or ID:* Each agent instance would have a unique name by which the agent could be located and addressed.
- *Owner:* User name, parent process name, or master agent name. IAs can have many owners. Humans can spawn agents, processes can spawn agents (such as stock brokerage processes using agents to monitor prices), or other IAs can spawn their own supporting agents.
- *Author:* Development owner, service, or master agent name. IAs may be created by people or processes and then supplied as templates for users to personalize.
- *Creation date and time:* Date and time of the request.
- *Account:* Billing information, electronic addresses. IAs must have an anchor to an owner's account and an address for billing purposes or as a pointer to their origin.
- *Goal:* Goal statements, measures for success. Crisp statements of successful agent task completion are necessary, as well as metrics for determining the task's point of completion and the value of the return. Measures of success may include simple completion of a transaction within the boundaries of the stated goal, or a more complex measure of returned information [such as a retrieved document partially [50%] fitting my goal need].
- *Duration:* Response requested by a certain date.
- *Moving parts or general services:* Used to perform the agent's tasks.
- *State:* A single or composite superstate value or values that represent the agent's important variable values and those of agents or objects with which it is in relationship. For example: running, waiting, shutdown, unloaded...
- *Authorization:* Does the agent need to interact with a user or host system and get some type of approval before consummating an action?
- *Resources needed:* A description of the resources the agent needs to complete a task or reach its goal, possibly in some canonical resource descriptor language. (Refer to Section 5, where we discuss a resource descriptor language as part of an architecture.)
- Knowledge of available *host functions* (the host API).
- *Initialization functions* (called by host).
- *Exit functions* (clean up).

- *Money or credit* needed to complete transactions. Does the agent need to deliver an (encrypted) account number to some other entity or agent to consummate a transaction?
- *Instructions for negotiation:* Rules, negotiation strategies, constraints on resources expended to achieve some goal.

Additional components may include:

- *Connections and relationships:* Used primarily in work-flow agents or in agents collaborating with other agents, these indicate where the information handled by one IA is to be routed, or where messages are to be sent.
- *Potential connections:* This is a collection of the types of entities with which an agent may connect.
- *Error address:* The address to which error messages are sent.
- *Return address:* The address from whence the agent came.
- *Process priority:* The requested execution priority of the process in which the agent executes.
- *Memory resources:* Memory needed to run adequately.
- *CPU resources:* Needed processing power.
- *Cache or checkpoint resources:* Disk space needed to store the agent/stat.
- *Applications or methods accessed:* Names of the methods the agent expects are supported at its destination host.
- *Rules protocol:* This is the logic the agent uses to carry out its goals.
- *Attach file:* Any files the agent may need to carry with it to the next destination.
- *Next destination:* Address where the agent goes next.
- *Trigger condition expression:* The conditions that must be true to launch the agent.
- *Termination condition expression:* The system conditions that must be true in order for the agent to cease operation.
- *Hibernation condition expression:* The system conditions that must be true in order for the agent to enter into a suspended await state.

6.8 Classifying Agents Based on Degree of Mobility

As discussed in Section 8, there are almost limitless applications for agents (intelligent or otherwise). While they may differ greatly in function, agent designs can be fit into one of two general patterns: stationary or fixed agents and mobile or itinerant agents. There are, of course, several other criteria upon which to base the major classification of your agents. This approach, based on the ability to move around, is but one. What you will find is that based on the huge amount of code necessary to differentially support either mobile or stationary agents, an OO-based classification based on that operational or behavioral criterion is at least *one* valid starting point. Of course, if the IAs that you are considering are definitely in one or the other camps, this is a moot point.

6.8.1 Fixed or stationary agents

Fixed or stationary agents execute as independent tasks on one processor (for example, client or server), but may spawn other tasks to execute on the same or other processors. Stationary agents work very well as information searchers and filters, data stream monitors, intelligent device controllers, E-mail triage agents, and advisors.

Fixed agents remain in a single location throughout the duration of their execution. Though there are several types of fixed agents, the common trait of immobility can simplify their design. For example, stationary agents don't have to interact with an infrastructure that supports agent transport over a network, the extensive security measures necessary for mobility, or complex agent-to-agent communication and synchronization.

One type of fixed agent, the interface agent, is principally involved in some aspect of interacting with the user. These agents may communicate to the user by using actors (anthropomorphic representations of the agent's activities or state), speech, or other mechanisms specific to the type of work being done by the agent. Examples of interface agents include advisory agents and task assistants (coaches and wizards are simple examples). All these agents have some interface to the user, and their domain of expertise and operation is the user interaction.

Another type of fixed agent is the functional agent. This type of agent can run on the workstation or on a server in the network, but its domain of expertise is not the interface, but rather some functional area. It may have an interface to the user, but that is not its principal focus. Examples of functional agents include the following:

- Control and monitoring agents
- E-mail handling agents and other automation agents
- Information retrieval agents
- Agents representing mobile users who are disconnected from the network
- Agents implementing work-flow management
- Agents in the network acting as servers for requests by other agents (for example, an agent to facilitate negotiations among other agents)

Most search agents are stationary, though they may spawn a number of tasks which search down different channels (such as on different servers) simultaneously. These agents use constantly updated index tables or navigate the web by accessing data and following links. When all searches are completed, the results are collated and duplicates removed. Even though a number of tasks work together, they all execute on the same server.

In a real-time process control context, nonmobile agents monitor conditions and make changes to fulfill system goals. Or they may alert human operators of alarm conditions, then, depending on the operator's action, undertake remedial activities. In a database, stationary agents can provide triggered retrieval capability, giving users the newest information in real time.

Some of the most promising types of stationary agents are those that exist to provide enhanced or intelligent help to users. Such agents generally help in two different ways. First, there is the coach paradigm. A coaching agent, (such as Microsoft's Bob) runs as a background task, ready to provide unsolicited advice on how to use an operating system or application more effectively. A help desk type agent (such as Ask.Me from Software Affiliates) provides answers to thousands of common user questions, all accessed through a natural-language query interface.

6.8.2 Mobile or itinerant agents*

Because we believe Telescript is agents done right (whatever its current or final status within the evolving agent development and execution environment universe), we present the following concepts about mobile agents within the context of the Telescript technology.

In other words, just as Smalltalk is the quintessential OO language to which all newcomers are compared, so, we believe, is Telescript the agent platform that all others might strive to emulate. If you plan on implementing mobile agents within other environments and with other languages, you could use whatever architectural and design concepts embedded in Telescript you deem appropriate for your problem.

A mobile agent can copy itself or be moved to execute on more than one processor, or to move about a network infrastructure in order to concentrate processing and use of other resources in the most advantageous way, or to provide processing flexibility. For example, a mobile retrieval agent may be sent from a client to execute on a server to put the processing near the data so that information need not be moved back and forth across the network. Since clients often have more graphical processing power available than servers, agents built as Java applets or servelets for example, may be sent from a server to a client to provide a faster and more impressive user interface. Downloading only an executable applet also avoids moving blocks of graphics across a network. Mobile agents can provide flexibility by copying and loading themselves into a variety of systems. To put new capabilities into network nodes or components

---

*Subsections 6.8.2.1 through 6.8.2.4 are courtesy of James White, Vice President, Telescript Technology at General Magic—used with permission.

of an embedded system, agents can be loaded to upgrade or change the focus of the system. This is often preferable to installing new software manually into many processing modules.

As a process or processes that may migrate through a computer network in order to satisfy user requests, a system of agents could perform many of the same functions traditionally performed by operating systems. Agents can make an especially attractive option in the face of a lack of true support in most operating systems for distributed computing. For example, an agent can find the best printer in a distributed system for your document, taking into consideration document format, printer location relative to where you are, printer speed, how busy each printer is, and print quality. Systems that support agent execution, communication, cooperation, and so on, are available, however, and it may be instructive to get an overview of such a system now.

Telescript is an OO remote programming language, environment, and agent architecture all rolled into one. It is a platform that enables the creation of active, distributed network applications. There are three simple concepts to the language: agents, places, and "go." Agents "go" to places, where they interact with other agents to get work done on a user's behalf. Agents are in fact mobile programs capable of transporting themselves from place to place in a Telescript network.

The language is implemented by the Telescript engine, a VM technology. The Telescript engine is a multitasking interpreter that integrates onto an operating system through a programming interface called the Telescript API. The Telescript engine is a server implementation. Together these techniques enable mobile agents.

6.8.2.1 Mobile agent paradigm.
The concept of a mobile agent sprang from a critical examination of how computers have communicated since the late 1970s. This section sketches the results of that examination and presents the case for mobile agents.

6.8.2.1.1 Current approach.
The central organizing principle of today's computer communication networks is RPC. Conceived in the 1970s, the RPC paradigm views computer-to-computer communication as enabling one computer to call procedures in another. As shown in Fig. 6.1, each message that the network transports either requests or acknowledges a procedure's performance. A request includes data that are the procedure's arguments. The response includes data that are its results. The procedure itself is internal to the computer that performs it.

Two computers whose communication follows the RPC paradigm agree in advance on the effects of each remotely accessible procedure and the types of its arguments and results. Their agreements constitute a protocol.

A user computer with work for a server to accomplish orchestrates the work with a series of RPCs. Each call involves a request sent from user to server and a response sent from server to user. For example, to delete from a file server all

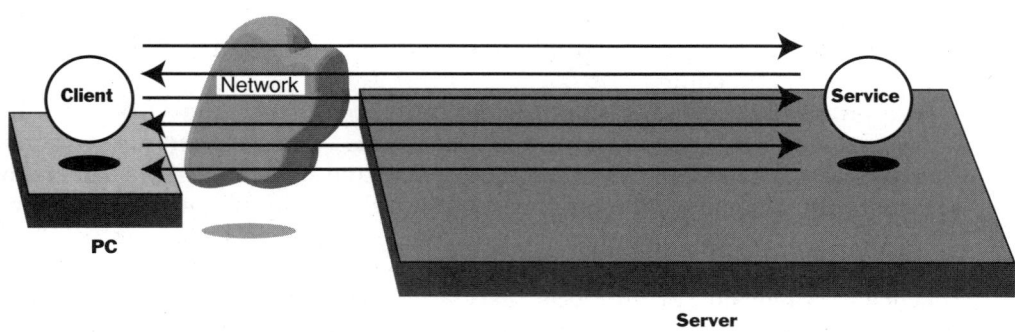

Figure 6.1 Current approach: RPC communications paradigm.

files at least two months old, a user computer might have to make one call to get the names and ages of the user's files and another for each file to be deleted. The analysis that decides which files are old enough to delete is done in the user computer. If it decides to delete n files, the user computer must send or receive a total of $2(n+1)$ messages.

The salient characteristic of RPC is that each interaction between the user computer and the server entails two acts of communication, one to ask the server to perform a procedure, and another to acknowledge that the server did so. Thus ongoing interaction requires ongoing communication.

6.8.2.1.2 New approach. An alternative to RPC is remote programming (RP). The RP paradigm, shown in Fig. 6.2, views computer-to-computer communication as enabling one computer not only to call procedures in another, but also to supply the procedures to be performed. Each message that the network transports comprises a procedure that the receiving computer is to perform and data that are its arguments. In an important refinement, the procedure is one whose performance the sending computer began or continued, but that the receiving computer is to continue; the data are the procedure's current state.

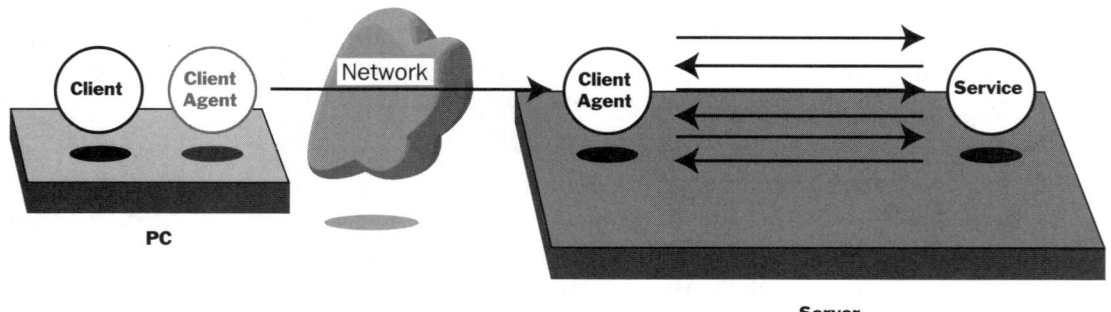

Figure 6.2 New approach: RP communications paradigm.

Two computers whose communication follows the RP paradigm agree in advance on the instructions that are allowed in a procedure and the types of data that are allowed in its state. Their agreements constitute a language. The language includes instructions that let the procedure make decisions, examine and modify its state, and, importantly, call procedures provided by the receiving computer. Such procedure calls are local rather than remote. The procedure and its state are termed a mobile agent to emphasize that they represent the sending computer even while they are in the receiving computer.

A user computer with work for a server to accomplish* sends to the server an agent whose procedure there makes the required requests of the server (for example, "delete") based on its state (for example, "two months"). Deleting the old files of the previous example—no matter how many—requires just the message that transports the agent between computers. The agent, not the user computer, orchestrates the work, deciding on site which files should be deleted.

The salient characteristic of RP is that a user computer and a server can interact without using the network, once the network has transported an agent between them. Thus ongoing interaction does not require ongoing communication. The implications of this fact are far reaching.

6.8.2.1.3 Tactical advantage.
RP has an important advantage over RPC. This advantage can be seen from two different perspectives, one quantitative and tactical, the other qualitative and strategic.

The tactical advantage of RP is performance. When a user computer has work for a server to do, rather than shouting commands across a network, it sends an agent to the server and thereby directs the work locally rather than remotely. The network is called upon to carry fewer messages. The more work to be done, the more messages RP avoids.

The performance advantage of RP depends in part on the network: the lower its throughput or availability, or the higher its latency or cost, the greater the advantage. The public telephone network presents a greater opportunity for the new paradigm than does an Ethernet. Today's wireless networks present greater opportunities still. RP is particularly well suited to personal communicators, whose networks are presently slower and more expensive than those of personal computers in an enterprise. It is also well suited to personal computers in the home, whose one telephone line is largely dedicated to the placement and receipt of voice telephone calls.

A home computer is an example of a user computer that is connected to a network occasionally rather than permanently. RP allows a user with such a computer to delegate a task—or a long sequence of tasks—to an agent. The computer must be connected to the network only long enough to send the agent

---

*The opportunity for RP (like that for remote procedure calling) is bidirectional. The example depicts a user's agent visiting a server, but a server's agent can visit a user's computer as well. In an electronic marketplace, if the user's agent is a shopper, the server's agent is a door-to-door salesperson.

on its way and, later, to welcome it home. The computer does not need to be connected while the agent carries out its assignment. Thus RP lets computers that are connected only occasionally do things that would be impractical with RPC.

6.8.2.1.4 Strategic advantage. The strategic advantage of RP is customization. Agents let manufacturers of user software extend the functionality offered by manufacturers of server software. Returning to the filing example, if the file server provides one procedure for listing a user's files and another for deleting a file by name, a user can add to that repertoire a procedure that deletes all files of a specified age. The new procedure, which takes the form of an agent, customizes the server for that user.

The RP paradigm changes not only the division of labor among software manufacturers but also the ease of installing the software they produce. Unlike the stand-alone applications that popularized the personal computer, the communicating applications that will popularize the personal communicator have components that must reside in servers. The server components of an RPC-based application must be statistically installed by the user. The server components of an RP-based application, on the other hand, are installed dynamically by the application itself. Each is an agent.

The advantage of RP is significant in an enterprise network but profound in a public network whose servers are owned and operated by public service providers like America Online. Introducing a new RPC-based application requires a business decision on the part of the service provider. For an RP-based application, all that's required is a buying decision on the part of an individual user. RP thus makes a public network, like a personal computer, a platform.

6.8.2.2 Mobile agent concepts.
The first commercial implementation of the mobile agent concept is General Magic's Telescript technology which, by means of mobile agents, allows automated as well as interactive access to a network of computers. The commercial focus of Telescript technology is the electronic marketplace, a public network that will let providers and consumers of goods and services find one another and transact business electronically. Although the electronic marketplace doesn't exist yet, its beginnings can be seen in the Internet.

Telescript technology implements the following principal concepts: places, agents, travel, meetings, connections, authorities, and permits. An overview of these concepts indicates how the RP paradigm provides the basis for a complete and cohesive RP technology.

6.8.2.2.1 Places. Telescript technology models a network of computers, however large, as a collection of places, as shown in Fig. 6.3. A place offers a service to the mobile agents that enter it. In the electronic marketplace, a mainframe computer might function as a shopping center. As shown in the illustration, a very small shopping center might house a ticket place where agents can pur-

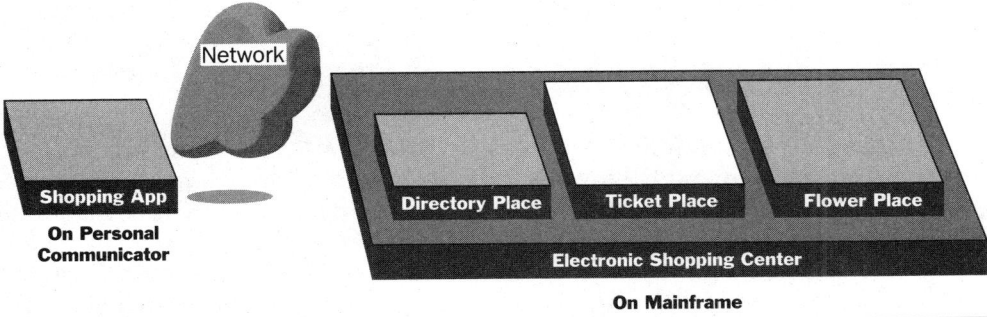

Figure 6.3 Computers as places.

chase tickets to theater and sporting events, a flower place where agents can order flowers, and a directory place where agents can learn about any place in the shopping center. The network might encompass many independently operated shopping centers, as well as many individually operated shops, many of the latter on personal computers.

Servers provide some places, and user computers provide others. For example, home place on a user's personal communicator might serve as the point of departure and return for agents that the user sends to server places.

6.8.2.2.2 Agents. Telescript technology models a communicating application as a collection of agents. Each agent occupies a particular place. However, an agent can move from one place to another, thus occupying different places at different times. Agents are independent in that their procedures are performed concurrently.

In the electronic marketplace, the typical place is permanently occupied by one distinguished agent. As shown in Fig. 6.4, this stationary agent represents the place and provides its service. For example, the ticketing agent provides

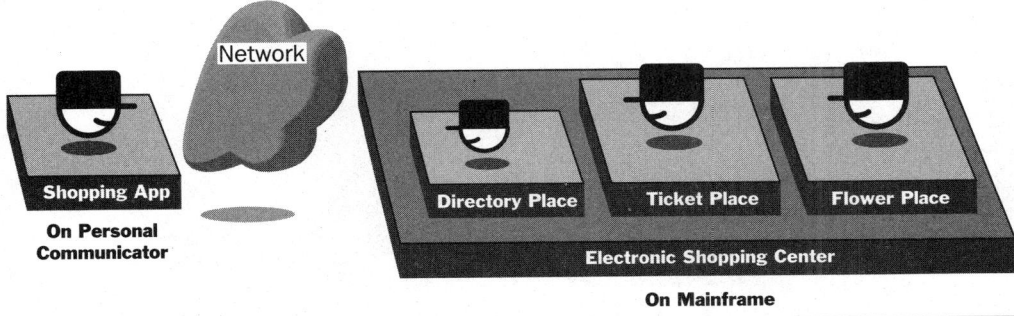

Figure 6.4 Agent representing a place.

information about events and sells tickets to them, the flower agent provides information about floral arrangements and arranges for their delivery, and the directory agent provides information about other places, including how to reach them.

6.8.2.2.3 Travel. Telescript technology lets an agent travel from one place to another, however distant. Travel is the hallmark of a RP system.

Travel lets an agent obtain a service offered remotely and then return to its starting place. A user's agent, for example, might travel from home to a ticketing place to obtain orchestra seats for *Phantom of the Opera*. Later the agent might travel home to describe to its user the tickets it obtained. Figure 6.5 illustrates this notion.

Moving software programs between computers by means of a network has been commonplace for 20 years or more. Using a LAN to download a program from the file server where it's stored to a personal computer where it must run is a familiar example. What's unusual is moving programs while they run, rather than before. A conventional program, written, for example, in C or C++, cannot be moved under these conditions because neither its procedure nor its state is portable. An agent can move from place to place throughout the performance of its procedure because the procedure is written in a language designed to permit this movement. The Telescript language in which agents are programmed lets a computer package an agent—its procedure and its state—so that it can be transported to another computer. The agent itself decides when such transportation is required.

To travel from one place to another an agent executes an instruction that is unique to the Telescript language, the go instruction. The instruction requires a ticket, data that specify the agent's destination, and the other terms of the trip (for example, the means by which it must be made and the time by which it must be completed). If the trip cannot be made (for example, because the means of travel cannot be provided or the trip takes too long), the go instruction fails and the agent handles the exception as it sees fit. However, if the trip

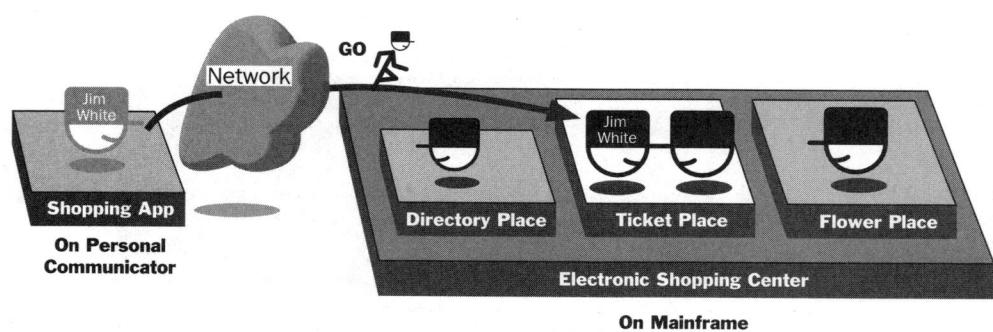

Figure 6.5 Traveling agent.

succeeds, the agent finds that its next instruction is executed at its destination. Thus in effect the Telescript language reduces networking to a single instruction.

In the electronic marketplace, the go instruction lets the agents of buyers and sellers colocate so that they can interact efficiently.

6.8.2.2.4 Meetings. Telescript technology lets two agents in the same place meet. A meeting lets agents in the same computer call one another's procedures.

Meetings are what motivate agents to travel. As Fig. 6.6 illustrates, an agent might travel to a place in a server to meet the stationary agent that provides the service the place offers. The agent in pursuit of theater tickets, for example, travels to and then meets with the ticket agent. Alternatively, two agents might travel to the same place to meet each other. Such meetings might be the norm in a place intended as a venue for buying and selling used cars.

To meet a colocated agent, an agent executes the Telescript language's meet instruction. The instruction requires a petition, data that specify the agent to be met and the other terms of the meeting, such as the time by which it must begin. If the meeting cannot be arranged (for example, because the agent to be met declines the meeting or arrives too late), the meet instruction fails and the agent handles the exception as it sees fit. However, if the meeting occurs, the two agents are placed in programmatic contact with one another.

In the electronic marketplace, the meet instruction lets the colocated agents of buyers and sellers exchange information and carry out transactions.

6.8.2.2.5 Connections.* Telescript technology lets two agents in different places make a connection between them. A connection lets agents in different computers communicate.

Connections are often made for the benefit of human users of interactive applications. The agent that travels in search of theater tickets, for example,

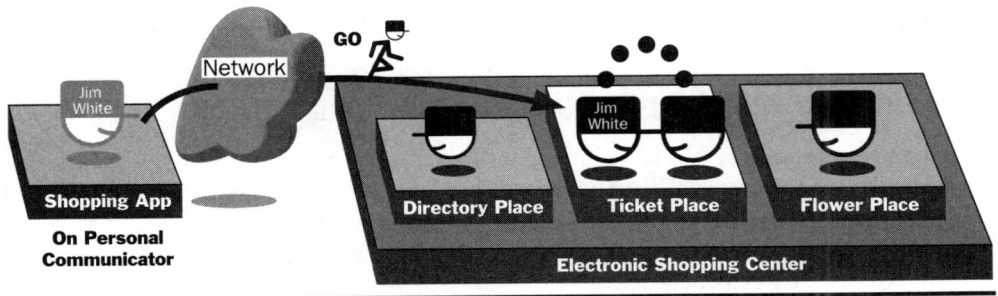

Figure 6.6 Meetings.

---

*Connections are not available in the present version of Telescript technology.

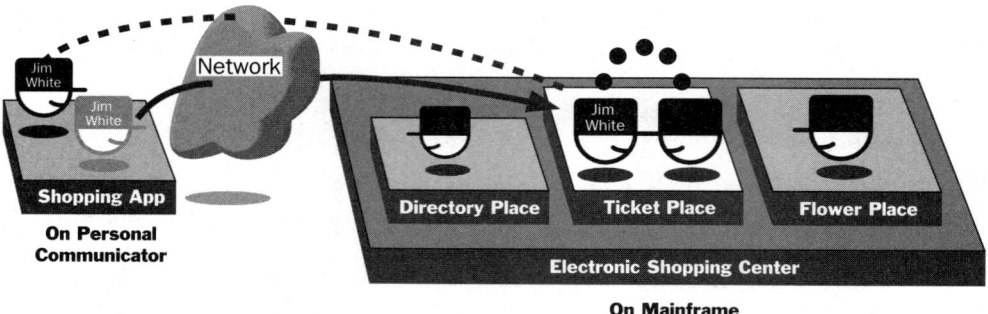

Figure 6.7 Agent connections.

might send to an agent at home a diagram of the theater showing the seats available. The agent at home might present the floor plan to the user and send to the agent on the road the locations of the seats the user selects. Figure 6.7 illustrates this notion.

To make a connection to a distant agent, an agent executes the Telescript language's connect instruction. This instruction requires a target and other data that specify the distant agent, the place where that agent resides, and the other terms of the connection, such as the time by which it must be made and the quality of service it must provide. If the connection cannot be made (for example, because the distant agent declines the connection or is not found in time or the quality of service cannot be provided), the connect instruction fails and the agent handles the exception as it sees fit. However, if the connection is made, the two agents are granted access to their respective ends of it.

In the electronic marketplace, the connect instruction lets the agents of buyers and sellers exchange information at a distance. Sometimes, as in the theater layout phase of the ticketing example, the two agents that make and use the connection are parts of the same communicating application. In such a situation, the protocol that governs the agents' use of the connection is of concern only to that one application's designer. It need not be standardized.

If agents are one of the newest communication paradigms, connections are one of the oldest. Telescript technology integrates the two.

6.8.2.2.6 Authorities. Telescript technology lets one agent or place discern the authority of another. The authority of an agent or place in the electronic world is the individual or organization in the physical world that it represents. Agents and places can discern, but neither withhold nor falsify their authorities. Anonymity is precluded.

Authority is important in any computer network. To control access to its files, a file server must know the authority of any procedure that asks it to list or delete files. The need is the same, whether the procedure is stationary or mobile. Telescript technology verifies the authority of an agent whenever it

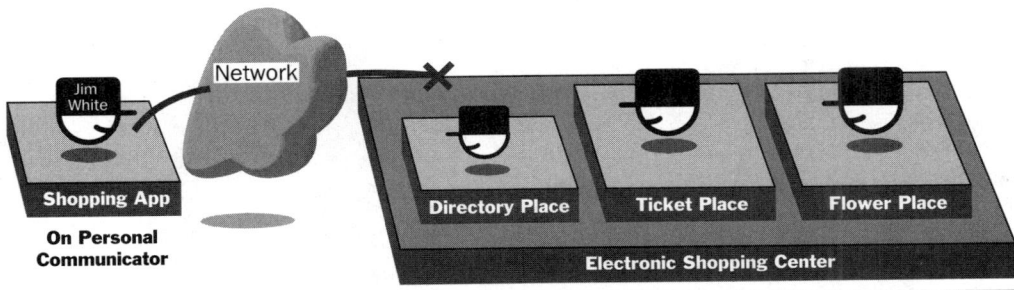

Figure 6.8 Agent authorities.

travels between regions of the network, as shown in Fig. 6.8. A region is a collection of places provided by computers that are all operated by the same authority. Unless the source region can prove the agent's authority to the satisfaction of the destination region, the latter denies the agent entry. In some cases, highly reliable, cryptographic forms of proof may be demanded.

To determine an agent's or a place's authority, an agent or place executes the Telescript language's name instruction. The instruction is applied to an agent or place within reach for one of the reasons discussed in the following paragraphs. The result of the instruction is a telename, data that denote the entity's identity as well as its authority. Identities distinguish agents or places of the same authority.

Authorities let agents and places interact with one another on the strength of their ties to the physical world. A place can discern the authority of any agent that attempts to enter it and can arrange to admit only agents of certain authorities. An agent can discern the authority of any place it visits and can arrange to visit only places of certain authorities. An agent can discern the authority of any agent with which it meets or to which it connects and can arrange only to meet with or connect to agents of certain authorities.

In the electronic marketplace, the name instruction lets programmatic transactions between agents and places stand for financial transactions between their authorities. A server agent's authority can bill a user agent's authority for services rendered. In addition, the server agent can provide personalized service to the user agent on the basis of its authority, or can deny it service altogether. More fundamentally, the lack of anonymity helps prevent viruses by denying agents that important characteristic of viruses.

6.8.2.2.7 Permits. Telescript technology lets authorities limit what agents and places can do by assigning permits to them. A permit is data that grant capabilities. An agent or place can discern its capabilities but cannot increase them.

Permits, as shown in Fig. 6.9, grant capabilities of two kinds. A permit can grant the right to execute a certain instruction. For example, an agent's permit

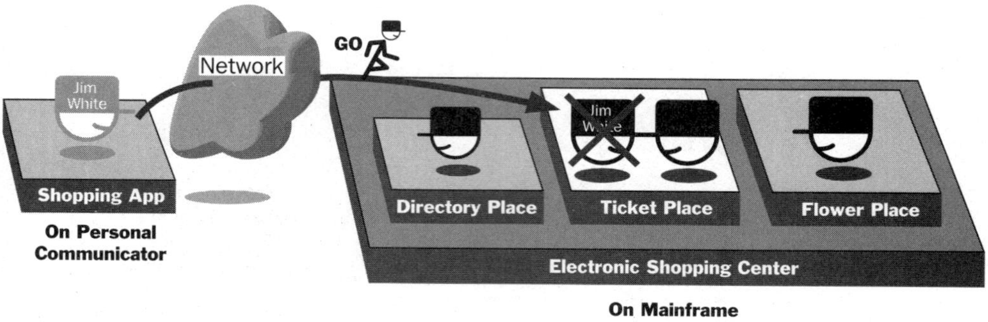

Figure 6.9 Agent permits.

can give it the right to create other agents.* An agent or place that tries to exceed one of these qualitative limits is simply prevented from doing so. A permit can also grant the right to use a certain resource in a certain amount. For example, an agent's permit can give it a maximum lifetime in seconds, a maximum size in bytes, or a maximum amount of computation, its allowance. An agent or place that tries to exceed one of these quantitative limits is destroyed.†

To determine an agent's or a place's permit, an agent or place executes the permit instruction of the Telescript language. This instruction is applied to an agent or place within reach for one of the reasons discussed in Subsection 6.8.2.2.6, "Authorities."

Permits protect authorities by limiting the effects of errant and malicious agents and places. Such an agent threatens not only its own authority but also those of the place and region it occupies. For this reason the technology lets each of these three authorities assign an agent a permit. The agent can exercise a particular capability only to the extent that all three of its permits grant that capability. Thus an agent's effective permit is renegotiated whenever the agent travels. To enter another place or region the agent must agree to its restrictions. When the agent exits that place or region, its restrictions are lifted, but those of another place or region are imposed.

In the electronic marketplace, the permit instruction and the capabilities it documents help guard against the unbridled consumption of resources by ill-programmed or ill-intentioned agents. Such protection is important because agents typically operate unattended in servers rather than in user computers, where their misdeeds might be more readily apparent to the human user.

---

*An agent can grant to any agents it creates only capabilities it has itself. Furthermore, it must share its allowance with them.

†An agent can impose temporary permits upon itself. The agent is notified, rather than destroyed, if it violates them. An agent can use this feature of the Telescript language to recover from its own misprogramming.

Agent-Design Considerations 243

6.8.2.3 Putting things together. An agent's travel is not restricted to a single round-trip. The power of mobile agents becomes fully apparent when one considers an agent that travels to several places in succession. Using the basic services of the places it visits, such an agent can provide a higher-level composite service.

Figure 6.10 shows that, in our example, traveling to the ticket place might be only the first of the agent's responsibilities. The second might be to travel to the flower place and there arrange for a dozen roses to be delivered to the user's companion on the day of the theater event. Note that the agent's interaction with the ticket agent can influence its interaction with the flower agent. For example, if instructed to get tickets for any available evening performance, the agent can order flowers for delivery on the day for which it obtains tickets. (Remember the Doonesbury cartoon.)

This simple example has far-reaching implications. The agent fashions from the concepts of tickets and flowers the concept of special occasions. In the example as presented, the agent does this for the benefit of an individual user. In a variation of the example, however, the agent takes up residence in a server and offers its special-occasion service to other agents. Thus agents can extend the functionality of the network; and thus the network is a platform.

6.8.2.4 Mobile agent technology. New communication paradigms beget new communication technologies. A technology for mobile agents is software—software that can ride atop a wide variety of computer and communications hardware, present and future.

Telescript technology implements the concepts of the previous section and others related to them. It has three major components: the language in which agents and places (or their "facades") are programmed; an engine, or interpreter, for that language; and communication protocols that let engines in different computers exchange agents in fulfillment of the go instruction.

6.8.2.4.1 Language. The Telescript programming language lets communicating application developers define the algorithms that agents follow and the

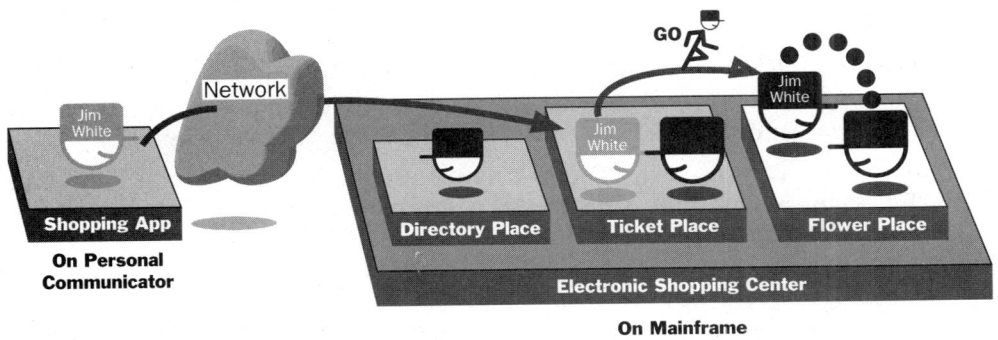

Figure 6.10 Agents with complex itineraries or goals.

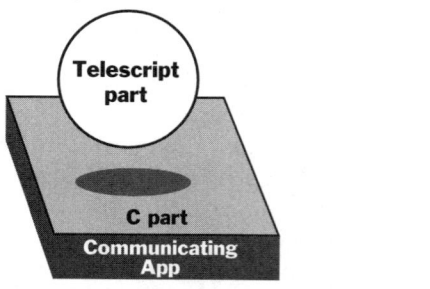

Figure 6.11 Telescript as agent definition.

information that agents carry as they travel the network (see Fig. 6.11).* It supplements system programming languages such as C and C++. Entire applications can be written in the Telescript language, but the typical application is written partly in C. The C parts include the stationary software in user computers that lets agents interact with users, and the stationary software in servers that lets places interact, for example, with databases. The agents and the "surfaces" of places to which they are exposed are written in the Telescript language.

The Telescript language has the following qualities that facilitate the development of communicating applications:

- *Complete:* Any algorithm can be expressed in the language. An agent can be programmed to make decisions; to handle exceptional conditions; and to gather, organize, analyze, create, and modify information.

- *Object-oriented:* The programmer defines classes of information, one class inheriting the features of others. Classes of a general nature, such as agent, are predefined by the language. Classes of a specialized nature, such as shopping agent, are defined by communicating application developers.

- *Dynamic:* An agent can carry an information object from a place in one computer to a place in another. Even if the object's class is unknown at the destination, the object continues to function—its class goes with it.

- *Persistent:* Wherever it goes, an agent and the information it carries, even the program countermarking its next instruction, are safely stored in nonvolatile memory. Thus the agent persists despite computer failures.

- *Portable and safe:* A computer executes an agent's instructions through a Telescript engine, not directly. An agent can execute in any computer in

---

*Despite its name, the Telescript language is not a scripting language. Its purpose is not to allow human users to create macros, or scripts, that direct applications written in "real" programming languages like C. Rather, it lets developers implement major components of communicating applications.

which an engine is installed, yet it cannot access directly its processor, memory, file system, or peripheral devices. This helps prevent viruses.
- *Communication-Centric:* Certain instructions in the language, several of which have been discussed, let an agent carry out complex networking tasks, such as transportation, navigation, authentication, access control, and so on.

A Telescript program takes different forms at different times. Developers deal with a high-level compiled language not unlike C++. Engines deal with a lower-level interpreted language. A compiler translates between the two.

6.8.2.4.2 Engine. The Telescript engine is a software program that implements the Telescript language by maintaining and executing places within its purview, as well as the agents that occupy those places. An engine in a user computer might house only a few places and agents. The engine in a server might house thousands (see Fig. 6.12).

At least conceptually, the engine draws upon the resources of its host computer through three APIs. A storage API lets the engine access the nonvolatile memory it requires to preserve places and agents in case of a computer failure. A transport API lets the engine access the communication media it requires to transport agents to and from other engines. An external applications API lets the parts of an application written in the Telescript language interact with those written in C.

6.8.2.4.3 Protocols. The Telescript protocol suite, shown in Fig. 6.13, enables two engines to communicate. Engines communicate in order to transport agents between them in response to the go instruction. The protocol suite can

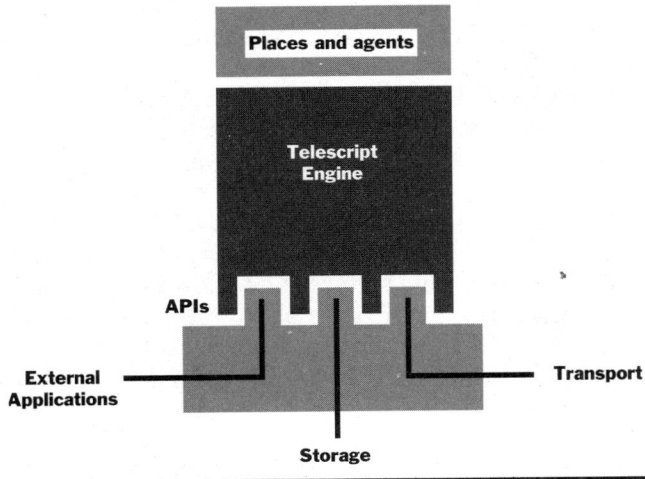

Figure 6.12 Telescript architecture—The Telescript engine.

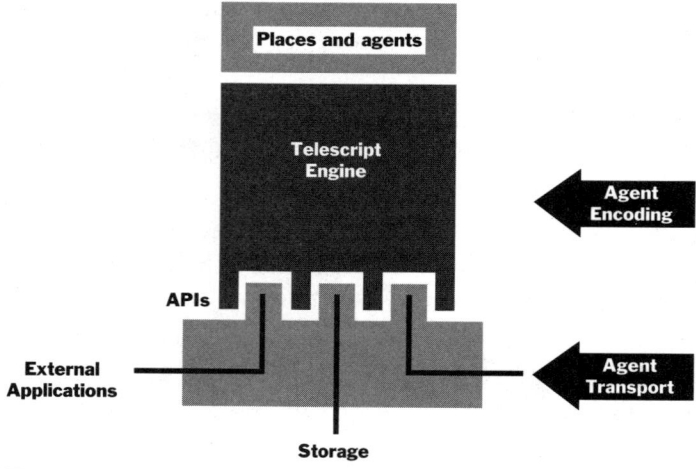

Figure 6.13 How Telescript engines communicate.

operate over a wide variety of transport networks, including those based on the TCP/IP protocols of the Internet, the X.25 interface of the telephone companies, or even electronic mail.

The Telescript protocols operate at two levels. The lower level governs the transport of agents, the higher level their encoding and decoding. Loosely speaking, the higher-level protocol occupies the presentation and application layers of the seven-layer OSI model.*

The Telescript encoding rules specify how an engine encodes an agent—its procedure and its state—as binary data and sometimes omits portions of it to optimize performance. Although engines are free to maintain agents in different formats for execution, they must employ a standard format for transport.

The Telescript platform interconnect protocol specifies how two engines first authenticate one another (for example, using public key cryptography) and then transfer an agent's encoding from one to the other. The protocol is a thin veneer of functionality over that of the underlying transport network.

6.8.2.5 Mobile agents—Conclusion. It should be evident that Telescript technology and architecture represent the best integrated agent development and execution environment available so far. They address all of the keys to agent success, and have all the infrastructure needed for successful agent execution. As you explore and evaluate agent design concepts, it's good to know that there exists today a system built explicitly for implementing good agent-based appli-

---

*The Telescript protocols are not OSI protocols. The OSI model is mentioned here merely to provide a frame of reference for readers who are acquainted with OSI.

cations. More detailed information about Telescript and other options for implementing agents will be discussed in Section 7.

We have learned so far that mobile agents are essentially software entities capable of replication and autonomous movement from one computing node to another, and of performing tasks based on information collected from throughout a network. In the ideal system, well-trained agents simply go about their tasks, collecting and processing information and waiting in the background for a set of events to occur.

Itinerant agents are dispatched from a source computer and roam among a set of networked servers until they are able to accomplish their tasks. Under this model, the agent can process and filter data extensively where they reside rather than moving them all to another location. Moreover, a dynamic itinerary would allow the agent to achieve its goal even when the capabilities of specific remote servers are not known ahead of time. In the patchwork quilt of today's interconnected public and private networks, this is a valuable asset. Examples of itinerant agents include information retrieval and processing agents.

Mobile agents can help answer the question of how to share processing and data across distributed nodes. As network bandwidth gets more and more crowded, having IAs do the work becomes increasingly attractive. For example, instead of an investor needing to check stock quotes at every spare moment in search of the right moment to buy, then trying to get an order through in time, he or she give rules to an agent, then turns the agent loose to find shares at a good price. When the agent finds a good deal, it contacts its owner for buying approval. After it receives approval, it purchases the shares, notifies its owner, and shuts down.

Mobile agents can do far more than buying and selling, however. Imagine a network of controllers in a processing plant. As production equipment changes, new monitoring software can be loaded into each node in the network via agent technology. IAs could ascertain the state of any piece of plant equipment simultaneously, and adjust parameters to achieve optimum efficiency.

Agents that move and interact in a distributed environment must have a portable representation which can execute on a number of VMs without recompiling, that is, byte codes or source code as opposed to machine language representation. One reason for this is that the execution environment needs to check each agent to make sure it will not try to execute illegal methods, access off-limits data, or forge and dereference pointers to wreak havoc on the host system.

Another reason that the agent will need to be compiled and executed dynamically is to take advantage of specific hardware features of the execution platform such as in lining and register usage. In addition, agents must not use pointers or programmer-directed memory allocation and deallocation. Agents must also support dynamic resource binding, since the execution environment will bind agent variables as appropriate.

Even with a dynamic, portable, clever agent, getting good results from a web-roving software entity is fraught with challenges. The phases in the life cycle of a mobile agent may be as follows:

- Typically, the agent's owner or user gives it rules or instructions and the resources (such as electronic cash, passwords, SQL scripts, data, API calls) needed to perform a task, then packages it up and sends it (via any number of means from copying to mime E-mail, to Telescript) to a host. An agent can also be launched, and even created, by another agent or process.

- The agent determines where it needs to go based on instructions or goals. The agent's instructions may include a list of nodes to visit in search of what your owner wants. The agent would also have a goal to accomplish, along with instructions for how to negotiate and what to trade for, if buying and selling are important.

- When the agent arrives at a host, and the host chooses to run the agent's code, the agent could then request that the host give it what it wants. The agent is then primarily dependent on calling functions supported on the host.

- If the agent is successful, results get returned to a designated sender address.

- When an agent's run was completed, it would then request that the host package it up again and send it to the next destination on its list (possibly back to your home server).

- If the new host can provide the resources the agent needs, and the agent has the correct privileges, the new host begins to execute the agent.

- Stopping the agent can be the hard part. How do you know where it is or what it's doing? A time limit is a good idea, as is periodic reporting back to the agent's owner. In the best of cases, an agent will shut itself down once its goals are reached. If there is a problem, the agent should detect it and send an error log to the user, then gracefully shut down, taking care to release any resources it holds and inform its owner.

6.9 Authentication, Exceptions, and Security

Making sure agents behave politely and responsibly can be almost as big a task as making them intelligent. The biggest danger of any networkwide system that allows IAs is that some of the agents will deliberately or accidentally run amok. Wayward agents can have a lot in common with viruses. Both are independently executing programs that can destroy or take control of resources on a foreign machine. Due to scores of different factors in an execution environment, mobile agents may also run into trouble while trying to accomplish a user's goals. Without proper error handling, such an agent could take an incorrect execution path and cause major headaches to many users, including the one who dispatched it.

Agents need almost exactly the same kind of security as any other program executing in cyberspace. Mobile agents are the main problem, since they may seek to access data and use resources on many network nodes, although stationary agents may also attempt to access data for which they have (or should have) no permission. The most basic security key is the same as always: set protection for files and passwords and user names for databases. For a desktop system, or even for a single server, this can be relatively simple: don't allow unauthorized users (or agents) to connect to your system. In the most extreme case, this means not connecting any ports for access from the outside world. Since most computers are eventually wired for communication to other computers, mimicking the overall interconnectivity of the world in general, a less radical approach is usually taken.

6.9.1 Handling exceptions

Even if all goes well and an agent legitimately gains access to a protected server, a number of factors may still go awry. In spite of safeguards, some force outside the agent's direct control often causes the agent to cease operation or operate incorrectly. These include down or inaccessible servers, databases with schemas that won't support a specific agent-delivered query, uncooperative security measures, lack of resources to run the agent on specific servers, and missing or incorrect APIs.

An agent (especially a mobile one) may be able to display messages on the host when necessary, and take input from various sources until its task is completed, but on the whole, agents are left by themselves to handle the problems that crop up. Most software is interactive with the user, but once an agent is unleashed, it's essentially on its own. It usually can't ask the user for more instructions. It must know enough to handle all expected (and some unexpected) conditions. Agents are different from other software because they are autonomous. Autonomy means responsibility in situations such as the following:

- What happens if one or more agents don't return results or status from their tasks or they themselves don't return from the location where they were executing?
- What if an agent cannot find what it came for; for example, a meeting-scheduler agent finds that no rooms are available?
- What does the agent do when systems or other resources are not available, or are too expensive, yet the agent has a high priority on fulfilling its goal?

The major agent development languages explored in this book all have excellent exception-handling facilities. However, in spite of how well an agent is programmed to degrade gracefully, handle specific error conditions, and do no harm to the host, things can and will go wrong. Security, privacy and error-handling issues must be supported in the infrastructure (for example the Source Sokets Layer (SSL) over COBRA's IIOP), and the host needs to keep an eye on the agent by monitoring resource use and data access. Agent hosts can maintain an agent execution audit trail, accounting, and monitoring, and even charging for what an agent does.

Limiting the use of resources by an agent, or accounting and charging for resource use are some of the reasons Telescript and other agent languages are interpreted or late-binding. With such systems it is easy for a host to insert instruction counters into an interpreter loop. Other resource control can be done with wrapped methods which, for example, count bytes written to or read from files or network sockets.

6.9.2 Security considerations—general

Security is one of the most obvious problems with both mobile and stationary agent technology—you need to guard against wayward agents that install viruses, compromise the host, or pilfer your database. The flip side, however, is that agents must be free to perform valuable tasks and must be interoperable among a variety of computing platforms. The ideal solution may lie in a new generation of operating environments that are extremely open and secure, such as Telescript, distributed Smalltalk, or some variation of Lotus Notes.

These systems have the capability to send not only distributed packets of data to remote machines, but packets of executable code as well. This code is much like any other program, with several specific differences. Since a network-navigating agent must execute on remote host systems, there must be provisions in the host execution environment for not only facilitating agent execution and communication, but also preventing the agent from accessing sensitive information (and vice versa) as well as protecting memory and limiting CPU cycles. In short, the incoming agent can do only what the host allows.

Autonomous mobile agents have some similarity to viruses, and indeed, without safeguards built into the infrastructure, agents could potentially trample memory or files needed by other programs. Privacy issues also emerge when software agents are on the prowl. While it may be convenient to have an agent sort through mail or news groups worldwide to retrieve information on your favorite topics, but you wouldn't necessarily want everyone (especially junk mail generators) in the world to know all your likes and dislikes. There are also agents that exist solely to spy on competitors and consumers, trolling for buying patterns, data transfers, and E-mail memos between employees at another company.

Ideally, a major difference between agents and viruses is the security administered by the system upon which an agent executes. In a well-managed system, an agent should be allowed access only to specific resources (memory, disk space, CPU cycles, and so on) within a network node, or at a meeting place for agents.

6.9.3 Security considerations—authentication and digital signatures

The topic of access control and information safety amounts to limiting the damage an ill-behaved, mischievous, or evil agent can do. The most dangerous things an agent can do are, of course, modifying and deleting data on the system when it's not supposed to leading to business failure, or worse a catastrophe measured in bodily harm to humans. Agents must generally have read-only access, along with a limited number of file accesses permitted, and limited memory and processing time. The main idea in this line of defense is verifying the source of an agent, and limiting facilities to allowed or trusted sources.

In the vast free-for-all of interconnectivity that is the Internet, there are many servers and intranetworks around which web masters have erected fire walls against invading users. While users demand access to and from the outside world, they also demand security. Intranet security thus resembles a walled city, with a limited number of gates or ports to the world outside. Fire walls are computers that stand guard over these ports by constantly running software to check the validity of incoming information. Anyone trying to access anything on a protected intranet must go through these fire walls—they're the only ones in the network wired for outside access. To get through, users (or users' agents) must have valid identification, and attempt to access only those ports that are acknowledged by the fire walls. In general, this identification consists of an Internet protocol (IP) address that is in the correct range and refers to an acceptable user name stored in a domain name server somewhere on the Internet.

Once an agent or user gets inside the fire wall, there are other protections, for example, passwords to various resources, such as databases, and encrypted data. In addition, systems which execute agents must check the agent's byte codes or source code to determine that it will not try to do anything harmful such as use more than its share of memory or fill a disk with garbage files. Some security problems can be solved by using a layer of encryption and cryptographic authentication to keep track of an agent's origin. Using this information, a fire wall could immediately answer the eternal questions: "Friend or foe?" and "What privileges (if any) do you have?" Encrypting files such as E-mail, as well as setting protections on them, provide yet another line of defense. A rogue agent may get in, may locate files, and may even be able to open them, but getting any meaningful information is another matter. Secure agent-to-agent communications is anabled by such service as the SSL standard.

6.10 Agents Programmed or Configured by the End User

Much of the foregoing concentrates almost exclusively on what you, the IA designer and developer, can do to create successful agents. But there is another way that combines your efforts with that of your most important ally, your users, or, more accurately, the users of your IAs. "Agents must be flexible enough to be tailored to each individual. The most flexible way to tailor a software entity is to program it. The problem is that programming is too difficult for most people today" (Smith, Cypher, and Spohrer, 1994).

If this ease of programming issue is not resolved, agents may whither on the vine. Targeting specific agent functionality to specific domains is fine. It is possible to create an agent builder that generates an agent script, especially if many agent-related utilities are already defined in the library. But as users have become burdened by application featuritis and system software configuration complexities, we see the possibility of the impending agent in every pot, so to speak, becoming another burden to already perplexed users.

Although this is an extreme scenario, the point of view has some merit. Many agent examples, prototypes, and concepts discussed in this book have relied on making the agents very flexible as a matter of architecture as they come off the developer's bench. One type of flexibility means coming complete with facilities for the user to customize the agent before launch.

This is very difficult to do, as the agents may become bloated resource hogs, or must take advantage of unknown infrastructural amenities (for example, network protocols, access to databases, or agent-aware helper applications). For certain definable domains, with assured infrastructure support, these approaches suffice; for example, enabling access to giant commonsense knowledge databases like Cyc, or enabling agents to enter into societies or federations from which flexible approaches to tasks may be garnered and experiences shared.

However, in other cases infrastructure assumptions (such as where your agent may live) or, in the absence of a known shared architecture guaranteeing interagent communications, another approach may be needed. One such promising approach is to develop your agents such that they can be programmed on the fly in some user-friendly fashion. Much research has been done along these lines of thought. For example, user-friendly scripting languages have been developed, like Apple's HyperTalk. Other techniques include visual interactive programming (VIP), enabling end-user programming by analogy, and letting users edit examples of agents that already do useful work (either in other domains or as embedded in a specific infrastructure).

Enabling easy end-user agent programming—or even just configuration—using the techniques just mentioned requires adherence to certain principles. Many of these principles are outgrowths of good user-interface design concepts promulgated since the advent of the Macintosh and earlier work at Xerox Parc. Tognazzini (1992) and Laurel (1993) are two good generic references.

Many good user-interface practices are followed in KIDSIM, a simulation-making environment for children. They include the following. (Note that adult users are capable of more. In reality, agent definition or, more accurately, configuration most likely resides in the domain of the power user.)

- Make everything that is important to the user and the agent visible and manipulable on the screen.

- Make it easy for the user to understand the consequences of actions; show feedback.
- Limit the use of modal interactions, that is, don't get the user in a situation where there is no way to back up, leave the agent-programming task for some other venue, or otherwise limit their input to your (and only your) interface.
- As mentioned earlier, include example agents that do useful things, and show the user (such as via tutorial) how to modify the agent for other tasks. This is easier than giving the user some language specification and saying "go to it."
- Make copious use of direct manipulation; this implies that you provide graphic objects that represent your agents and the various aspects of your agents, together with any supporting architectural and infrastructural element the agents are to interoperate with.
- Don't use computerese when describing how to program their agents. Use concrete terms and concepts that relate to concrete domains and tasks.
- Try to make the result of the user's programming efforts map into a one-to-one relationship with the internals, or data structures (that is, objects) of the finished agent. This is good OO design and development practice. Although not always practical when generating code fragments and hidden or proprietary mechanisms, you could make use of OO principles such as encapsulation to help out here. For example, say the user is teaching an agent to recognize certain graphic objects, and your end result is to be a neural net embodiment of that task. At the conclusion, let the user see some graphical object encapsulating the net's innards, rather than attempting to educate the user in the intricacies of that representation.
- Provide an environment where users can engage their agents in simulations to test them before commissioning them to a real environment.

Some of the other ideas embodied in KIDSIM are extrapolable to any end user of agents.

- Allow users to make their own representations of their agents. This implies providing a graphical tool kit or enabling a commercial graphics package's output structure to dovetail with your agent's design.
- Allow users to define their own agent's characteristics. Provide root agent classes that can be specialized via derivation, or agent objects that can be customized via variable configuration.
- Allow users to define (to some extent) the behavior of their agents. This implies a behavior editing tool. For those users who you believe are computer literate, this implies use of a scripting or other standard language, or to make it easier, a VIP language. Again, allowing users to derive their

agents from developer-supplied root or standard agents with default behaviors allows users to program by extension and editing, rather than from scratch.

This last idea is embodied in KIDSIM by innovatively combining two ideas:
First is the use of *graphical rewrite rules* that enable the transformation of a region of a simulation environment from a before part or state to an after part or state. A rewrite rule is similar in structure to an if-then-else or production-system-based environment. "A graphical rewrite rule consists of a (possibly generalized) visual image and zero or more property tests. In order for a rule to match, its visual image must conform to a situation on the game board [in the KIDSIM environment], and all of its property tests must evaluate to true" (Smith, Cypher, and Spohrer, 1994).

Saying this another way, using graphical rewrite rules involves allowing a user to specify, within an agent's simulated environment, a before and after state where the definitions of such states are expressible with attributes that, in turn, affect some graphic content visible to the user. The user is telling the agent to transform the first or current (leftmost) state into the second or goal (rightmost) state (left to right being the standard production-rule form of expressing a transformation).

A practical example might be showing the graph of a network in a troubled state and a smoothly operating state. Troubles could be shown by nodes whose backed-up traffic is causing the node's display graphic to turn red and blink. When an agent fixes the traffic problem, the state's graphic could turn a steady blue. The mechanism of how the agent does this is to understand how to perform this transformation (that is, do its network troubleshooting task). This is the subject of the next idea.

Second is *programming by demonstration*. "Programming by demonstration (PBD) is a technique in which the user puts a system in 'record mode,' then continues to operate the system in the ordinary way, and the system records the user's actions in an executable program" (Smith, Cypher, and Spohrer, 1994).

Recording the "how" in this fashion implies that you have graphically expressed all the steps an agent can take, such that a user can choose the necessary ones and specify the sequence necessary to carry out a specific task. In the example of network troubleshooting, you may want to lead the user to a more explicit rendering of the attributes within a node that affect its performance. Then, in record mode, let the user change the substates of the node (graphically or via text entry) that affect its return to a normal state. What is required here is a mechanism to translate the user's actions to an executable that is then made part of the network troubleshooting agent (probably as one or more methods).

Combining these techniques within an agent is even better: "...combining graphical rewrite rules and programming by demonstration result in a system that is stronger than either. Graphical rewrite rules solve the programming by demonstration representation problem, and programming by demonstration solves the rule-semantics problem" (Smith, Cypher, and Spohrer, 1994). While we do not know of any commercial systems that employ both ideas in a single generalized programming environment, there are examples of both ideas out there.

For example, typical macrorecording software exemplifies PBD. Certain simulation languages environments (such as MODSIM) employ user-driven graphical programming. Data-flow language environments allow various levels of graphical programming or VIP. In addition, applications that allow systems to be specified graphically as objects and relationships, such as STELLA for the Macintosh, with the effects each object has on others specified from a selection menu, are examples of what agent developers should be looking to provide end users in the way of programming ease.

6.11 Agent Communications

Interprocess communication is normally a straightforward matter. Processes can send messages to each other regarding their states so that, for example, one process won't try to access a result before it is calculated, and another won't perform a calculation until all the needed data have been collected.

IAs are essentially independently executing processes, but since IAs deal in knowledge, coordinating activities among them can be much more complex. Like all independent tasks, IAs must communicate with each other and with users. There are, of course, a multitude of languages in which you can encode messages for interagent communications. Especially interesting are those languages that have been developed for the typical lightweight communications needs of agents. These include Telescript, Java, Limbo, TcL, Python, and several others. Of course, the capabilities of these languages are combined with infrastructural capabilities (such as CORBA) to form a complete communications mechanism. However, something more than just the agent's state and simple messages may need to be passed around. Since IAs act on rules and ontologies, which are much more complex than the average message, protocols have been defined for knowledge interchange. For example, the Cyc system uses a language called CycL.

The following is an example of a protocol or language designed explicitly for agent-to-agent communications. Section 5 discusses the use of such a protocol in the context of agent architectures.

6.11.1 Knowledge query and manipulation language

The KQML is an all-purpose agent communication and query language which conforms to the KIF specification (see Subsection 6.11.2). An advanced query protocol, KQML allows diverse agents to communicate since it does not force agents into a specific structure or implementation. KQML is a message format as well as a message-handling protocol. This supports not only simple state communication, but knowledge sharing among agents. Through a language such as KQML, agents can share knowledge and information to cooperate with each other to perform complex tasks.

A good ACL has many needs, some of which are in competition. KQML is a new communication language that addresses many—although not all—of these needs. The following information is taken, with permission, from the KQML website.*

> KQML—Knowledge Query and Manipulation Language—a high-level agent-communication language. KQML is a language and protocol for exchanging information and knowledge. KQML can be used as a language for an application program to interact with an intelligent system or for two or more intelligent systems to share knowledge in support of cooperative problem solving. It focuses on an extensible set of performatives, which defines the permissible operations that agents may attempt on each other's knowledge and goal stores.
>
> The performatives comprise a substrate on which to develop higher-level models of inter-agent interaction such as contract networks and negotiation. In addition, KQML provides a basic architecture for knowledge sharing through a special class of agents called communication facilitators which coordinate the interactions of other agents. The ideas which underlie the evolving design of KQML are currently being explored through experimental prototype systems which are being used to support several testbeds in such areas as concurrent engineering, intelligent design and intelligent planning and scheduling (Finin, McKay, and Fritzson, 1992).

6.11.2 Knowledge interchange format

KIF is a standard protocol for knowledge sharing and communication among diverse agents. It defines the capability for declaring reasoning rules and expressions, and for creating arbitrary sentences in the first-order predicate calculus. KIF also provides the capability to define objects, functions, and relations related to knowledge representation.

KIF can be considered an interlingua for knowledge sharing and communication among heterogeneous agents. It is a computer-oriented language for the interchange of knowledge among disparate programs. It has declarative semantics (that is, the meaning of expressions in the representation can be understood without appeal to an interpreter for manipulating those expressions); it is logically comprehensive (it provides for the expression of arbitrary sentences in the first-order predicate calculus); it provides for the representation of knowledge; it provides for the representation of nonmonotonic reasoning rules; and it provides for the definition of objects, functions, and relations (Finin, McKay, and Fritzson, 1992).

6.11.3 Agent communications example

The following is an example of how agents communicate to get a job done. This example is based on an information-distribution application being developed

---

*Navigate to http://www.cs.umbc.edu/kqml for in-depth information on KQML.

Agent-Design Considerations 257

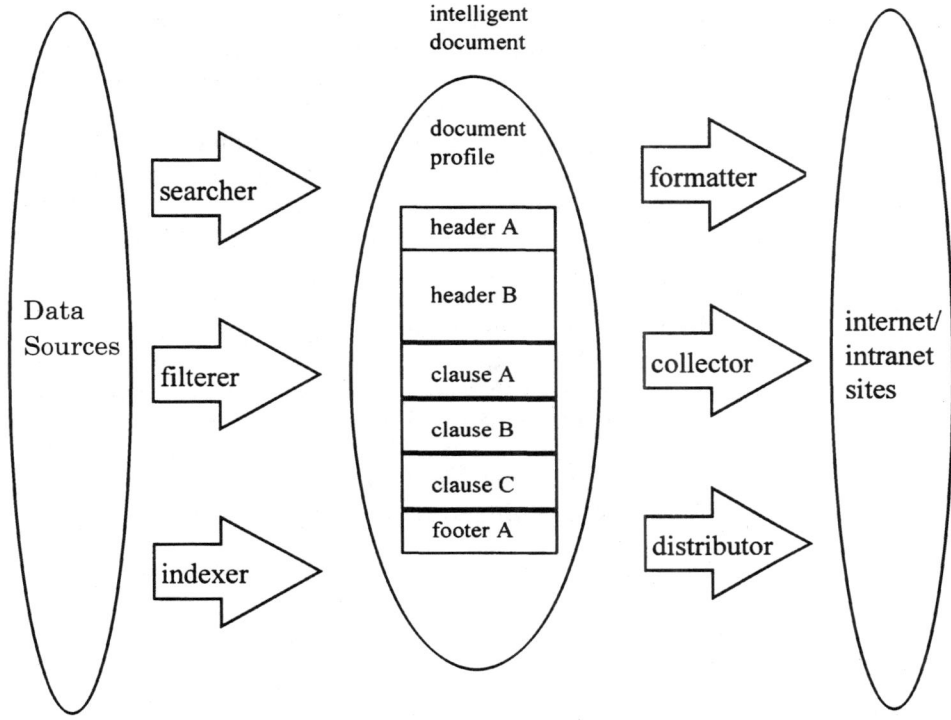

Figure 6.14 Agent collaboration on WWW documents.

for an on-line software support system. In this system agents work together to dynamically build and display intelligent web documents (see Fig. 6.14).

Each document consists of processing agents and components. Components may be controls (ActiveX, Java, or Java script), images, formatted text, or links to other sites. Processing agents include text searchers, data collectors, formatters, filterers, indexers, and distributors.

When a web page component wants to request a page update, it may first ask a text searcher for pages related to a specific user request, such as: "Find all information about installing NetWare." The searcher then locates all related database information, ranking it by number of search-term occurrences, and highlighting the search terms in all text fields found.

The searcher then sends a message telling a data collector to gather all information found from various sources into a "page frame" corresponding to a user-generated "page profile." This consists of clauses, a header, a footer, and other component specifications. In this process the collector may invoke filterers and indexers to select only the most relevant items from all sources, and to index all items into a database for fast text-based retrieval.

The distributor caches this information, along with other recently gathered page frames. The distributor is responsible for intelligent maintenance of the page cache. A data gatherer may also be triggered by changes to specified data entities, or by an update timer.

As requested by web pages and users, the distributor either sends information unformatted for automatically updating screen components, or sends a message to a formatter. This is how dynamic web sites are created. After creation, they may be updated as long as users are interested, and deleted or left in suspension otherwise.

The formatter takes filled-in page frames and formats them according to corresponding page profiles, adding images, links to related pages, and text marked up in a language such as HTML, Java applets, existing JavaScript components, and ActiveX controls. The distributor then sends these to appropriate locations, automatically building a new web page as a user navigates to it. After the page is built, the gatherer and the distributor constantly update it with new data as long as users are interested. The agents in this system are written in distributed Smalltalk and VisualWave.

Section

7

Developing Intelligent Agents Now

7.1 Introduction

In this section we provide an overview of the major development environments that we feel are especially conducive to agent building. For the most part these tools are available commercially, but as of this writing, some aren't quite there yet. Of course, theoretically, you could build agents in assembly language, C, or COBOL. Indeed, many database management systems provide procedural languages and triggers with which you could build *simple* "agents." But creating the kind of agent system that will fulfill the promise of ubiquity, intelligence, and autonomy mentioned throughout this book requires sophisticated design and development approaches. These in turn require the tools and technologies to match. So far in this book we have tried to give you an overview of *the types of* technologies that we feel you should consider incorporating into your agents or agent systems, as follows:

- What agents are and the roles an agent can play in giving users improved access to distributed systems, and how an agent can easily automate routine tasks.
- How OO "thinking" and development are a natural match for the agent creation tasks ahead of you.
- Various AI-driven techniques, including soft-computing technologies such as neural networks and fuzzy logic that can be incorporated into agents to make them (more) "intelligent."

- Certain infrastructures or middleware that you can use to help bridge the environment and network gaps between agents in distributed systems.
- Several agent architectures incorporating concepts of collaboration, communications, and organization that can help guide your own design toward truly effective agent systems.
- General design considerations that you should be aware of. These considerations relate both to the agents you will be building and to the environments in which they will run.

Many of those technologies and concepts will help provide the "intelligence" or help solve the distributed-agent-design problems you will face. Nothing beats reuse. For example, the use of a CORBA-compliant infrastructure will save you time and the headaches involved in getting agents to communicate with each other at the lowest levels of your system. This frees your valuable time to concentrate on what you want your agents to do, rather than the minutiae of cross-platform communications.

Eventually, however, to build and launch agents successfully, agent-building tools, effective execution environments, and, usually, an end-user interface to the agents are necessary. Today there are a number of environments and tools that have great potential to develop and run truly useful agents. In this section we'll explore Smalltalk, Telescript, and Java, all technologies that incorporate an OO language and development environment, with a VM-based environment in which your agents can execute. We'll also briefly look at other types of tools, such as VisualBasic and IBM/Lotus Notes, and discuss where they fit into the agent technology picture.

7.2 Building Simple Agents

The simplest agents are, for example, UNIX shell scripts launched with the UNIX cron utility or some other timer or system event to start up at specific times every day. Simple agents can also be built using database triggers, such as provided in Sybase. Sending an SQL query string from a client program to execute on a database server is an example of simple agent behavior.

Database triggers and stored procedures can be used to turn formatted reports into on-line text search queries, or a report formatter can be timed to run daily. A printer demon can manage print queues, triggered when a job is sent, or a postscript file can be sent to a printer where it is interpreted into a formatted document. These simple agents have very little in the way of autonomy or intelligence, but are the first steps in making agent technology useful.

7.2.1 Agents (almost) without programmers

There is no shortage of relatively simple agent builders, such as MicroStrategy's DSS Agent, Verity agent servers, or IBM/Lotus Notes. All of these have their place, but their usefulness is limited to searching databases, formatting complex reports, and providing data views.

As an example of a relatively simple agent builder, IBM/Lotus Notes, release 4.x and above, has some interesting capabilities. Notes, for those of you living on another planet, is an environment that supports a document-centric model of collaborative groupware. You can design a database composed of Notes documents that can incorporate multiple user-designed forms and reports. Forms can "inherit" design characteristics from other forms that are designated as templates. (A Notes template is very roughly akin to a class in the OO world.)

Forms can include all the familiar field types, as well as pointers to, or embedded documents created with, other applications (such as a Microsoft Word document). When a change is made to a Notes document, any place where that document is replicated on other nodes, or servers, is also updated via a replication process. Forms and reports can have related stored procedures that fire when some event occurs.

What is interesting from the agent builder's point of view is that this package includes an explicit agent template with which you can specialize specific types of agents. Notes agents are an encapsulation of complex behavior you want to occur when triggers or events happen. Notes' agent building involves specifying the events that will launch the agent, together with the (Notes-specific) code and attributes, where you specify what the agent is and does.

7.2.2 Power-user environments for building agents

Another step up the development-environment complexity and flexibility hierarchy, and we find ourselves using 4GLs together with the appropriate infrastructural components to build slightly more elaborate agents. So-called power users generally have the expertise to write code in any of several user-scripting environments, including VisualBasic (VB), HyperCard, and others. Our presentation will outline what the power user, already adept at building VB applications, would have to do in addition to build a small, relatively simple agent system using off-the-shelf components and VB.

You can use OLE Automation's ability to support dynamic invocation of interfaces of COM objects from scripting languages to invoke CORBA objects. For example, VB, PowerBuilder, Delphi, FoxPro, Excel, and Word can access OLE Automation objects. One specific example of this would be using VB, that ubiquitous language and tool kit, together with OLE and CORBA to build agents that monitor conditions (possibly remote) and execute VB scripts when conditions so dictate (Fig. 7.1). [The basic outline of this technique was generated by Roy and Ewald (1995)]. Generically, this sort of scenario would proceed using the following technologies:

1. *The OLE Automation client* (a controller): This is a client of OLE Automation objects. A VB controller can be used to create OLE Automation objects and then access their properties and methods. This is done via an

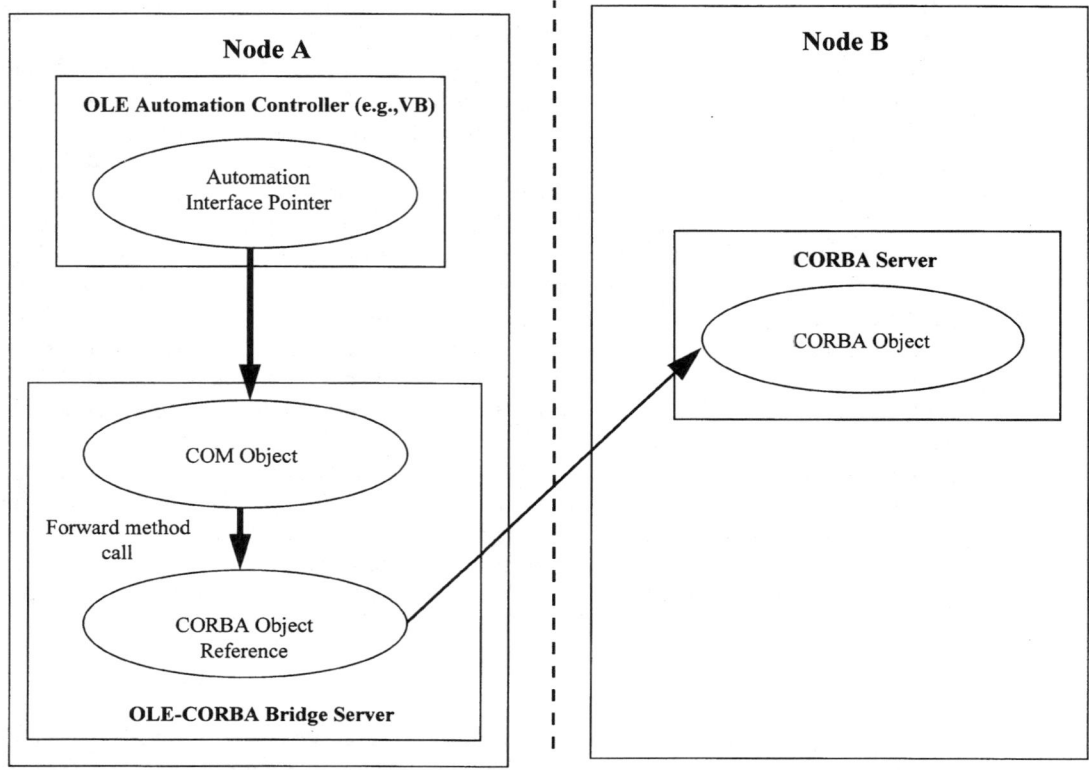

Figure 7.1 Simple agents using VB, OLE, and CORBA. (*Roy and Ewald, 1995.*)

    interpreter that translates script requests into invocations on OLE Automation servers.
2. *An OLE-to-CORBA bridge server:* The bridge is both an OLE Automation server and a CORBA client. A bridge can be in the same process as the client (via DLL) or in a separate process on the same machine (a local server). The OLE Automation objects forward invocations to matching CORBA object references (that is, messages to the agent).
3. *A CORBA server:* A standard implementation as specified in the Object Management Group's IDL (OMG IDL).

    You can view this scenario and the components as comprising one agent, or you can design it as a VB-based agent collaborating with other types of agents that need to be accessed over some network.

    One of the first things to consider is that when you develop agents that execute on different platforms, it is ideal if you can use middleware that serves as the bridging functions. In Section 4 we discussed CORBA-based middleware. In our present example, you also need to map the OMG IDL to Microsoft's ODL. A company called IONA makes Orbix, a Windows-based CORBA-compliant bridge between CORBA and OLE Automation objects.

To extend the example into a real domain, let's now consider a VB-based agent (VB is the OLE Automation controller), executing on a Windows machine that collaborates with a remote, CORBA-based stock Transaction agent. Together, they track the price of a stock and transact a quantity of that stock when the price reaches a certain point. The VB-based agent does this by calling the CORBA-based agent representing a generic stock transaction object (ostensibly, set up on some broker's server to track some stock's activity).

Here are some code fragments of the OMG IDL needed to define what the remote CORBA-based broker's agent (the agent or program hosted by the broker's CORBA-based server that tracks a stock's price and records transactions) has to offer a potential client agent:

```
interface stockTransaction {
  void stockName (in string stockSymbol );
  void performTransaction (in string transactionType
    in float buyPrice, in float sellPrice
    in long sharesToTransact,
    in string accountNumber);
  readonly attribute float currentBuyPrice
  readonly attribute float currentSellPrice
  readonly attribute int commissionPercent
  ...
    };
```

Obviously, there would be other codes to register the transaction with the client's account file and to actually place the transaction with the broker's on-line stock trading agent. Also, some code to send an acknowledgment back to the transaction-requesting agent and for exception handling (for example, a void HasException routine) would be needed.

For an OLE Automation agent (such as the VB-based agent) to use the CORBA-based resource or agent, the complete IDL (including the preceding code) is compiled to produce, among other things, an ODL specification and a makefile. One target of the makefile is a bridge server between the VB agent (as the OLE Automation server piece) and an Orbix-constructed agent. Some of the code produced as the ODL, which is a one-to-one mapping of the OMG IDL to ODL, is as follows:

```
library stockTransaction {
...
  interface IT_IstockTransaction : IUnknown {
  short IT_BindMarkedHost([in]date, [in] time);
  ...
  void IT_Release();
  ...
  void stockName([in] BSTR stockSymbol);
```

```
      void performTransaction ([in] BSTR transactionType,
    [in] float buyPrice,
    [in] float sellPrice,
        [in] long shareToTransact, [in] BSTR accountNumber);
        [propget] float currentBuyPrice();
        [propget] float currentSellPrice();
        [propget] int commissionPercent();
...
    };
    ...
};
```

As before, other codes would be included in the ODL reflecting the extra code in the IDL that handles exceptions, and so on (an IT_HasException routine).

The ODL contains some lines not in the CORBA IDL definition— IT_BindMarkedHost binds the VB agent to a CORBA agent, and IT_Release tells VB when to release the CORBA agent for use by another agent.

Now let's look at some of the VB-based agent code:

```
Sub Check_Stock ()
DIM stockTrans as object
Set stockTrans = CreateObject
    ("MyTransaction.stockTransaction")
stockTrans. IT_BindMarkedHost("011196", "10:30am")
If stockTrans.IT_HasException Then
...
End IF
```

Now, a loop or timer can be set up to periodically execute the following code:

```
If stockTrans.currentBuyPrice <= 35.00 Then
    buyItFor = stockTrans.currentBuyPrice
    stockTrans.performTransaction ("buy", buyItFor, nil, 10,
        "286 56 6578")
End IF
End Sub
```

Since this is a VB subroutine (it could be a function), some main VB program would have to initiate the event Check_Stock, so that the subroutine executes. The main program, the Check_Stock sub, and other subroutines and functions (for example, to enter the transaction automatically into your personal finance software package (such as an OLE-enabled package like Microsoft's Excel or Money) would make for a robust and useful "agent."

In addition you could expand either the VB-based agent or the remote agents with more intelligence. For example, a fuzzy logic DLL call from VB can process several more financial parameters, including those that may be relatively uncertain, like expected company earnings for some time period, and the general market direction based on several measures. Or you could develop an expert system that encapsulates an expert broker's buy/sell heuristics using, for example, M.4 VB from Teknowledge Corp. This package includes the mechanisms necessary to incorporate rule-based, procedural control and OO programming constructs into VB.

7.3 Serious Agent Tool Kits and Development Environments—Introduction

To create the types of IA systems we envision in this book now and in the future, you'll have to use what we term a serious agent tool kit. This tool kit would be based on a major, modern programming language which, in line with our previous contentions, would be OO. Most OO programming languages have one or more fully featured development environments.

Aside from the standard visual tool sets and facilities for user-interface building and database access, a serious agent builder or language provides the ability to access utilities, infrastructural components and facilities (such as CORBA), other applications, and functions defined in the execution environment, as well as creating the agent definitions themselves. The language environment should also provide the ability to test the agent safely before it is commissioned into the target system.

When it comes to OO languages, three stand out as particularly conducive to agent development: Java, Telescript, and Smalltalk. One of the major reasons we like these languages, aside from their relatively pure OO nature, is the fact that these languages are intimately tied to a dynamic, portable execution environment, called a virtual machine (VM). This means your agents, once developed, can do what IAs in a distributed environment must do—move to nodes (sporting the VM, of course) that have the resources necessary and relevant to the fulfillment of an agent's goals.

In addition, each of these languages has enjoyed success in meeting certain specific needs of the agent developer. Even in Java's case, itself a relative newcomer, institutions such as Stanford University have developed extensions to the language specific to agent development (see Subsections 7.4.5 and 7.4.6). As we discuss each of these languages and their relative development environments, we'll point out the important agent-related features and show how these language environments can be used to build a successful agent system.

7.3.1 Smalltalk agents. Smalltalk is probably the best known pure OO language. Some say that once you've programmed in Smalltalk (which is really a matter of extending the behavior of the existing Smalltalk objects to suit your purpose), you'll never go back to whatever you were using. After using several other languages over the last 20 years, both of us have become seasoned Smalltalk programmers, so we can say there is more than a little truth in that notion. Although the programming-by-extension paradigm is not unique to Smalltalk, it has an advantage of a mature, well-thought-out class library, wonderful development environments, and exciting new developments coming in the near future (for example, a standard declarative model for Smalltalk, new, very fast VM-based interpreters, interoperation with Java, and more). And with the introduction and integration of (formally) Hewlett-Packard's implementation of Distributed Smalltalk (DST) into major vendors' Smalltalk environments, you'll have a CORBA-compatible Smalltalk-based ORB as a ready agent-communication-enabling mechanism. More on Distrib-uted Smalltalk follows.

In this subsection, although you probably won't get any in-depth Smalltalk programming examples, we hope that some of the flavor of Smalltalk will show through. What you really need to do though (unless you are a Smalltalk programmer already, or have chosen another language), is get your hands on one of the good Smalltalk books available, and get a demonstration of one of the Smalltalk environments (such as VisualAge, VisualWorks, and VisualWave). This is the only way to evaluate the potential of Smalltalk for use in your general and agent development efforts.

Before we continue, we want to give you a brief list of some of the more important characteristics of Smalltalk, from the practical developer's point of view:

- Language and development and execution environments are intertwined. You craft applications by reusing and extending existing classes. Delivered applications are "stripped" of development environment and other unused classes.
- VM-based interpreter allows fast programming and debugging; no compile-link-load-run cycles.
- Single-inheritance class hierarchy with class object at the top; class libraries extensive; rich and mature libraries available.
- Hooks and handles to other languages, databases, and applications available.
- Model-view-controller (MVC) application structure is a classical way to separate concerns:
  - Model represents the application logic.
  - View is the user interface to the model; N views per model.
  - Controller glues the model and its views together.
- Built-in update-change dependence mechanism. Changes to variables are propagated to those objects who register their interest.
- Message passing hierarchy. Message float follows class hierarchy path.
- Lots of third-party tools available, just like in any other major language, from configuration control and team development tools to graphics, database access, and communications objects that support all the major protocols.
- Last but not least, the VM, by virtue of its portability, takes care of a substantial portion of the cross-platform concerns for you. Develop on the Mac, and your application runs on Windows 95, Windows NT, or UNIX. And the Smalltalk vendors maintain that platform portability for you.

7.3.1.1 Smalltalk development environments. ParcPlace/Digitalk VisualWave and VisualWorks, and IBM's VisualAge are ideal OO development environments for making agents that can be delivered and run just about everywhere, since they provide binary portability via VMs on at least 14 different platforms. Once a VM is installed on a supported platform, a Smalltalk image created on any platform will execute on any other. This system also makes it relatively simple to move any ParcPlace/Digitalk Smalltalk application image onto the WWW since, in general, images that run on any supported platform will also execute on a VisualWave web server (which includes a VM) with little or no modification.

Right out of the box, good Smalltalk environments such as ParcPlace/ Digitalk VisualWorks and VisualWave support many of the facilities needed for agent development and execution. The Smalltalk development environment has a built-in, integrated class, a variable and method browser, a work space to test chunks of code, a resource manager, GUI builder, and an incremental JustInTime (JIT) compiler which makes the language appear interpreted. Smalltalk is fully OO, including encapsulation, inheritance, polymorphism, dynamic binding, exception handling, and automatic memory management. Using Smalltalk, a root agent class can be built, then specialized for different types or subclasses of agents.

Since the Smalltalk VM is layered over the native execution environment, agents can transport themselves and execute on a known layer, even in foreign territory. Through this VM strategy, Smalltalk can execute on almost any platform. The VM is small.

In addition, Smalltalk provides a standard, widely known development language that is suitable for most applications. Smalltalk development environments need at least 16 Mb of memory to run acceptably, and most Smalltalk applications need 8–16 Mb. Smalltalk applications can execute in both 16- and 32-bit environments without recompiling. In the early years of the personal computer revolution, Smalltalk applications were relatively rare because most machines did not have enough memory to execute them. As memory becomes ever cheaper, Smalltalk will continue to grow in popularity.

Smalltalk provides a complete software development environment and a layered VM as its operating environment for agent execution. It also includes all the tools needed for developing distributed client/server applications. Smalltalk code developed on other systems can also be imported. In addition, Smalltalk environments are source-level modifiable, facilitated by included source code for all objects in the hierarchy. Smalltalk also includes an huge and expandable library of classes.

Smalltalk contains all of the basic features necessary for agent development, including multithreaded execution, no need or support for pointers, automatic memory management, late binding, and OO construction. Smalltalk also supports a number of dynamic information structures such as lists, ordered collections, and strings.

7.3.1.2 Basic components of a Smalltalk-based agent.

Developing software in Smalltalk consists in defining classes, instantiating them into objects, and sending messages to the objects. Autonomous Smalltalk objects respond to messages and perform actions just as agents do. For example, in Smalltalk you could define a root, or abstract, class called agent. The class could then be subclassed to provide the specific functionality for a specific domain or problem, as the following class hierarchy demonstrates:

```
Object
  Agent
    SearchAgent
    BiddingAgent
    AuctioneerAgent
    DocumentManagementAgent
    ExportAgent
    ImportAgent
    FormatAgent
```

Objects could be instantiated as:

```
aFormatAgent := FormatAgent new.
anImportAgent := ImportAgent new.
```

Messages could be sent to the objects as follows:

```
aFormatAgent start
aFormatAgent stop
```

The basic functionality of an agent might be defined by first creating a root agent class as follows:

```
Object subclass: #Agent
  instanceVariableNames: 'name owner returnAddress nextAddress
  triggerExpression executionRules goalHierarchy status
  processPriority terminationExpression'
  classVariableNames: ''
  poolDictionaries: ''
  category: 'Agents'
```

The instance variables defined in the foregoing include the following:

1. The `name` of the agent is a unique identifier in the execution name space, for example, "Agent007."
2. The agent's `owner` is the user name of the agent creator and/or launcher.
3. The `E-mail returnAddress` is where the agent is to return results.
4. `nextAddress` is where the agent is to be sent, if possible.
5. The `triggerEexpression` defines when the agent is to be launched; for example, "when the price of mutual fund x reaches 40."

6. The `executionRules` for the agent are the set of facts and inference rules the agent uses in order to complete its mission without causing problems for the rest of the execution environment. These are also known as *ontologies* (see Section 3). For any nontrivial agent, this would be complex. One simple rule could be something like: "If stock x drops to price z, buy 100 shares."

7. The agent's `goalHierarchy` is what the agent is out to achieve for its owner. The highest-level goal might be "make a 25% profit on my portfolio."

8. The agent's `status` indicates whether or not the agent is in a good state or in an error condition.

9. `ProcessPriority` is the priority level at which an agent is executing.

10. When `terminationExpression` evaluates to true, the agent does whatever is necessary for shutdown.

7.3.1.3 Smalltalk agents by example—an agent auction.

One of the main uses for agents in the future will be to buy and sell products on their owners' behalf. One of the emerging venues for electronic buying and selling is the on-line auction, which seems to be the perfect place for agents. In the simple example we outline here, agents representing auctioneers manage an inventory of items to auction, and notify potential buyers when they decide to hold an auction. Agents may then be sent to the auction automatically on behalf of interested users.

Users give their bidding agents the goals and rules by which to bid. The goals could consist of simple directives such as: "buy a copy of software x for under $40," or be rather more complex, involving several subgoals for the agent to meet on behalf of its owner. The rules an owner encodes in his or her agent tell it what to do in specific circumstances, as will be explained.

A simple agent auction system consists of several parts:

- An *auction queue,* through which processes can safely share information.
- The *bid* class, which contains the data and methods needed to create a bid to place on the queue.
- The *auctioneer* agent is a process that monitors the E-mail for incoming agents, validates the agents (checks source code), and launches bidding agents as they arrive, then notifies the winner of the bid.
- The agents themselves run as independent processes, checking the items offered, determining a reasonable bidding limit, checking the current high bid, and sending a new bid to the queue as warranted. E-mail from the auctioneer can cause agents to be launched, or launching could be triggered by some other means, such as time or date or a need (inventory is low, for example).

The auctioneer agent objects are the central arbitrators of an auction system and are responsible for initiating the auction, validating bidding agents, receiving bidders, determining the items to be offered, setting minimum bids, and so on. When the auctioneer receives an agent (possibly via E-mail), it will first check to make sure that the agent doesn't reference nonexistent variables, that it has no statements that may result in infinite loops, and that the bidder has enough money to participate in the auction, and it will give the agent the auctioneer's rules to execute.

Although the auctioneer does much of the work in running the auction, truly effective agents would have the intelligence to determine what items to bid on, what their bidding limits should be, and how to implement a user's expressed preferences. Bidding agents can increase their maximum expected utility by employing AI capabilities based on the technologies explained in Section 3. For instance, agents can decide what items to bid on and how much to bid with the help of fuzzy logic. Users can also employ genetic algorithms to determine which agents provide maximum excepted utility bases on various success factors.

7.3.1.4 Specializing Smalltalk agents for particular tasks.

To create an auctioneer or bidder, you could derive or specialize the Smalltalk root agent as follows:

```
Object
  Agent
    BiddingAgent
    AutioneerAgent
```

7.3.1.4.1 Auction agent.

The hypothetical behavior of both a bidder and an auctioneer is captured with a short scenario, as follows:

1. Go to the auction site.

   ```
   myBuyer := BiddingAgent new.
   myBuyer moveToAuction: someAuction.
   ```

2. Get verified. This means that the source code is checked to ensure that it will not cause problems while executing in the system.

   ```
   theAuctioneer verify: myBuyer.
   ```

3. Request resources. In general, resources can include files, databases, memory, disk space, access to money in the user's account, CPU cycles, other agents and other systems. In our auction case, the agent need only ask a for a seat at the auction. If a seat is granted, the agent is given access to its owner's funds and is put on the auctioneer's list of bidders.

   ```
   myBuyer requestSeat: theAuctioneer.
   ```

4. A bidding agent must bid according to the rules given by its owner in so far as those rules do not conflict with the auctioneer's rules. For example, if

agent1 bids higher than agent2 and agent2 has rules dictating that it attempt to outbid all other agents by an owner-specified amount, then agent2 will submit a bid. However, if agent2 does not have sufficient funds on account to cover the bid, or the agent entered the bid too late, the auctioneer agent rules will disallow the bid.

```
myBuyer submitBid: bidAmount to: theAuctioneer.
```

5. Pay for the goods if the agent's bidding is successful.

```
myBuyer pay: anAmount to: theAuctioneer.
```

6. Send the results back to the owner.

```
myBuyer sendResults: purchases to: returnAddress.
```

7.3.1.4.2 Auctioneer agent. In general, the duties of an Auctioneer Agent are as follows:

1. Notify users concerning auction details.
2. Monitor incoming mail for arriving agents.
3. Verify that agents received will execute and do not contain potentially harmful instructions.
4. Launch agents as independently executing tasks at the bidding-agent priority.
5. Terminate tasks when an auction is over.
6. Provide requested resources as appropriate.
7. Manage items upon which agents will bid.
8. Make sure agents bid according to rules.
9. Repackage agents and send them on to another destination if necessary.
10. Begin and end auctions.
11. Arrange for shipment of items to bidders.
12. Handle inventory and accounting functions.

7.3.1.5 Smalltalk support for agents—Some details. As stated previously, Smalltalk contains a number of features to facilitate agent construction and execution. These include independent task execution, interprocess communication, automatic memory management, execution via VM, support for accessing utilities in the execution environment, the ability to determine what resources are available, and provisions for exception handling.

Many of the capabilities illustrated next correspond to the agent auction example.

7.3.1.5.1 Independent task execution.
Smalltalk, and all good candidate languages for agent development can launch, perform, and manage independently executing tasks. An executable task can be launched via a forkAt: message to an executable block. The task then executes at the priority provided as an argument to the forkAt:.

```
[ dataBlock clear.
dataBlock := readDataPipe.
evaluateRulesFor: dataBlock.
repeat.]forkAt: reasonablePriority.
```

7.3.1.5.2 Interprocess communication via a shared queue.
To handle bids from multiple, independently executing agents, an auctioneer could use a shared queue, which would facilitate independent, asynchronous entries from multiple agents.

The following illustrates the methods necessary to support a shared bidding queue.

```
AuctionQueue class methodsFor: 'access'
next
    ^BidList next.
nextPut: item
    BidList nextPut: item
peek
    ^BidList peek.
!AuctionQueue class methodsFor: 'queue'
initialize
    BidList := SharedQueue new.
AuctionQueue initialize
```

To run an auction, the auctioneer could employ a shared queue along with "bid" objects, as shown in this simple example:

```
|aBid highBid|
aBid := Bid new.
highBid := Bid new.
aBid address: 'jmjohnson'.
aBid name: 'Jay Johnson'.
aBid amount: 10.
10 timesRepeat: [
  highBid := AuctionQueue peek.
  (((highBid amount) + 5) <= 30)
  ifTrue: [
    aBid name ~= highBid name
    ifTrue: [
      aBid raiseBid: (highBid amount) + 5.
      AuctionQueue nextPut: aBid
    ].
  ].
].
```

7.3.1.5.3 Automatic memory management.
As all good agent languages should, Smalltalk supports automatic garbage collection. In Smalltalk everything is an

object. To create a new object in Smalltalk, it is only necessary to send a message like this:

```
myAgent := Agent new.
```

When an object is not using myAgent anymore, the memory it occupied will be automatically returned to the memory pool for future use. There is no need to deallocate it explicitly. Although garbage collection is automatic, programmers can also direct it, which may be necessary in some cases for performance reasons.

7.3.1.5.4 The virtual machine. Because of the Smalltalk VM, a piece of Smalltalk code can be sent to another server, where it can execute as part of a Smalltalk image:

```
runAuction
    AuctionQueue initialize.
  [(Compiler evaluate:(('c:\visual\image\agent1.st') asFilename)
    contentsOfEntireFile)
 value.] fork.
    Transcript show: (AuctionQueue next amount) printString.
```

In this example we loaded in and executed an agent source file, but a Smalltalk image consists of byte codes, which execute on the VM.

7.3.1.5.5 Support for accessing a system utility. The following is a simple mail monitoring agent in Smalltalk using Novell GroupWise with the standard mail API (MAPI) interface:

```
delay := Delay forMilliseconds: 100.
|mailList| GroupWiseMAPI logon.
[
  [
    mailList := GroupWiseMAPI allUnreadMailMessages.
    mailList do: [ :eachMessage |
      eachMessage subject = 'Status Requested'
      ifTrue: [ sendMailTo: eachMessage sender
        ('status.rpt') asFilename contentsOfEntireFile].
    delay wait. ]
  repeat] forkAt: Processor userBackgroundPriority.
  GroupWiseMAPI logoff.
```

Here we log onto the E-mail system, then launch a task to check continuously for new messages. If one of the messages requests a status report, we automatically send the latest report to the requesting user.

7.3.1.5.6 Requesting available resources. An agent system really comes in several pieces including a host and the agent itself. The agent is sent to a host, either from the user or from another host. Once it arrives, the host checks the ASCII file to determine that the message is in fact an agent. The host must also check that the agent has the right encrypted password and access codes. The host checks the resources the agent is requesting, and the variables it will access to make sure there are no conflicts. The agent knows the definition of these methods, so it can call with the correct arguments. Using VisualWave Smalltalk from ParcPlace/Digitalk, the agent can ask the host if it understands the message it wants to send, and vice versa.

```
|aBid |
   aBid := Bid new.
aBid respondsTo: #open:.
```

If the host or another agent does not know how to respond to a message, the asking agent can request that the responding agent load in the new method. In a sense, this means that one agent could learn what another agent knows.

7.3.1.5.7 Exception handling. An executing agent system consists of two parts: the agent itself and the environment in which it runs (which may include a manager or agent arbiter, such as the auctioneer). The agent, if developed in a good agent language such as Smalltalk, Telescript, or Java, can only perform the methods provided in a limited environment. The agent cannot write to or delete files if such functionality is not provided in the environment, or if all of the files in the server environment were read, write, and delete protected such that only the owners had those privileges.

If things still go wrong in spite of the checking and limitations imposed on executing agents, Smalltalk has extensive support for error handling. In Smalltalk, error handling is facilitated by a large, user-extensible hierarchy of signals. When a signal is raised inside a handler block (as the result of a divide-by-zero error, for example), it creates an exception. The exception provides information about the error condition and the current execution state. The exception can then be used to either abort the operation, proceed, or restart.

7.3.2 Distributed Smalltalk

Distributed Smalltalk provides a flexible environment for creating multitiered distributed applications. It provides an integrated set of OO frameworks for the development and deployment of multiuser, enterprisewide distributed applications. Fully integrated with VisualWorks, Distributed Smalltalk lets developers build multitiered applications that are compliant with CORBA. Applications built with Distributed Smalltalk make the best use of existing resources for increased performance, scalability, security, and easy maintenance.

7.3.2.1 Interactive environment. Distributed Smalltalk provides an interactive environment for creating highly portable applications and a robust set of classes to simplify the process of building large, complex applications. By raising the level of abstraction, Distributed Smalltalk enables developers to become more productive when developing and deploying their applications. For example, you do not need to write a communications infrastructure, a process that typically is tedious and error-prone.

7.3.2.2 Open architecture.
Distributed Smalltalk provides an open architecture for building and deploying enterprisewide client/server/web solutions. The product allows you to build reconfigurable applications that are dynamic collections of client and object agent implementation components. You can access these components regardless of the location and can move them to more powerful machines as their requirements change.

Components may be located in different departments and divisions and may come from multiple vendors or be developed in house. Distributed Smalltalk's well-defined, reusable component interfaces allow you to increase productivity and improve code.

7.3.2.3 Standards-based object services—Distributed Smalltalk frameworks.
Distributed Smalltalk provides basic application frameworks based on the OMG CORBA 2.0 standard. This standard enables objects and applications from multiple vendors to interoperate with other OMG CORBA standard applications, regardless of the language or platform. Developers can leverage the following services as well as additional Distributed Smalltalk frameworks to create robust, enterprisewide, distributed applications:

- *Naming service:* Assigns a meaningful name to an object or agent within a naming context
- *Event service:* Enables agents to notify each other asynchronously of interesting events anywhere in the enterprise by using standard event protocols
- *Life-cycle service:* Allows programmers to create and initialize, delete, copy, and move agents independent of location or platform
- *Transaction service:* Defines and completes full two-phase commit transactions among distributed agents
- *Concurrency control service:* Enables multiple clients to coordinate access to shared resources

Extended Distributed Smalltalk frameworks include:

- *Compound life cycle:* Adds to the life cycle service to operate on more complex compound objects or agents
- *Properties and property management:* Part of an agent's external interface (owner, creation date, modification date, version, access control list, and so on)
- *Relationships—containment (or composition) agents and links:* Allows networked relationships among agents

- *Presentation/semantic split:* Provides an efficient architecture for distributed agent applications and a reusable framework based on an advanced application architecture
- *Application objects and their assistants:* Relatively large-grained objects that are used by end users (for example, file folder or order entry form)

Distributed Smalltalk solves large, enterprisewide problems with standards-based frameworks. Using off-the-shelf components, you can create applications with a CORBA interface, eliminating the need to reconstruct applications. You can reuse existing components to save application development time, and they can partition their applications to run anywhere in the distributed architecture.

Distributed Smalltalk lets you browse the shared interface repository graphically. As Fig. 7.2 illustrates, the Distributed Smalltalk architecture provides full two-way communication between computing components, and connectivity to different applications and agents that run on different platforms. Unlike

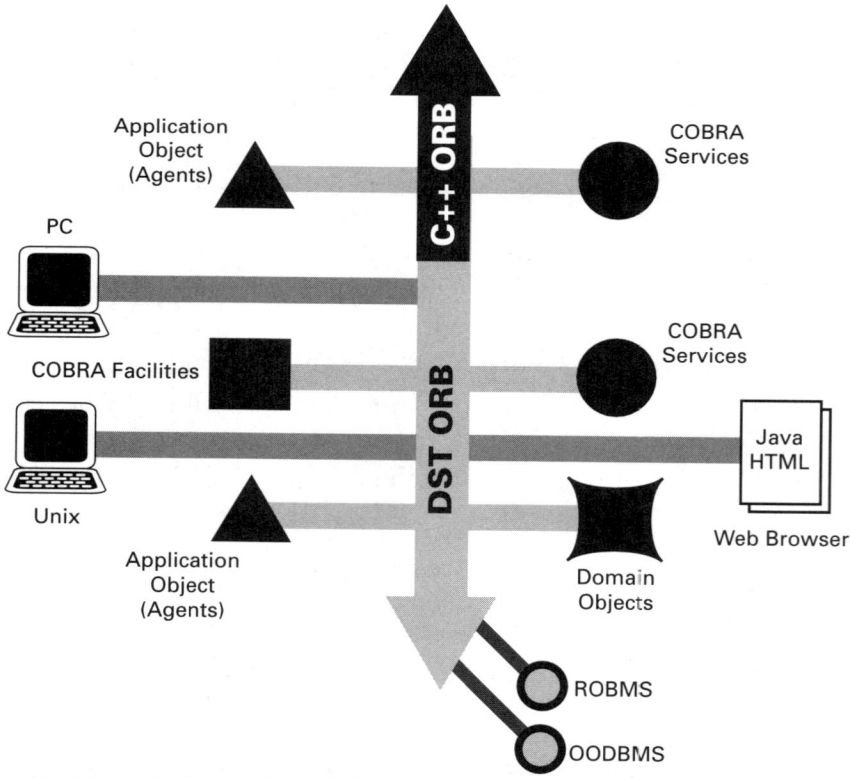

Figure 7.2 Distributed Smalltalk reaches out. (*Frost, 1996.*)

existing client/server applications, you can readily change and evolve systems as needs and requirements change.

With Distributed Smalltalk, you can make the best use of computer resources by allowing distributed objects and agents to reside on the machine that makes the most sense. For example, you developers you can isolate business objects on a central server or assign a computationally intensive task to another more appropriate server.

A number of sophisticated tools are included for testing, tuning, and delivering distributed applications. Using Distributed Smalltalk, you can find problems anywhere in the computing environment, delivering distributed systems with ease. Tools and facilities include the following:

- *Remote object debugger:* Debugs distributed processes across multiple Distributed Smalltalk images
- *Interface repository browsers:* Views and edits IDL interfaces for local and remote interface repositories
- *IDL interface generator:* Generates an IDL interface from an existing Smalltalk class
- *Local RPC tester:* Tests objects and agents in a simulated distributed environment
- *ORB monitor:* Provides a dynamic view of inter-ORB communications traffic
- *IDL interfaces:* Can be generated from existing Smalltalk classes

7.3.2.4 How Distributed Smalltalk enables agent interaction. Figure 7.3 illustrates how an agent running in a Smalltalk image on one machine could communicate with a different Smalltalk (or C++) agent running in a different image on another machine.

The mechanisms that Distributed Smalltalk relies on to get the work done are encapsulated as Distributed Smalltalk classes that you can use to generate your own agent-agent communications architectures. At a high level, the mechanisms or classes include the following:

- *IP* is the internet protocol that routes packets to a machine and a port on the machine.
- The *transmission control protocol* (TCP) is the session-based mechanism that maintains the connection for the delivery of sequenced packets.
- The *IIOP* maintains connections between ORBs across the Internet.
- The *ORB* itself is the broker that agents and objects rely on to ensure that the message sent gets routed to the correct receiver, and maps the message from a specific language to some neutral form.
- A *socket* is the way the ORB talks to the TCP/IP mechanisms.

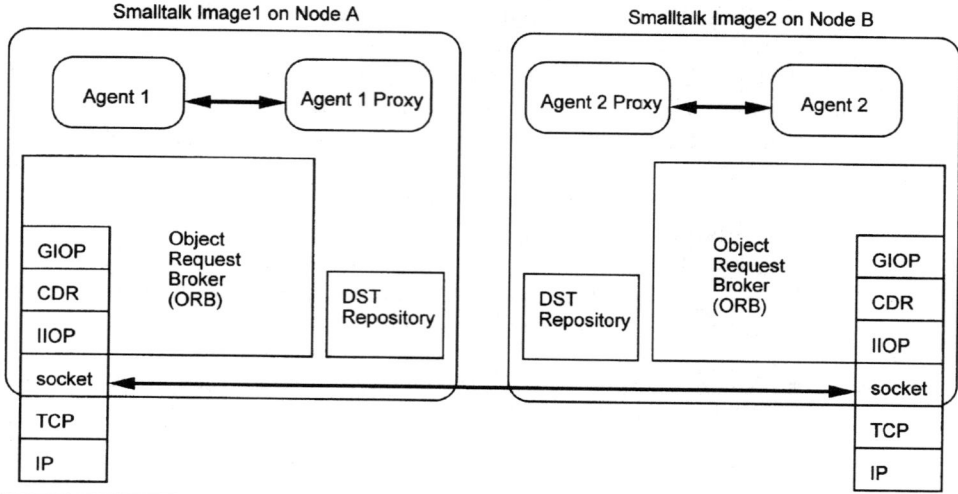

Figure 7.3 Distributed Smalltalk–enabled agent collaboration.

- The *common data representation* (CDR) maps the parameters within a message to a standard format suitable for transmission.
- The *general inter-ORB protocol* (GIOP) describes the transfer and message formats used in ORB-to-ORB interactions.
- The agents (agentA and agentB) are set up as a client/server pair. The requester or sender of the initial message is the client, and the recipient and provider of a response is the server. Of course, any agent can be both a client and a server.
- The *proxy* for each agent is the way in which transparent forwarding is achieved. When you design an agent, you want it to appear as though you were sending a message to another agent without regard to its location. When agentA sends a message, through its proxy, to agentB (received through its proxy), the sending agent's thread of execution is blocked until agentB replies (kind of like an RPC). The proxy or forwarding machinery is implemented within the DSTObjRef class (actually, subclasses DSTObjRefRemote—the real stuff—and DSTObjRefLocal, which wraps local objects so that you can test without a remote machine or image).
- Lastly, the *DST repository* contains the textual representations of each agent's interfaces, and the methods that convert those representations to and from an IDL, which is like an Esperanto for agents and objects written in different languages. ORBs use IDL-specified interfaces to map requests from agents and objects, through the proxy, into a language the ORB knows the intended recipient's proxy will understand. Several other detailed mechanisms provide the IDL-to-implementation-language lookup and delivery.

7.4 Java

One hope for agent standardization on the WWW, the Internet, as well as more private intranets is a language developed by Sun Microsystems—Java. In the course of a relatively short time, Java has generated a huge amount of interest. Software companies, including Microsoft, are falling all over themselves to license Java technology and develop Java programs, including software entities called applets and servelets. One of the reasons for this success is the lack of a powerful, standard network-based software development language. Microsoft, Borland, ParcPlace, and a number of other software companies are vying to make their network-development products popular, but most have admitted that their products must be Java-compatible.

The main reason for Java's initial popularity is that it can help bring otherwise bland and static web pages to life. When a Java-enabled browser such as the Netscape Navigator is used to access HTML pages on a WWW server, applets defined on the web pages will be downloaded and will execute on the user's workstation when the user clicks on the appropriate part of the page. While there are other means of doing this, Java offers an OO solution that is already integrated into the emerging universal client—web browsers.

Java was designed to look like C++ in order to take advantage of the relatively large C/C++ developer base. In many cases, however, the similarities are only skin deep. Using C++ as a point of departure, let's look at how Java and C++ are similar and how they differ. C++, while a good system programming language, is unsuitable for developing agents to execute in a distributed environment. The reasons are primarily its support for programmer-directed memory allocation and deallocation, pointers and pointer arithmetic, multiple inheritance, and procedural features such as functions, structures, unions, typedefs, defines, and preprocessor directives, including macros. Java has abandoned these C++ features in favor of automatic memory allocation and garbage collection, dynamic binding, and other features designed to make applets more robust, portable, and secure. Though it has many similarities with C++, in more than one way, Java is actually a cross between C++ and Smalltalk.

Programmer-directed direct memory access via pointers may be handy in certain applications, but it is too dangerous and confusing to use for distributed or mobile agents. Java does not support memory access via programmer-defined pointers. It disallows pointers, while C and C++ are generally known as pointer-centric languages. It is true that if you took pointers away from C/C++, it would not even look the same, nor would many of the idioms inherent in the languages be relevant. On this point alone, one would have to classify Java as a cousin of C/C++, not a direct descendant.

7.4.1 Agent applications in Java

Java is actually two separate languages. Java applets are used almost exclusively to spice up user interfaces on the WWW, since applets execute within a secure browser environment and have (courtesy of the browser, the loader, and the byte-code verifier) only severely limited access to memory, files, and databases. Java can also be used as an application development language that is not burdened by any of the applet restrictions.

Java may or may not make significant inroads into the corporate IS (Information Systems) application development, since COBOL and 4GL packages are well entrenched there, but with an emerging set of tools and standards [such as the newly defined Java database connectivity interface (JDBC)], it is beginning to emerge as a major WWW/intranet application development option.

Java still faces difficult challenges in becoming a viable mission-critical application language. It does not have a long track record of success in the client/server arena, and it still lacks advanced optimization for time and memory space, although this is improving. We are only beginning to see the first generation of integrated development environments, and at this time Java has only limited and immature library support, although this will certainly improve with time if Java continues to grow in popularity (Corcoran, 1996).

7.4.2 Java as an OO language

Java is fully OO, including dynamic binding and an interpretive execution paradigm. Java also has no support for pointers or references, and does not allow programmer-directed memory allocation and deallocation. Java does not support friend functions and supplies a final directive for classes, which means no further subclassing can take place. In addition, Java does not support operator overloading as does C++. These language characteristics serve to make programming simpler (Polese, 1996) and to increase the robustness of Java code, since it must execute on a variety of platforms.

As opposed to C++, Java supports only single inheritance and provides dynamic binding to support true polymorphism. Java forces the programmer to develop the code in an OO fashion—there is little support for lapsing back into a low-level procedural language like C and ignoring the object paradigm. You can't even sneak in a *struct,* which is found in many C++ programs. Java does not rely on simple C-style *#include* file statements and makefiles, but opts for a more sophisticated *import* package statement similar to Ada, where a package consists of a class hierarchy.

Java is primarily used as a compiled-then-interpreted language via Java applets. The Java compiler translates source code into assemblylike byte codes, which are then run on a platform-independent VM. Java is essentially fully OO, whereas C++ is not, although Java is not quite a pure OO language since it has no support for metaclasses, and the very lowest level entities in the language (integer, character, and so on) are not objects but simply types. Everything considered, the Java development and execution paradigm is actually closer to a pure OO language such as Smalltalk than it is to a hybrid like C++.

Smalltalk and Java are both compiled incrementally into byte codes that execute on platform-independent VMs, a revolutionary concept in OO programming, which happens to fit perfectly into the multiplatform WWW. Both feature dynamic binding and single inheritance. Both force the user to use the OO paradigm, although Smalltalk is a pure OO language since in Smalltalk everything is an object, even integers and classes. The syntax for both languages consists almost exclusively of an object designator followed by a method to be executed.

In Java,

```
myClass.doSomething();
```

In Smalltalk,

```
myClass doSomething.
```

Both Java and Smalltalk support exception handling, and neither supports pointers, references, or programmer-directed memory allocation and deallocation. Both rely on automatic garbage collection to manage memory. Both languages provide the same control structures, with the exception of a case statement (Java has one, Smalltalk does not), which is not really needed in a pure OO language.

Aside from Java's superficial resemblance to C++, Smalltalk and Java differ in that Smalltalk has no direct support for private methods as Java does. In addition, Smalltalk does not support a "final" designation for classes (prohibiting further subclassing). Other important differences relate to the respective maturity of the two languages. Smalltalk has been around for over 25 years, and has been by far the most widely used pure OO language. As such, Smalltalk has a variety of mature integrated development environments as well as excellent database support from all major vendors, and a huge, easily user-extended class hierarchy featuring extensive GUI, collection, conversion, and network support. These environments have been refined to support heavy-duty code optimization, giving Smalltalk an edge in performance.

7.4.3 Java tools

As of this writing, however, Java tools are going through some major evolution, and integrated GUI/visual development environments are beginning to emerge, such as Symantec Café, and Microsoft's J++, and a number of other vendors are making promises for bigger and better IDEs (Integrated Development Environment). Due to Java's enormous potential, a plethora of other tools and libraries is being developed.

For example, Marimba Inc., Active Software, and Callico Technology are three companies preparing Java-related tools as this book is being written. Marimba, whose workers include four of the original team from Sun that developed Java itself, is developing a framework that developers can use to construct Java-based clones of PointCast,' the Web interface that consists of multiple windows displaying news, graphics, and a stock ticker. This framework will be composed of Java widgets that are network aware. Developers can use these components by calling them from within their own code. So your agents can therefore make use of these facilities as well. One of the anticipated uses we envision in the agent world is to enable your agent to go to multiple sources on the Internet in order to get different types of information, then sift through that information, packaging it up for the user as knowledge.

Active Software is working on a technology called ActiveWeb, which will be very similar to ParcPlace/Digitalk's VisualWave in that it is a software communications platform that enables databases and browsers to exchange information across intranets and the Internet. The product enables companies to convert their databases into content that can be accessed via a web browser.

One of the uses of this kind of technology within an agent system is as an extrapolation of the preceding example. A knowledge-retrieval agent could use ActiveWeb to mine databases for relevant information, which it then uses to update and broadcast those important pieces of information to other databases or agents. Indeed, as users find information sources proliferating, they will desire their own personal "gofer" that learns what they need, knows where to go and get recent information, and which can travel to those locations and interface with the sources, travel back, and integrate the acquired information into a personalized database for the user.

Calico is building Java tools that enable its sales software to work on the web, effectively enabling electronic commerce. Calico provides order-processing systems that reach across the enterprise. This kind of technology is, for example, important to a fully integrated, agent-based work-flow application. Agents can take a customer's order and interface with the relevant software at each point in the flow of product ordering, routing, accounting, and reporting.

7.4.4 Applets and servelets as agents

As far as agent technology is concerned, Java has tremendous potential due its fast growing and already widespread use on the web and its suitability as an agent development language. How well Java can adapt to other distributed execution environments, however, remains to be seen.

A Java VM is included in Netscape (as well as other browsers and OSs), which is fast becoming the standard web browser, and a "universal client" for intranets. Sometimes data access agents can be run most advantageously on the server, and sometimes on the client. Where data searches are involved, the server side is the obvious choice—agents go from the client and execute on the server, or from server to server, eliminating the necessity for transferring massive amounts of data across a network.

When it comes to user interface, however, animation, film clips, music, sound, and so on, are best executed on the client machine where memory and CPU resources can be dedicated exclusively. Only a byte-code applet needs to be transferred, not a block of graphics. The applet is then interpreted on the client.

In the Java execution paradigm, applets (somewhat akin to agents) migrate from server (web page) to client. These applets execute on the client, freeing up the server and providing far more resources for applet execution, enabling more complex web page elements by focusing GUI execution on the client.

In addition to the Java applet, there is another Java-based entity, called the servelet, which lets a user upload an executable program to the network. It's similar to the remote-agent concept in General Magic's Telescript. With servelets, a client user or application could launch a servelet-based agent to search the network or other clients for information and respond automatically or give periodic updates.

Java has potential in more real-time systems as well. Because Java runs on a VM, Java-based agents can be loaded into embedded applications such as telephone switching networks without fighting hardware-specific battles. Indeed, the language was originally designed with embedded systems in mind. Embedded Java programs can report the status of hardware elements and report faults, as well as provide new control algorithms. This means that new agents could simply be loaded into existing systems, without redesigning and rebuilding the control software.

7.4.5 Agile agents in Java*

The impetus behind Java tool kit add-ons, such as the aglet (agile agents) system from IBM, is to focus on the agent developer's tasks. The Aglet Workbench provides facilities on top of, or in addition to, various other tool kits that extend Java to specific domains or problems, without requiring the developer to delve deeply into the myriad aspects of Java.

In a nutshell, the Aglet Workbench, available (as of this writing) on Windows NT, Windows 95, and Solaris operating systems, is a visual environment that gives the agent developer access to the complete aglet framework:

- The *Java agent application programming interface* (JAAPI) (more detail later).
- *Tahiti,* a user interface that will dispatch, monitor, and manage agents.
- *Tazza,* an agent builder that uses a drag-and-drop metaphor for composing JAAPI agents.
- *Fiji,* the aglet web launcher built as a Java applet, that enables a web browser to launch an aglet.
- The *aglet* itself.

---

*This section is taken from M.D.K.'s notes and personal memory of a presentation by Yamamoto and Chang (1996).

- The *agent transfer protocol,* independent of any particular agent implementation, provides application-level facilities to transfer agents (aglets that are Java byte codes) between networked computers using universal resource locators (URLs). Aglets can be halted, packaged with their current state in another object, dispatched to another environment, and resume execution. At the time of this writing it appears that execution starts "from the top," so to speak. The transferred state does not include any local (dynamically declared) variables or the current program counter (PC) that would let it be "reentrant." (Telescript agents, on the other hand, can be transferred completely, allowing a Telescript agent to resume execution where it left off.)
- *Packages* that enable aglet access to databases.

The following is a summary of some of the important classes and interfaces available via the workbench:

- *Aglet* is an abstract class from which you derive your aglets, overriding those properties and methods necessary to empower your agent to achieve its goals.
- The *aglet context* is an interface an aglet uses to gain knowledge about its environment. The aglet context incorporates security features that you can use to protect AGLETS from each other and to protect the environment from aglets.
- The *aglet proxy* is a class that encapsulates the real aglet, thus protecting against direct access to the aglet's public interface.
- The *aglet identifier* is a class that, when instantiated, is the unique identifier for an aglet.
- *Itinerary* is an abstract class from which you derive a class representing the routing or travel plans for an aglet.
- *Message* is a class that, when used with *future* and *arguments* classes, enables aglet-to-aglet communications. Messages can be sent both asynchronously and synchronously.

Several simple aglet scenarios were demonstrated at OOPSLA 96, including an aglet that was dispatched to say "hello" on a remote machine, and an aglet that was dispatched to a remote machine to find a certain file and return with it. Of course, more complex scenarios are limited only by your problem domain and imagination. Indeed, IBM has also supplied common usage patterns for agent collaboration that you can use to design your own agent system. These include typical relationships between agents such as master-slave, notifier-notification, and messenger-receiver.*

In addition, according to Steven Farley of Aurora Simulation, Inc., IBM's MAF (Mobile Agent Facility) Specification (largely based on Aglets) is one of

---

*The Aglets Workbench can be downloaded freely from http://www.ibm.co.jp/trl/aglets.

the top contenders to become the standard for Java-based agents. IBM has responded to the OMG (Object Management Group) Common Facilities RFP3 by submitting the MAF spec, which is also a leading candidate for standardization within CORBA (Farley, 1997).

7.4.6 The Java agent template

Java is rapidly becoming the language of the web and will be used widely on both the WWW and the intranets. While it may never replace the languages being used for mission-critical applications in the corporate data center, it will probably become an integral part of any application interfacing with web technology.

This recently embraced technology is heavily based on the distributed computing paradigm. The processing and resources to fulfill user requirements will be distributed across the web, not local to one server or client machine. While there are a number of languages and development environments that focus on a specific client (such as Windows 95) or server (database servers from Oracle, Sybase, and so on), there are few if any that are as fully dedicated to web programming as is Java. In addition, Java has all the language features to support agent building and execution.

Since it is within the Internet and intranet paradigm that agents will probably make the greatest impact, the signals are clear: serious Java agents are probably just over the horizon. As Java's popularity spreads, there will no doubt emerge several successful, widespread Java agent systems. A good start toward that future is the *Java Agent Template* (JAT) from Rob Frost at CrossRoute Software Inc. The JAT, along with other components, allows Java programmers to develop agents that can communicate.

Each agent created from the template is an autonomous task, understands KQML (refer to Section 3), and can thus share knowledge with other JAT agents. The agent carries out its mission using a user-supplied set of facts and rules of inference, or *ontologies* (see Section 3), and other resources, which can be loaded from an initialization file specified at creation time. Each agent also inherits a basic ontology general to all template-created agents.*

7.4.6.1 Using the Java agent template. The following is a description of a raw, or canonical, JAT. Section 8 contains an example of a JAT-based application.

The JAT consists of, among other things, an agent class which represents a multithreaded, autonomous software agent. Each agent has an associated CommInterface, which knows how to exchange KQML messages asynchronously with other agents. The agent must be created, along with the CommInterface and MessageOutput objects, in an external class, such as AgentFrame, which implements the AgentContext interface.

---

*We are grateful to Rob Frost for allowing us to include information from the JAT web site in this section. This section contains only a portion of the template and related classes. The complete JAT is downloadable from http://cdr.stanford.edu.

The agent communicates with its external class using a MessageOutput object, also created in the external class. This external class may contain a GUI.

Each agent, when created, loads a set of initialization messages from an initialization file. One of these messages will provide the address of an agent name server (ANS), to which the agent must send its name and address. Each agent automatically "knows" the KQML language (KQMLmessage class) and the agent ontology (AgentOntology class). When created, agents must be supplied with the URL location of any local classes the agent wants to share with other agents (for example, special languages, ontologies, or classes).

The internal knowledge of an agent is contained in a set of resource classes. Access to the data within these classes is synchronized to prevent race conditions. A special subclass of resource, RetrievalResource, provides the storage for addresses, languages, ontologies, and remotely loaded classes. When an agent attempts to access a nonexistent element from a RetrievalResource instance, the access thread is blocked and a message is sent to obtain the desired object. When the object is received, the access thread is restarted. (See the RetrievalResource class for more information.) This action is taken for unknown addresses, languages, ontologies, and general classes. For the retrieval of remote code, a hash table of network class loaders is maintained.

When a message is received by an agent, the following steps occur:

1. CommInterface receives a string message, parses it to create a KQML message, and calls the agent's receiveMessage method.

2. The KQML message is passed to the interpretMessage method of an instance of the specified ontology. (In other words, a specific instance of the AgentOntology class, or a subclass of the AgentOntology class.) If the agent does not possess the ontology, a request for that ontology subclass is sent, per the RetrievalResource functionality, to the agent who sent the message.

3. For each ontology, the message is interpreted according to the syntax (language) for the content. Again, if the agent does not possess the specified language subclass, a request is sent to the message sender.

Agent communications regarding addressing, code retrieval, and other organizational or architectural issues (not related to domain knowledge) are done using KQML for the message content language and ontology "agent."

7.4.6.2 The Java agent template application programming interface. This subsection contains only a small portion of the complete JAT API to illustrate some of the main capabilities of a JAT agent.* The code

```
public Agent(String n,
    MessageOutput mo,
```

---

*The complete JAT can be downloaded from http://cdr.stanford.edu/ABE/JavaAgent.html.

```
    URL init_url,
    URL shared)
```

constructs an agent with an output, name, default addresses file containing the locations of ontologies, languages, and so on, and the URLs of shared classes. It initializes the classes, addresses, ontologies, and language storage objects and automatically adds the agent ontology and KQML language to the ontology and language buffers. Finally, this constuctor loads the initialization file and processes all initialization messages. The parameters are:

n	String identifier for agent
mo	Output for system messages
init_url	URL for initialization file
shared	URL for shared classes

The code

```
    public void loadInitFile(URL file_url)
```

loads a set of KQML messages from a file at the specified URL. This file must be a text file containing one KQML message per line. The language and ontology fields for the messages are limited to "KQML" and "agent," respectively. At a minimum this file must contain a KQML message informing the agent of the address of the ANS. This message will have the following structure: (evaluate :sender file :receiver: agent :language KQML :ontology agent :content (tell-address :name <name> :host <host> :port <port>)). The parameters are:

file_url	URL for initialization file

The code:

```
    public void loadResource(String type,
       String name,
       String class_name,
       String url)
```

loads a given resource. If the resource currently exists, it does not overwrite. The parameters are:

type	Type of resource, either language, ontology, or class
name	Identifier for resource
class	Class name
url	Code base for class

The code:

```
    public synchronized void addSystemMessage(String message)
```

pipes a system message to the agent's MessageOutput object. The parameters are:

message   System message to output

The code:

```
public void sendMessage(KQMLmessage message)
```

is called to send a given KQML message. It calls the sendMessage method of the associated CommInterface. The parameters are:

message   KQML message to send

The code:

```
public void sendResult(KQMLmessage message,
  boolean status,
  String reason)
```

is called by the associated CommInterface to report the success of a sendMessage call. The parameters are:

message   KQML message which CommInterface was asked to send
status   True if message transmission was successful, false otherwise
reason   String with detailed reason.

The code:

```
public void receiveMessage(KQMLmessage message)
```

is called by the associated CommInterface when a KQML message is received. The message is handled based on the ontology for the message content. Content language is considered within the ontology handler. The parameters are:

message   KQML message that was received

The code:

```
public Class loadClass(URL code_url,
  String name)
```

is called to load a remote class into the current run-time environment. First check to see if a loader for the URL already exists. Only create a new loader if one doesn't currently exist. Returns null if no code is found. The parameters are:

code_url   Code base URL
name   Class name

Returns:

The loaded class
The code:

```
public void terminate()
```

is called by CommInterface when the last termination message has been sent. The code:

```
public void initiateTermination()
```

is called when the executable class which contains the agent terminates. The agent sends a remove-address message to the ANS.

For more information on the JAT refer to the JAT-based example in Section 8 and to the web site mentioned earlier.

7.4.7 Agent-support facilities in Java

Agents are often required to access and extract information from web pages and listen to communication ports via sockets. Fortunately one of the standard Java packages (java.net) provides these capabilities.

To access information from a web page, the first step is to get a URL as follows:

```
String urlString = "http://www.altavista.com/*"
aURL = new URL ( urlString );
```

The new instance of URL can open a stream to the contents of the web page:

```
String line;
InputStream aWebPage;
DataInputStream aWebPageContents;
aWebPage = aURL.openStream();
aWebPageContents = new DataInputStream ( new BufferedInputStream (
aWebPage ) );
```

Now you can set up a *while* statement to read and parse the contents of the web page:

```
while ( (line = aWebPageContents.readLine() ) != null ) {
  // insert code to parse string input, searching for links to other
  pages etc.
}
```

To monitor a port via a socket, create a new socket and set it to a port:

```
Socket monitorPort = new Socket ( hostname, portNumber );
```

Then create an input stream on the socket:

```
DataInputStream inStream = new DataInputStream ( new BufferedInputStream
( monitorPort.getInputStream() ) )
```

To provide robust access, this code should be wrapped in a *try...catch* statement to handle execution errors.

For additional information on Java-based agent architectures, refer to Farley, 1997. This is a good introduction to some of the technical issues involved in implementing an agent architecture in Java, including simple examples. The text of that article is available at www.sigs.com.

7.5 Telescript: The Complete Mobile Agent Environment

While Smalltalk and Java have tremendous potential as agent languages, and we are definitely seeing progress toward direct support for agent building and execution in these languages, they are not yet ready for prime time when it comes to providing a standardized agent execution environment and architecture. As we saw in Section 6, the concept of mobile agents and the necessary infrastructure for building and executing a system of Agents is best illustrated by Telescript from General Magic. In this section we explore examples of Telescript mobile Agents and present the details of developing useful Agents using Telescript.*

7.5.1 Mobile agent technology

New communication paradigms beget new communication technologies. A technology for mobile Agents is software—software that can ride atop a wide variety of computer and communication hardware, present and future.

Telescript technology implements the Telescript concepts presented in Sec. 6 and others related to them. It has three major components: the language in which Agents and places (or their "facades") are programmed; an engine, or interpreter, for that language; and communication protocols that let engines in different computers exchange Agents in fulfillment of the go instruction.

7.5.1.1 Language. The Telescript programming language (Fig. 7.4) lets communicating application developers define the algorithms that Agents follow and the information that Agents carry as they travel the network.† It supplements sys-

---

*The authors wish to thank General Magic and James White for what follows, which is based on White (1995). Additionally, we want to reiterate that, whatever the current or final status of Telescript, we believe the work that has gone into Telescript serves *as an example* of agents done right.

†Despite its name, the Telescript language is not a scripting language. Its purpose is not to allow human users to create macros, or scripts, that direct applications written in "real" programming languages like C. Rather, it lets developers implement major components of communicating applications.

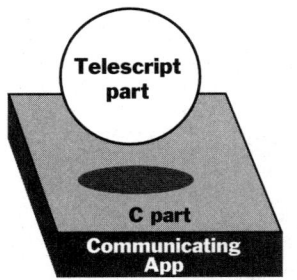

Figure 7.4 Telescript-based agent.

tems programming languages such as C and C++. Entire applications can be written in the Telescript language, but the typical application is written partly in C. The C parts include the stationary software in user computers that lets Agents interact with users, and the stationary software in servers that lets places interact, for example, with databases. The Agents and the "surfaces" of places to which they are exposed are written in the Telescript language.

The Telescript language has the following qualities that facilitate the development of communicating applications:

- *Complete:* Any algorithm can be expressed in the language. An agent can be programmed to make decisions; to handle exceptional conditions; and to gather, organize, analyze, create, and modify information.

- *Object-oriented:* The programmer defines classes of information, one class inheriting the features of others. Classes of a general nature, such as agent, are predefined by the language. Classes of a specialized nature, such as shopping agent, are defined by communicating application developers.

- *Dynamic:* An agent can carry an information object from a place in one computer to a place in another. Even if the object's class is unknown at the destination, the object continues to function: its class goes with it.

- *Persistent:* Wherever it goes, an agent and the information it carries, even the program counter marking its next instruction, are safely stored in non-volatile memory. Thus the agent persists despite computer failures.

- *Portable and safe:* A computer executes an agent's instructions through a Telescript engine, not directly. An agent can execute in any computer in which an engine is installed, yet it cannot access directly its processor, memory, file system, or peripheral devices. This helps prevent viruses.

- *Communication-centric:* Certain instructions in the language, several of which have been discussed, let an agent carry out complex networking tasks, such as transportation, navigation, authentication, access control, and so on.

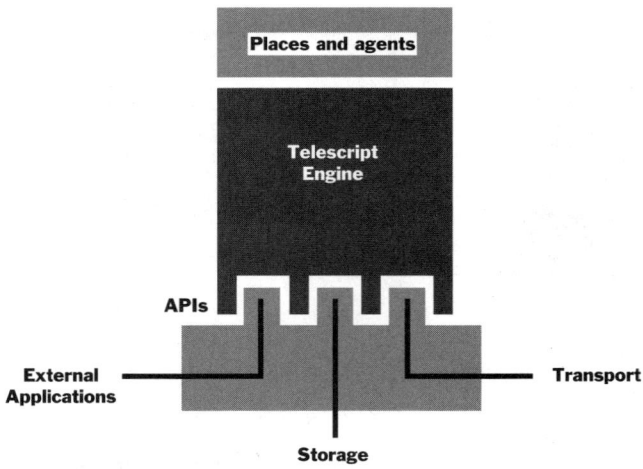

Figure 7.5 Telescript engine.

A Telescript program takes different forms at different times. Developers deal with a high-level compiled language not unlike C++. Engines deal with a lower-level, interpreted language. A compiler translates between the two.

7.5.1.2 Engine. The Telescript engine (Fig. 7.5) is a software program that implements the Telescript language by maintaining and executing places within its purview, as well as the agents that occupy those places. An engine in a user computer might house only a few places and agents. The engine in a server might house thousands.

At least conceptually, the engine draws upon the resources of its host computer through three APIs. A storage API lets the engine access the nonvolatile memory it requires to preserve places and agents in case of a computer failure. A transport API lets the engine access the communication medium it requires to transport agents to and from other engines. An external API lets the parts of an application written in the Telescript language interact with those written in C.

7.5.1.3 Protocols. The Telescript protocol suite (Fig. 7.6) enables two engines to communicate. Engines communicate in order to transport agents between them in response to the go instruction. The protocol suite can operate over a wide variety of transport networks, including those based on the TCP/IP protocols of the Internet, the X.25 interface of the telephone companies, or eve-electronic mail.

The Telescript protocols operate at two levels. The lower level governs the transport of agents, the higher level their encoding and decoding. Loosely

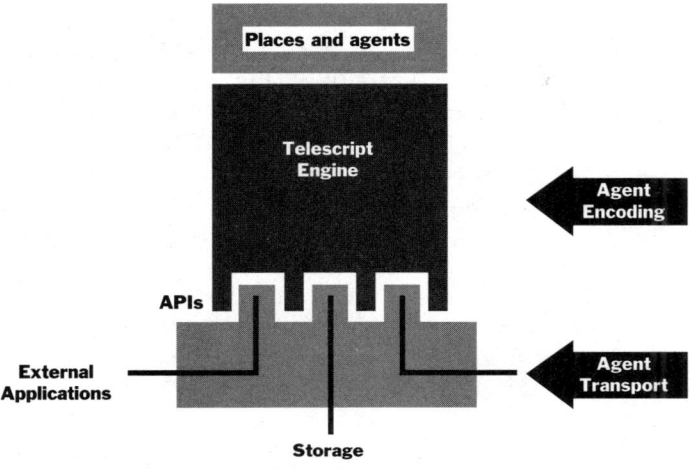

Figure 7.6 Telescript protocols.

speaking, the higher-level protocol occupies the presentation and application layers of the seven-layer OSI model.*

The Telescript *encoding rules* specify how an engine encodes an agent—its procedure and its state—as binary data and sometimes omits portions of it to optimize performance. Although engines are free to maintain agents in different formats for execution, they must employ a standard format for transport.

The Telescript *platform interconnect protocol* specifies how two engines first authenticate one another (for example, using public key cryptography) and then transfer an agent's encoding from one to the other. The protocol is a thin veneer of functionality over the underlying transport network.

7.5.2 Programming mobile agents—The Telescript object model

This section explains how a communicating application works. It does this by implementing, in the Telescript language, both an agent and a place. This section begins with a discussion of the low-level concepts and terminology that underlie the example, with particular attention to the Telescript object model, which governs how either an agent or a place is constructed from its component parts.

7.5.2.1 Object structure. Like the first OO language, SmallTalk, the Telescript language treats every piece of information, however small, as an object. An object has both an external interface and an internal implementation.

---

*The Telescript protocols are not OSI protocols. The OSI model is mentioned here merely to provide a frame of reference for readers who are acquainted with OSI.

An object's interface consists of attributes and operations. An attribute, which is an object itself, is one of an object's externally visible characteristics. An object can get or set its own attributes and the public, but not the private, attributes of other objects. An operation is a task that an object performs. An object can request its own operations and the public, but not the private, operations of other objects. An operation can accept objects as arguments and can return a single object as its result. An operation can throw an exception rather than return. The exception, an object, can be caught at a higher level of the agent's or place's procedure, to which control is thereby transferred.

An object's implementation consists of properties and methods. A property, an object itself, is one of an object's internal characteristics. Collectively, an object's properties constitute its dynamic state. An object can directly get or set its own properties, but not those of other objects. A method is a procedure that performs an operation or that gets or sets an attribute. A method can have variables, objects that constitute the dynamic state of the method.

7.5.2.2 Object classification. Like many OO programming languages, the Telescript language focuses on classes. A *class* is a "slice" of an object's interface combined with a related slice of its implementation. An object is an *instance* of a class.

The Telescript programmer defines his or her communicating application as a collection of classes. To support such *user-defined* classes, the language provides many *predefined* classes, a variety of which are used by every application. The example application presented in this section consists of several user-defined classes that use various predefined classes.*

Classes form a hierarchy† whose root is object, a predefined class. Classes other than the object class inherit the interface and implementation slices of their superclasses. The *superclasses*‡ of a class are the root and the classes that stand between the class and the root. A class is a subclass of each of its superclasses. An object is a member of its class and each of its superclasses.

A class can both define an operation or attribute and provide a method for it. A subclass can provide an overriding method unless the class *seals* the operation. The overriding method can invoke the overridden method by *escalating* the operation using the language's ^ construct. The overriding method selects and supplies arguments to the overridden method.

One operation, which the object class defines, is subject to a special escala-

---

*The example uses the predefined classes: Agent, Class, Class Name, Dictionary, Event Process, Exception, Integer, Meeting Place, Nil, Object, Part Event, Permit, Petition, Place, Resource, String, Teleaddress, Telename, Ticket, and Time and various subclasses of exception, such as Key Invalid.

†The language permits a limited form of multiple inheritance by allowing other classes that extend the hierarchy to a directed graph.

‡The language permits a class to have implementation superclasses that differ from its interface superclasses. Such classes are rare in practice.

tion rule. InitializeOperation is requested of each new object. Each method for the operation initializes the properties of the object defined by the class that provides the method. Each method for this operation must escalate it so that all methods are invoked and all properties initialized.

7.5.2.3 Object manipulation.
The Telescript language requires a method to have references to the objects it would manipulate. References serve the purpose of pointers in languages like C, but avoid the "dangling-pointer" problem shared by such languages. References can be replicated, so there can be several references to an object.

A method receives references to the objects it creates, the arguments of the operation it implements, and the results of the operations it requests. It can also obtain references to the properties of the object it manipulates.

With a reference to an object in hand, a method can get one of the object's attributes or request one of the object's operations. It accomplishes these simple tasks with two frequently used language constructs, such as the following:

```
file.length
file.add("isEmployed", true)
```

The example application makes use of the predefined dictionary class. A dictionary holds pairs of objects, its *keys* and *values*. Assuming that file denotes a dictionary, the first program fragment, file.length, obtains the number of key-value pairs in that dictionary, whereas the second program fragment adds a new pair to it. If these were fragments of a method provided by the dictionary class itself, file would be replaced by *, which denotes the object being manipulated.

References are of two kinds, *protected* and *unprotected*. A method cannot modify an object to which it has only a protected reference. The engine intervenes by throwing a member of the predefined reference protected class.

7.5.3 Programming a place

The agent and the place of the example (Fig. 7.7) enable this scene from the electronic marketplace. A shopping agent, acting for a client, travels to a warehouse place, checks the price of a product of interest to its client, waits if necessary for the price to fall to a client-specified level, and returns when either the price reaches that level or a client-specified period of time has elapsed. The construction of the warehouse place and the client's construction and eventual debriefing of the shopping agent are beyond the scope of the example .

The warehouse place and its artifacts are implemented by three user-defined classes, discussed in the following subsections.

7.5.3.1 The catalog entry class.
The user-defined catalog entry class implements each entry of the warehouse's catalog, which lists the products the ware-

296 Section Seven

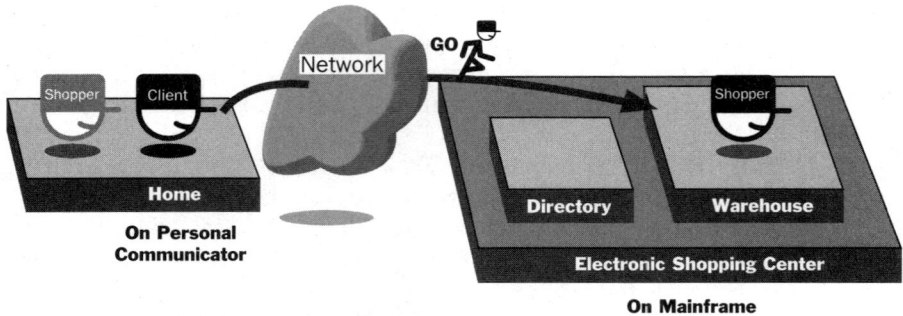

Figure 7.7 Telescript agent programming.

house place offers for sale. Implicitly, in this example, this class is a subclass of the predefined object class.

A catalog entry has two public attributes and two public operations. The product attribute is the name of the product the catalog entry describes, the price attribute is its price. The two operations are discussed after the following example.

```
CatalogEntry: class =
(
  public
    product: String;
    price: Integer; // cents
    see initialize
    see adjustPrice
  property
    lock: Resource;
);
```

The special initialize operation initializes the three properties of a new catalog entry. The product and price properties, implicitly set to the operation's arguments, serve as the product and price attributes. The lock property, set by the method to a new resource, is discussed next.

```
initialize: op (
  product: String;
  price: Integer /* cents */ ) =
{
  ^();
  lock = Resource()
};
```

A catalog entry uses a resource to serialize price modifications made using its adjustPrice operation. A Telescript resource enables what some languages call *critical conditional regions*. Here the resource is used to prevent the warehouse place and an agent of the same authority, for example, from changing a product's price simultaneously and, as a consequence, incorrectly.

The public adjustPrice operation adjusts the product's price by the percentage supplied as the operation's argument. A positive percentage represents a price increase, a negative percentage a price decrease.

```
adjustPrice: op (percentage: Integer)
throws ReferenceProtected =
{
  use lock
  {
    price = price + (price*percentage).quotient(100)
  }
};
```

A catalog entry, as mentioned earlier, uses a resource to serialize price modifications. Here the language's use construct excludes one agent or place from the block of instructions in braces, as long as another is executing them.

The operation may throw an exception. If the catalog entry is accessed using a protected reference, the engine throws a member of the predefined reference protected class. For example, if the shopping agent rather than the warehouse place tried to change the price, this would be the consequence.

7.5.3.2 The warehouse class.
The user-defined warehouse class implements the warehouse place itself. This class is a subclass of the predefined place and event process classes. A warehouse has three public operations, as shown in the following example:

```
Warehouse: class (Place, EventProcess) =
(
  public
    see initialize
    see live
    see getCatalog
  property
    catalog: Dictionary[String, CatalogEntry];
);
```

The special initialize operation initializes the one property of a new warehouse place. The catalog property, implicitly set to the operation's argument, is the warehouse place's catalog. Each key of this dictionary is assumed to equal the product attribute of the associated catalog entry.

```
initialize: op (
  catalog: owned Dictionary[String, CatalogEntry]) =
{
  ^()
};
```

A region can prevent a place from being constructed in that region the same way it prevents an agent from traveling there (see Subsection 6.8.2.2.7). Thus a region can either prevent or allow warehouse places and can control their number.

The special live operation operates the warehouse place on an ongoing basis. The operation is special because the engine itself requests it of each new place. The operation gives the place autonomy. The place *sponsors* the operation, that is, performs it under its authority and subject to its permit. The operation never finishes; if it did, the engine would terminate the place.

```
live: sponsored op (cause: Exception|Nil) =
{
  loop {
    // await the first day of the month
    time: = Time();
    calendarTime: = time.asCalendarTime();
    calendarTime.month = calendarTime.month + 1;
    calendarTime.day = 1;
    *.wait(calendarTime.asTime().interval(time));
    // reduce all prices by 5%
    for product: String in catalog
    {
      try { catalog[product].adjustPrice(-5) }
      catch KeyInvalid { }
    };
    // make known the price reductions
    *.signalEvent(PriceReduction(), 'occupants)
  }
};
```

On the first of each month, unbeknownst to its customers, the warehouse place reduces by 5% the price of each product in its catalog. It signals this event to any agents present at the time. A Telescript *event* is an object with which one agent or place reports an incident or condition to another.

The public getCatalog operation gets the warehouse's catalog, that is, returns a reference to it. If the agent requesting the operation has the authority of the warehouse place itself, the reference is an unprotected reference. If the shopping agent requests the operation, however, the reference is protected.

```
getCatalog: op () Dictionary[String, CatalogEntry] =
{
  if sponsor.name.authority = = *.name.authority {catalog}
  else {catalog.protect()@}
};
```

As mentioned earlier, one agent or place can discern the authority of another. Using the language's sponsor construct, the warehouse place obtains a reference to the agent under whose authority the catalog is requested. The place decides whether to return to the agent a protected or an unprotected reference to the catalog by comparing their name attributes.

Of course, it is possible to make E-mail a little too clever, like person X who programmed his mailbot to reply to a list of people, including person Y, who might send a message while X was on vacation. X's message said: "Sorry I missed your message. I promise to read it as soon as I come in on Monday." Meanwhile person Y had programmed his mailbot to do essentially the same thing. Both X and Y then went on vacation at the same time, and each returned to find a mailbox full of "Sorry I missed your…" messages. This could have been avoided if the mail agent were programmed not to send the same message to the same address twice.

8.14 Database Access via the Web

An agent can periodically query a set of databases, locating records which meet a user's interest criteria. Those records may then be translated for display in various venues, such as the Internet, the WWW, CompuServe, Usenet, or Lotus Notes. Thus web pages can be linked to live databases via agents. Results can be displayed with highlighted search terms, hypertext, virtual-reality markup language (VRML), and so on.

8.15 Agents in Industrial Automation and Control Domains

In industrial automation and control applications, agents can serve as partial proxies, helpers, and assistants to those humans (process operators, process engineers) that are responsible for various aspects of a production line. For example, process operators and maintenance technicians usually must gather and rely on large amounts of data and activities occurring within a process-control system. Now monolithic optimization programs monitor process variables and other system parameters (environment, memory usage, processing "units" in use, and so on) within or across process units to:

- Smooth out complex feedback-loop responses (eliminate ringing or hysteresis)
- Ensure optimal resource usage (feedstocks and raw and intermediate materials, as well as use of equipment and compute power)
- Ferret out problems caused by the interaction of many variables
- Propagate alarm and preventative information to appropriate user-interface devices
- Act, in a closed-loop fashion, to change process variable set points in order to achieve desired production goals in both continuous and discontinuous (batch) processes.

What IAs potentially bring to the table is the ability to further remove the human operators and technicians from the loops and put in IA-based surrogates and proxies. Again, trust in the technology is one of the big issues in a domain that is potentially dangerous to not just bits, but atoms—human atoms, that is. But put that aside for a moment (after all we *do* trust machines with our lives every day, so why should agents be any different).

Are there any things a process operator, a maintenance technician, or even a process manager does that an agent can do (either better or for less cost)? Of course, in any domain we can use agents to run around the net, collecting, formatting, and delivering customized reports on conditions and activities in that domain. However, we want something more substantial.

The key is in the understanding of what this book means by IAs, as opposed to any other computer program: autonomous, encapsulated intelligence or expertise, possibly mobile software. In production control, much of the expertise is in the heads of process operators and process engineers, and the best ones are not on the job 24 hours a day. IAs that encapsulate a process engineer's knowledge of the process chemistry, or a process operator's knowledge about what valve affects what process variable can be built using AI techniques. An IA can be built that just knows standard operating procedures so that in case of a process upset, an operator's assistant agent could guide the operator through difficult times.

An IA with autonomy and maybe some mobility (to visit nodes within the distributed process control system) could serve *as* the operator in certain control situations. It could bring processes up faster, shut them down more safely during process upsets, and find potential production and maintenance problems faster than human counterparts.

8.16 Governmental Agents

How will the activities of governmental bodies be affected by the use of agents? Agents can be extremely useful in searching legislation, regulations, and generally mining the information veins in local and national government databases.

Governments at every level are prodigious consumers and producers of information. It is used for decision support, policy making, resource allocation, and deflection and diffusion of responsibility. Since it is not always the most effective use of humans' time to sift, sort, catalog, and store data, most governments have automated information management and retrieval of the information they collect, namely, census, taxation, and so on.

As data repositories or "warehouses" have become ever more massive and complex, paralleling the increase in population and the complexity of government, complex data organization and mining strategies are now necessary. Online analysis and processing (OLAP) agents are being used to view data from many perspectives in order to make sense of the data. These agents can spot trends in data values, then "drill down" into the data, finding clues for why the trends are occurring. (Refer to the DSS agents in this section.) Eventually humans or expert systems can use the clues found to explain the trends and help formulate future policies.

It used to be that worldwide distribution of information and ideas was subject to numerous gatekeepers to the mass media such as book and magazine publishing. Access to reputable mass-media channels was jealously guarded. To gain widespread publication of any information, an author had to gain the attention and approval of gatekeepers such as reporters and editors. Then, if deemed profitable, the author's ideas would be edited, checked (at least to some extent) for accuracy, and distributed in a perishable form, such as radio, television, magazines, or newspapers.

Now, thanks to the Internet and other networks, anyone can shoot from the hip and publish a web page or add his or her comments to a news group. Those words immediately gain worldwide distribution, and may carry the same weight as a carefully researched article in a reputable magazine, with the added advantage that they may be freely altered, electronically copied and E-mailed, as well as attributed to any source the author or subsequent readers can dream up.

The worldwide flood of information characterized by the WWW has caused and increasingly will cause problems for government agencies. When monitoring on-line sources, how does one determine the validity of the information? When publishing to on-line channels, how do governments make sure information is understood as official, and is not maliciously altered? What about secret communications? How can a government keep secrets when civil servants can anonymously and instantly E-mail or publish sensitive information? While one government's cyberagents monitor information sources, hostile governments can flood channels with bogus messages as well as monitoring. Therefore government A's agents must check for hostile agent activity from government B, besides encrypting and routing secret information and possibly sending decoys of its own.

IAs could help increase security and get real information out of a mass of data. By monitoring channels for incoming and outgoing messages, providing intelligent gateways in a government fire wall, agents can, at the risk of becoming Big Brother's eyes and ears, make sure that no one dealing with sensitive information can distribute it on cyberchannels. Information could be sent as autocorrecting agents, much like parity bits and Hamming codes, which only receiving agents could decode. Data mining can provide timely, useful information for decision and policy making support with the need for an army of information analysis bureaucrats. In addition, agents could automatically trace the source of any potentially important information, verifying attribution before a message is accepted. This may need to be done sparingly, however, since it could consume an unacceptable amount of time and resources.

8.17 Medical Agents

Agents can be used to monitor and track patients, assist in diagnosis (via encapsulation of or reference to appropriate expert systems), and inform appropriate medical personnel of patient conditions, especially those patients in intensive care. Agents can facilitate efficient medical treatment by monitoring patient conditions and reporting anomalies to the appropriate personnel (for example, within the realms of in-home baby monitoring, elderly care, and in-patient monitoring scenarios as agents are embedded into monitoring equipment).

8.18 Military Agents

Military commanders are always interested in finding new and better ways to give effective orders. Agents could provide the means to decode and encode, receive, route, and validate information nearly instantaneously. Agents could coordinate transportation, logistics, even battlefield maneuvers. Agents could read satellite and air-surveillance images to understand the big picture, while coordinating with other intelligence-gathering agents to make decisions about tactical situations. Battlefield command, control, and communications agents could give an army a great advantage in speed and intelligence, but with lives at stake, agents would probably need to be limited in scope and authority.

8.19 Computer-Aided Design Assistants

Much of product design (especially of digital systems) consists of knowing what off-the-shelf components exist and how they could fit into a design. With many vendors offering thousands of components, an effective component search can be extremely difficult. Krakatoa, a parametric search engine, allows a designer to enter parameter values of the devices desired, and information will come back on specific parts with a URL to a web page describing more. Krakatoa, written in Java, provides a designer with the ability to search on-line databooks from any vendor. Cadis has put information from National Semiconductors into Adobe's PDF (Portable Document Format) format, which is an SGML-Krakatoa-searchable format, and Tektronix and others will offer the facility.

8.20 JAT-Based Agents for Interactive, Collaborative, Concurrent Design and Engineering

The JAT we discussed in Section 7 is being used in a design engineering context to aid in communicating design over the Internet. These engineering agents encapsulate engineering software. Using JAT-based agents that perform remote simulations, download code and design specifications, and notify, via broadcast, other designers of design changes, this paradigm shift in the way interactive design is done will certainly upset some monolithic applecarts (those applications that are all-in-one approaches to design). Of course, any collaboration done over the Internet using JAT could also take place more securely over a company's intranet. Agents can be programmed to find out what each collaborator needs to do a particular part of the job and locate and provide those resources to the collaborator.

We see this use of agents spreading to other domains in the near future, now that a *content-independent* architecture for JAT-based collaborative engineering has been developed. Figure 8.2 shows that architecture.

Indeed, as you can see, the architecture can be used for agent-based concurrent collaboration in domains other than engineering, such as financial planning, company operations, troubleshooting in almost any domain, and so on. Notice that the agent contains its needed resources, including ontologies (domain-specific knowledge) and any specialized languages that can be translated to a standard ACL based on, for example, KQML. This would allow agents based on different languages to interoperate.

Another way this technology can be used is to integrate many different stand-alone applications in a work-flow scenario. For example, as you design and produce a new product, JAT-based agents could get parts information from parts suppliers and integrate these data into your own internal and user documentation. They could interface with manufacturing agents to provide custom configuration information, and when a product is ready to ship, agents can be dispensed to update all the relevant accounting systems and as-built documentation.*

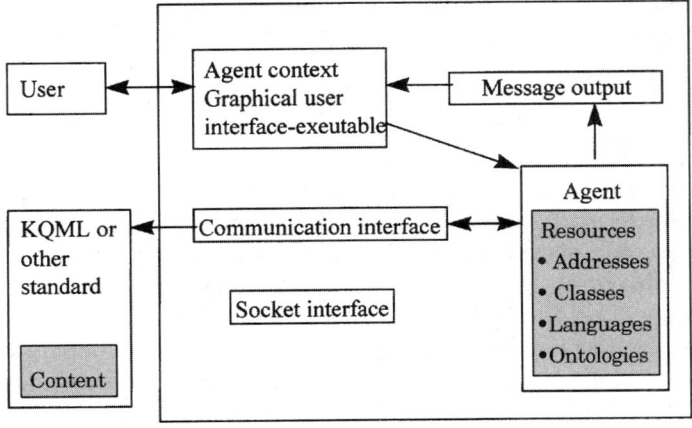

Figure 8.2 Distributed collaboration framework using JAT. (*From Center for Design Research, Stanford University, Stanford, CA.*)

---

*You can download JAT, which currently runs under Windows NT and UNIX, from the Stanford web-based site: http://cdr.stanford.edu/ABE/JavaAgent.html.

8.21 Technical Assistance Agent

Agents can assemble an HTML web page to solve a specific user problem. Since you don't want a whole set of manuals dumped on you, an agent can look up all reported problems related to your problem, then give the top 30 to you, ranked in order or relevance. Of course, an agent can also automatically download a self-extracting patch file to fix your software.

Let's say you are using an application, and suddenly the performance of the application takes a nosedive. You access the home page of the vendor, then fill out a form, including a text description of the problem and information about your configuration. A help-desk agent:

- Provides help on a variety of supported products
- Understands natural-language queries
- References "solutions" database of known problems
- Is constantly updated
- Can download new, self-extractable, and self-installing versions and patches
- May be able to monitor the client system via remote agent

An automated troubleshooting agent working over the web, reporting relevant information to a browser, might work as follows. A user fills out a form, including some text describing the problem. The agent then extracts keywords and dispatches topic agents to the database. TIDs (Technical Information Documents) are found which match product, version number, keywords, or general information. Then HTML pages are built dynamically to display the information found.

8.22 Decision Support Agents

One of the things a DSS agent does is dynamically create the SQL necessary to produce the defined reports without the user needing to formulate complex queries involving operations such as table joins, group by, where clauses, and so on. A DSS agent, like work-flow agents, can define multiple distribution paths involving E-mail, fax, DSS agents, CompuServe, Internet, and so on.

Only the results need to be moved, so it is highly economical to send a DSS agent to a data warehouse on the Internet, create the reports, and distribute the results. Using DSS agents, a user can define reports using templates and filters, then drag and drop them into an agent and define a trigger condition for the agent.

Agents are a bit like background batch jobs, but they run in distributed environments, are end-user programmable, can communicate, and behave with some degree of intelligence. They can be triggered on events, not just periodically.

The results from DSS agents are retained in a buffer where they can be shared, so the same query won't need to be repeated. Turning users loose with

SQL can bring a server to its knees, not to mention the possibility of corrupting data. An agent can remedy this by coordinating its activities with the current loading of the system and higher-level programs or agent.

For example, the DSS agent from MicroStrategy allows users to define alerts that are triggered by a set of conditions being met. The conditions are defined by selecting variables from a database and incorporating them into logical expressions. Various analyses can be performed on the collected data and alerts. For example, inventory agents can monitor orders, slow sellers, and to-date totals. Performance agents can interact with inventory agents to determine profitability by market segment, region, or product. Ranking agents can keep track of what is doing the best and what is doing the worst. And time-period agents can collect and collate data based on various time periods. All of these agents could then feed their data to graphing and reporting functions, or could feed higher-level decision-support software or agents.*

8.23 "Bots"

The word "bot," as it is often used on the Internet, refers to virtually every kind of computer program from a television listing generator to human impersonator. While very few bots are artificially intelligent, they are autonomous, meaning that they can monitor and take action relative to their environment without human direction. Like serious agents, bots can carry on and perform their duties tirelessly while their creators relax or cruise the web.

Bots are usually written in a language such as C or Perl to run on UNIX, and usually execute on Internet relay chat (IRC) channels. Bots can be programmed to "chat" a little like human users by repeating phrases geared to specific situations. IRC bots are really just curiosities or "toys," however. To really accomplish useful work, or serious mischief, a bot must execute outside of the IRC boundaries.

Some bots have caused problems by hogging resources on servers that are not adequately protected by fire walls, but in general they've been pretty benign. This is not to say that in the future agents, or bots, may not do some serious damage (either accidentally or on purpose) as they become more intelligent and autonomous.

8.24 Author's Assistant

Agent assistants will eventually be more than administrators. In the future they may become creative associates. In the near future an agent could search through some (pointed-to) document and pull out all references to books, magazines, or persons (presumably an author of some works) and search on-line literature databases for the correct source information (author, title, date,

---

*Navigate to: http://www.strategy.com/ for more information.

publisher, page, and so on). The agent could then autogenerate a bibliography or references section in the format selected, for example, a designated WordPerfect template. (Oh, do we wish we had one of these while writing this book; hey, that gives us an idea....)

The agent could reference a format designation as well as a sorting order (such as order as cited or alpha order) and for those references not located, just add them in proper order as above and flag for the author to edit them. The agent could also autogenerate the permission-request forms for each reference selected (using a designated template) and print them, along with the address label to the publisher. Publisher information could be retrieved from the Internet.

For this type of agent to work, there would have to be some user-supplied connection between the context in which the agent is expected to execute and the agent. For example, the author's assistant could be a "plug-in" or "helper application" to WordPerfect. When an author runs WordPerfect, a configuration would be set up by the author that tells the agent under what conditions to execute, and to notify or not to notify the author concerning a range of problems (for example, can't connect to a literature database).

8.25 Anthropomorphic Agents: Firefly

Firefly is a personal software agent capable of communicating with other users and recommending music that it knows the user will enjoy. Firefly automates the word-of-mouth process, learning about the user and his or her opinions, and leveraging that information to best serve the user's needs. Firefly uses the tastes, opinions, preferences, and idiosyncrasies of those most similar to you (your "nearest neighbors") in order to suggest new music that you might like too. The more the user trains its agent, the more useful and accurate it gets. The more other people use the system, the smarter the Firefly community becomes.*

8.26 Big Brother Agents

Of course, with all the capabilities mentioned in this book, some will be tempted to use IA technology to monitor (and thence control) other humans. For example, a performance review agent that runs in the background could automatically document activities and creative product production for your annual performance review, including the following:

- E-mail sent
- Software changes made
- Design documents created or changed
- Keystrokes
- Reports generated

---

*For more information, navigate to http://agents.www.media.mit.edu/groups/agents/.

Indeed, some of this "Big Brother is watching" technology is already in use. For example, telemarketing centers already have background tasks that record keystrokes and orders taken per unit time. As agents become more adept at monitoring various aspects of users' proclivities, production "quotients," and other measurable, quantifiable items, computer users will have increasingly more data about them made available to others.

Imagine a personal profiling agent that follows you around, recording your every move in cyberspace. The partial underpinnings of such a scenario already exist in the way your purchases are recorded, especially if paid for by credit or debit card. As the cashless society becomes reality, those who choose to participate (there will always be those "outlaws"—cyberluddites, isolationists, the paranoid, criminals, and so on—that refuse to so participate), can look forward to detailed dossiers being composed on them, on the fly so to speak.

IA technology (especially the autonomy and mobility aspects) extends this capability by enabling much more information and analysis to be done on the cyberactivities of people wherever they go. For example, as we start ordering pay-per-view television from digital satellite systems and from video servers, agents can travel to your "web-TV" box collect your viewing preferences and send these data to product manufacturers, whose agents in turn will send you pitches on related products.

8.27 Agents as Meeting Facilitators

A good deal of what we expect IAs to be doing is helping us manage the vast amount of information coming to us from all quarters. One place where an IA would be particularly helpful in this regard is to help recall and organize the salient concepts, ideas, agreements, and other information that result from a meeting. Indeed, meetings, especially brainstorming meetings, tend to ramble in all directions and produce lots of information that may not be directly or sequentially connected—call it "hyperspeak."

These kinds of meetings usually have a facilitator, (hmmm, sounds like a name for an agent!) a person who directs the discussion toward fruitful avenues, restores order among the inevitable chaos, and sometimes plays the role of scribe as well. The facilitator, and maybe others, are then charged with boiling all that down into a report or presentation. What if an IA could help with most of these tasks? Well, some researchers have developed an IA that does just that.

Chen, Houston, Nunamaker, and Yen (1996) have built a C++-based agent that mines and manipulates information from GroupSystems electronic meeting system. The agent incorporates:

- *Automatic indexing:* Looks for terms or phrases within each comment made in the meeting and logs how many times those terms appear elsewhere in the session.

- *Cluster analysis:* Weights each term or phrases. Weights relate the relative importance of the term, the relationship to other terms and to the whole document representing the meeting session.

- *Hopfield neural net–based term* classification organizes terms into a single-layer network, each neuron of which represents a term, and the weighted connections represent the term's relationship to other terms. When the net is run, a term is input into the net's single-input neuron, and a Hopfield algorithm activates the neuron's neighbors. The output is a list of all concepts related to the input term. Repeated for each term, this process yields clusters of related ideas from the meeting session.

The agent was used to collect all comments made in the meeting session, and generated categories and other helpful data about a meeting, essentially acting as an electronic scribe. In addition:

- The agent effectively dealt with the information overload by collating and categorizing meeting ideas.
- The agent generated a collaborative vocabulary—a set of common ideas that all participants could agree were common and salient.
- The agent does not miss anything; even sensitive or off-color comments and ideas are captured.
- The agent does this faster than a human (of course).

One issue was that the agent generated too many categories of concepts—some of which were either too specific or too general to be useful. Another was the lack of generalization capability. Both of these problems can be fixed, and the researchers are hard at work doing just that, and making other enhancements to this IA. They are also working on similar IAs for common groupware applications such as Lotus Notes.

We see this use of IAs as particularly indicative of where this technology can have a big impact. After all, information, knowledge, and wisdom are being valued as heavily as material capital in an ongoing enterprise. And capturing, analyzing, and transforming data and information into knowledge and wisdom can be delegated with success to IAs. The faster you can do this, the faster that knowledge and wisdom can be made operational.

Section

9

Agent Futures

"In the future, everyone will be famous for 15 minutes."
ANDY WARHOL

"In the future, everyone will have an agent."
JAY JOHNSON

In this section we speculate as to where agent technology may be headed. We present some examples of projects and noncommercialized (yet) software that could portend what to expect in the near future. We also speculate as to where a longer-term view of AI and IAs may take us.

9.1 Introduction

Today relatively simple-minded search agents are used widely, and larger, monolithic special-purpose agents, such as real-estate or travel agent applications, lurk on the web. Small, mobile agents, such as Telescript agents, are also emerging slowly, although at this point these are far from common. The future will see systems of small agents cooperating together to produce an intelligent result. These agents must adhere to a common standard of communication and actions, and each may be specialized to perform one function.

Compared to human history, computing is essentially in the middle ages today. We have begun to develop a cybercivilization, but conditions are still mostly chaotic. Agents will help bring efficiency and order to cyberspace by providing uniform access to web resources, making it possible to get timely information from the Internet without dedicating your life to cybersurfing, making it practical to use the Internet for commerce, and performing many tasks to improve the lives of individual users.

The vast decrease in the costs of computer memory and computer power in terms of millions of instructions per second (MIPS) is well known. Because of this, desktop systems powerful enough to run broadly useful operating systems and applications are becoming the rule instead of an exception reserved only for "power users." While powerful desktop workstations are becoming pervasive, truly "pervasive computing" will occur only (on the fast-approaching day) when the main source of computing power is the network, not the desktop. One of the major contributing factors to the future success of agents in distributed systems is the orders-of-magnitude decrease in the cost of storing and pumping bits around.

As this cost goes down, not only desktop computers but all intelligent devices, including embedded systems in automobiles, fax machines, printers, and industrial control equipment, can become part of a worldwide pervasive computing network.

Like environments and languages such as Java, with its ability to run in embedded environments using, for example, the picoJava microprocessor, IAs will also be able to exist in these environments. This means that IAs can gain access to many more types of resources and information than is currently the case. For example, an IA that functions as a personal assistant—an electronic butler or secretary if you will—can serve much more effectively if it knows about and can communicate with the other devices in your life.

For example, if an agent can get on your in-home intranet, it can query your automobile (which at home is, of course, plugged into your intranet, as well as the internet for daily remote diagnostics and "recalibration") to see whether it is ready for the long road trip you told the agent to plan for you. The agent senses the important variable within the auto's network, or talks to the auto's computer directly to discover if you will have to stop for gas on the way out, or get your tires filled to specification with air.

The popularity of the Internet seems to increase dramatically every year, overloading the bandwidth of the system and making it highly frustrating if not impossible to access many servers. Enter agent technology. Servers running IAs, which can increase the efficiency of access by orders of magnitude, along with the decreasing expense of moving bits around, will eventually make the line of separation between systems disappear. Today it can be a complex matter for users to execute programs on several different servers, using data distributed across still others. Thanks to IAs, users will be able to go on about their work without having to know or care where the computing and data resources they use reside, or what is needed to access them.

Pervasive computing will extend not only to desktops, servers, and intelligent devices, but to wireless communication as well. Imagine, and this is not far away, an in-car navigation system using an IA to access a global network in order to assemble detailed local-area maps as needed. These maps, along with an intelligent device providing real-time vocal directions to any location (specified via a hand-held global positioning system device), would mean a driver need never get lost, even in a foreign city. Agents could also constantly diagnose the condition of the vehicle, reporting to the driver as necessary. If something went wrong, a "repair agent" could find the nearest garage with the parts and facilities necessary to make the repair, also requesting a tow if necessary.

Similarly, IAs could provide real-time navigation for Internet WWW surfers. Your navigation agent could perform dynamic queries and build detailed information maps to lead you to the next interesting web site on your quest. Such a "guide" agent would figure out the most important links to follow, and do what was necessary for you to access those sites (Henry, 1995).

The future will see the increasingly pervasive use, not only of simple agents to handle E-mail, network communications, and so on, but of advanced agents, which will play an increasingly important role in decision support. As databases grow in both the amount of data stored and the complexity of the relationships between data elements, and as distributed databases become ever more prevalent, agents continue to gain in importance. Agents will be able to monitor and correlate trends in real time, pinpoint the reasons for the trends, and offer options for remedying or profiting from them.

Except in highly specialized cases, observing, reproducing, and improving the way users perform tasks usually amounts to simply tracking and replicating keystrokes. Until an agent can see the bigger picture, and understand a user's true goals and how it can find shortcuts to achieving them, it will be of very limited use. In the future, agents will be able to piece together clues about how a user goes about his or her business. These agents will be able to figure out their masters' goals, and come up with ways to achieve them.

Agents that can perform more natural-language-related topic searching may be on their way in a few years, but a human's ability to know what kind of information he or she is interested in is difficult to automate. Agents will nonetheless be useful in cataloging nonmission-critical information, such as clippings from on-line periodicals. If an agent makes a mistake there, it will probably be lost in the shuffle.

9.2 The Future of Network Management

In the future, computing will become pervasive will network nodes interconnected globally. Agents will need to provide constantly updated network maps and could be instructed to manage a variety of systems. As networks continue to sprawl and link with other networks, monitoring and managing network nodes will become ever more crucial. Into this potential chaos steps the IA. Agents can be installed on every node in the network, monitoring conditions and alerting the network management system before trouble starts. Ideally, IAs can do more than this. If an agent is "smart" enough to find a defective or potentially defective component, it should be able to either fix it or prevent it as well.

For example, today's agents can detect a bad router port, but cannot patch around it. They can only summon a technician to do the job—not very practical when the problem is miles away in the wee hours of the morning. The current standard for network agents is not up to that intelligence level, however.

Agents can also be used to change the behavior of a network or remote node, based on user reconfiguration needs. For example, new sets of alarms need to be generated based on different combinations of timing and limits. As networks grow beyond sizes manageable by humans, IAs will roam LANs and WANs searching for and fixing problems, reconfiguring and updating users as nodes are added, and adjusting the system for maximum performance.

9.3 No Surfing

As the sheer sizes of networks make human-interactive surfing for information, negotiating for resources, and so on, impractical, agents or avatars will represent humans or organizations so that they can be in more than one place at a time to represent their interests. Systems such as General Magic's Telescript are already in place, as are a number of cyber meeting places for human surrogates or servants. These will become far more widespread as time goes by.

9.4 Commercial Agents

Research has shown that when it comes to purchasing any item, most people want as broad a selection as possible, but they want this selection "edited down" for them, as per their preferences. Agents are perfect for this, and many companies are working on prototypes. One of the first (mentioned in Sec. 8) is Andersen Consulting's BargainFinder* which shops for the best price for CDs from known on-line stores. While this is a noble effort, it immediately ran into several problems, including speed and the lack of any retained knowledge.

Most importantly, it proved that in order for a purchasing agent to be successful, there has to be a benefit for retailers who do not want to be judged on price alone. Some retailers are actually blocking BargainFinder from their sites, even though they may lose cybercustomers. In the future, vendors on the web will conclude that it is more profitable to be included in purchasing services than to preserve outdated, though once profitable ways of doing business.

Sellers will simply have to adjust to a new way of doing business if they want to stay alive. While purchasing agents "pull" information and products from the net, selling agents can be "pushing" products and information into users' domains. This will give rise to user agents to defend against sales agents, which will lead to smarter and more ubiquitous sales agents, and so the battle for consumer dollars will escalate as web computing becomes pervasive.

Another aspect of future commerce facilitated by IAs is the probability that they will replace the middleman: distributors, retailers, and others that profit from and, some would say, unnecessarily hinder the free exchange of product

---

*Navigate to http://bf.cstar.ac.com/bf/.

from manufacturer directly to consumer. For decades, commerce has been facilitated by intermediaries. Many manufacturers saw the opportunities for greater sales if they sold direct to the public, either by opening their own factory outlets or through direct phone or catalog sales. This trend, called disintermediation, is continuing today.

But at the same time, within the world of electronic commerce, specifically the Internet, several companies have sprung up that offer services traditional within the purview of retailers and distributors: the gathering in one place of multiple product offerings, knowledge and answering of questions about such products, and provision of the best price obtainable for a chosen product (see discussion of BargainFinder in Section 8). This service amounts to reintermediation and may be the trend for the *immediate* future.

However, it is our contention that all the services provided by these new, virtual, or electronic-based intermediaries can be performed by increasingly sophisticated IAs. Thus we will perhaps witness the final closing of the circle or loop of commerce, with "redisintermediation": your IA will take your preferences for a new product and interact directly with—not retailers, not distributors—but the direct manufacturers or providers (in the case of services) of the product. Indeed, this trend will enable product makers to keep close, unbiased, and instantaneous tabs on consumer desires, as well as enable specialty and niche manufacturers and service providers to expand their market shares.

Another commerce-related question affecting the future of agents is; where to put the intelligence? The majority of agents are used to monitor and retrieve information from a network, and most of the "intelligence" has been put into the gathering and not the interpretation of the results.

In the future, agent designs may let the information or resource provider control the gathering, while the agents' "intelligence" can be concentrated into the interpretation of information, giving both providers and consumers new tools for electronic commerce.

For example, buying cars and trucks may be getting a lot easier thanks to on-line auto purchasing agents. An Internet-based service such as Auto-By-Tel*, or a research aid such as the Edmunds guides on-line† have made it possible to purchase a vehicle sans haggling with a determined salesperson. People have long been looking for a better way to buy vehicles, so the impact on car and truck salespeople is potentially larger than that felt by real estate or travel agents. In the near future, it may become commonplace to not only research a car or truck on the web to find the features and price you need, but commission a cyber agent to find and purchase a vehicle according to your specifications at the best price (Greenwald, 1996). This could be the ideal task for an intelligent agent: to use the plethora of on-line vehicle information sources and recommend a short list of makes and models to fit your needs, then search for the

---

*Navigate to http://www.autobytel.com/.

†Navigate to http:www.enews.com/magazines/edmunds.

right price at a local dealer. Of course, you'll still need to go to a car dealership to take a test drive, but thanks to technologies like VRML and Java, even this may change.

A personal auto-buying agent could collect your requirements for the vehicle, then search sources such as the Edmunds guides or *Consumer Reports* on-line to find the best candidates. Depending upon your input at that point, it could then start looking for the lowest price on the best vehicle in your price range (new or used) with the options you want. Of course, you would need to approve the final purchase, but your agent could save you weeks of research and negotiation. This could work for any major purchase, from a computer system to a pair of skis.

With the initial success of on-line purchasing agents such as BargainFinder and Auto-By-Tel, users will certainly be able to define their own personal shopping agents, containing price range, priority, needs, personal preferences, and so on. These agents, coupled with the other commerce agents roaming various networks, will undoubtedly revolutionize the way humans buy and sell commodity from information to power tools.

9.5 Net Searching and Information Mining

Agent-based searching will go far beyond literal text matching. For example, we envision agents that can do patent and copyright searches. This would involve sophisticated pattern extraction, context-sensitive searches, and some degree of natural language understanding. Intelligent searching and data-stream monitoring will also involve relating sets of ideas and even graphical pattern matching. This kind of facility is being worked on in other contexts as well, as discussed in Section 8, where agent-facilitated meetings employ similar strategies for extracting, analyzing, and relating ideas and concepts from multiple sources.*

9.6 Agents to Infiltrate Applications

Users can expect to see vendors applying smart agent technology to more types of resources and applications. Rick Kreyser, vice president of marketing from OpenVision Technologies Inc., Pleasanton, California, is of the opinion that "the next step in agent technology is to use smart agents to collect information about databases, systems, operating systems, and business applications." This requires embedding manageability and monitoring capabilities right into an application when it is written by using a set of application programming interfaces that are typically packaged in software development kits.

Just as GUIs were initially introduced as a "value-added" feature in some applications and are now virtually indispensable, eventually, software packages that don't include agent support won't be salable.

---

*For more on the potential of this facility, navigate to http://www.spo.eds.com/patent.html, and //www.uspto.gov.

9.7 Military Agents

When it comes to military communication, command, control, and intelligence, agents as "cybersoldiers" have great potential. The idea that military operations can be carried out by robotic entities that communicate and coordinate with each other and obey instantly is very appealing. Using cybersoldiers can preserve the lives of flesh-and-blood soldiers.

Since orders, in corporate hierarchies as well as in military organizations, are given via (hopefully) secure computer networks, hostile entities could benefit greatly by using smart agents to infiltrate communications channels. Agents will therefore be needed to monitor communications networks and guard against information-pilfering bots.

9.8 Database Agents

There are tons of good information available on the Internet, but most of the important data reside in corporate databases. Most database packages support triggers and stored procedures for launching agents, but in the future, agents may automate all database access and updating, as well as checking the validity of the data and performing natural-language (topic) queries rather than simple keywords and phrases in text searches. Database agents will also coordinate application execution among distributed databases. Real-time text searching with relevance ranking and search-term highlighting is already available via Fulcrum, Folio, and other products.

To be truly useful as a data gatherer and processor, an agent, whether mobile or stationary, will need to be able to access a variety of databases in a uniform way. An effective agent will need a flexible interface that will allow it to query a broad spectrum of relational and OO databases. Since the agent itself will probably be OO, a vendor-independent object-to-relational/relational-to-object interface such as ObjectSelect* will be required.

Agents will also be used to increase database performance by buffering, distributing, and triaging incoming requests. These agents can also support data security requirements, maintain referential integrity, and help keep databases clear of obsolete, incorrect, and redundant data.

9.9 "Big Supplier" Is Watching You

Retailers don't want to have to be monitored constantly by purchasing agents looking for the lowest price. Business will take on an entirely different dimension when individual purchasing agents can cruise through hyperspace and haggle for the lowest possible price. The agent could find out that you're overstocked and haven't sold one of a specific item in a month, so you would have

---

*Available from Bob Capel and associates at www.objectselect.com.

to reprice goods constantly. You could also use some antiagent strategies, like sending your own agents to confuse them in their search.

Meanwhile, the buyer agents that retailers use to stock their cybershelves would have to be smarter and faster than ever to outwit their suppliers. Sales agents will know about you, and they will be much more persistent than a mere E-mail message. Agents have no ethics, emotions, fear, or embarrassment. They may hide secrets, but they probably know some dirt about the other side. This could be high-speed cutthroat capitalism at its worst.

9.10 Trust

There seems to be a hidden element of distrust of AI, and of software in general. AI seems to be defined as "whatever computers can't do." People may never believe software can be made smart enough to be of much use. In spite of the widespread use of investment expert systems, AI is expensive to develop and seems to have a limited appeal. In addition, people are reluctant to rely on AI for anything more than providing a challenging chess partner or reading characters via an optical scanner.

People in general feel comfortable with agents finding resources and helping with research, but when it comes to making important decisions based on the data agents gather and analyze, humans may never trust them.

Agents lack ethics. How do we program software to "understand" what we mean by "morality?" Of course there are always methods to ensure an agent does no wrong. You could codify Asimov's laws of robotics, and the various corollaries that have been developed. Perhaps a good way of ensuring that all software adheres to the basic tenets of morality and doing no wrong, is to make a Cyc-based ethics ontology, and make it mandatory that all agents have access to it.

9.11 Information Agents and Cooperative Information Systems

In the future, agents will be able to learn from and cooperate with each other, and they'll be able to learn from their environments. An information agent is an agent that has access to at least one, and potentially many, information sources, and is able to collate and manipulate information obtained from these sources in order to answer queries posed by users and other information agents. A network of interoperating information sources is often referred to as intelligent and cooperative information systems.

The information sources may be of many types, including traditional databases as well as other information agents. Finding a solution to a query might involve an agent accessing information sources over a network. A typical scenario is that of a user who has heard about research being done concerning a specific topic. The agent is asked to investigate, and after a careful search of various FTP sites, returns with an appropriate technical report, as well as the name and contact details of the researchers involved.

In the future agents will locate loosely specified items from a range of document repositories. Teams of agents (ala webCrawler) will be able to search web sites and heterogeneous databases and work together to answer queries that are outside the scope of any one of the individual databases. These agents will start searching at a variety of entry points, and coordination or agents or duplicate removal could be handled either by a supervising agent, or via communication between searching agents. By searching in parallel, adding more computing power can greatly speed up the search, giving rise to services consisting of clusters of search servers.

While an intelligent network, functioning as an execution environment for autonomous entities, can replace less efficient and less flexible execution models, it also has potential to work well for modeling the real world, since human beings and organizations of humans operate as autonomous agents. For example, multiple agents can represent different real-world parties and are therefore self-interested. Two competing agent-interaction approaches are arbitration versus contracted self-interest. One approach is to carry out exchanges without enforcement or arbitration, depending on self-interests balanced through predefined interaction contracts. The arbitration approach involves a central authority or system to act as a referee and perform functions on behalf of agents, thereby helping to ensure that everything will run smoothly. This is the approach most commonly used today, but principles of independent competition, cooperation, negotiation, enforcement, and so on, may come into play in the agent interactions of the future.

9.12 Future Agent Builders

In this subsection what we mean by agent builders is the interface the end user will see and use that allows agent configuration, launch, and monitoring. This is not the same kind of builder as the agent development environments that agent developers will use to create the agents in the first place. Indeed, the end-user builder will also be created, if necessary, by the agent developers.

For the end user, developers will provide many ways for the user to interact with agents—to configure them for tasks, give them access to resources (or place limits on them), and set up monitoring facilities. The agents will be trainable, programmable by example, graphically programmable, or programmable via goal statements or even natural language. Although languages such as Smalltalk, Telescript, and Java will be important, these will only interest the software developer community. Agent technology will not become widespread until end users can configure their own agents with relative ease, and be assured of their safe and effective execution.

For example, if a user wants to configure his or her own insurance purchasing agent, the user interface must be relatively simple and intuitive. A form, filled out with the help of a "wizard" or intelligent interface agent, could be ideal. Then the purchasing agent could be given the task of going out over the net and finding a policy that fits the criteria on the form, returning with pricing alternatives from several companies.

The last things future agent users will want to worry about are the syntax of a language, debugging algorithms, or the details of network protocols. The user will then simply want the agent to perform its tasks effectively and without negative consequences.

All we can effectively do today is develop single-purpose agents. But in the future, agents will be called upon to perform a more general range of functions. Today, for example, we can build stationary Internet or database search agents. The future will demand sophisticated information agents that can search content from many sources, intelligently filtering and formatting the information in a variety of ways. End-user agent builders will probably accomplish this through combining with other agents and infrastructure utilities, along with AI constructs. Future agent builders must therefore be not only more user-friendly, but much more intelligent and flexible as well.

9.13 Social Issues Pertaining to Agent Technology

Human agents who work primarily as information brokers, from bank tellers to travel agents to newscasters, will eventually need to rethink their professions. Why would you sit and listen to a news anchor read a few news stories at 6 P.M. when your agents could filter a number of news feeds and give you only the stories (condensed and integrated if desired) in which you had an interest, long before the six-o'clock broadcast. In fact, the bases for such services are now being laid. Providers like PointCast allow you to customize a news feed to your desktop. It is only a matter of time until some entrepreneur will provide a similar facility, but allowing *you* to choose from an array of sources, not just those contracted by the PointCast-like provider, and sans the advertising.

Why would you go into a bank to do business when your agent can negotiate the best interest rates with hundreds of banks, and take care of all your banking business on-line? Why would you give a human travel agent a commission when there are dozens of good on-line travel services your software agents can work with to design the perfect vacation or business trip? Why would you ever set foot inside a grocery, hardware, or department store when your agents can always find just what you need at wholesale prices—and have it delivered to your home?

To survive, human agents will have to offer value-added features your agents can't match, such as insightful and entertaining local commentaries on the news, highly personalized service, specialized inventories, greater product expertise, and package deals that beat what your agent can find. These human agents will also need to get software agents of their own to find out what their customers want, get it faster and cheaper, and advertise the fact.

Certainly, the best and brightest human agents will survive, as will others who find out how to specialize in service niches and combinations where software agents can't compete. Sadly, this could drive human-delivered news even farther toward entertainment, violence, gossip, and sensationalism.

Agents will not only be able to search the world to find the lowest price, but could access unbiased product reviews to find the best buys. To combat this, retail stores will need to develop drawing power beyond brand or name loyalty, and come up with sales techniques more effective than the standard high-pressure pitchman. To stay in business, retailers will need to present their wares in elaborate three-dimensional virtual stores on the web and offer free computer art and software downloads to get customers interested. Retailers will also need to pay browsers and search services to funnel business into them. Manufacturers will advertise and sell their wares at gigantic on-line markets. If they can't or won't participate in the markets, they'll lose business to the competition.

It is in such a world of on-line commerce that Telescript was built to thrive. Telescript agents are designed to go to defined "places" and conduct business on behalf of a user. Right now, on-line agents are relatively difficult to create and debug. Before such agents become popular with the average web user, truly graphical agent builders will need to come on-line. Such builders will never be as flexible as a programming language, but in the future, the best agent development systems will support both approaches.

In the case of television channels, competition will become brutal when most viewers come to rely on agents to filter through thousands of channel offerings in lieu of channel surfing. The Nielsen ratings system will be strangely skewed when agents can monitor all channels at once and present viewers with only those segments of programs that interest them, such as a single local news story from three different points of view, the most exciting parts of ten different sporting events, or the last half of a movie you missed earlier. All of these could be stored automatically and viewed at any time.

Of course, your agents could cut out every advertisement, switching you to another interesting channel or stored program during commercial breaks (also true of on-line magazines and newspapers). This will force advertisers to find more innovative ways of integrating their product pitches in with entertainment and news programs.

9.14 Replacing Humans with Software

Automating activities performed by humans is the raison d'être of computers. Replacing those humans with computers is the nasty side effect, a trend that has been accelerating over the last 25 years. In our opinion, humans that now provide information-based services are just like the product middleman, who is the intermediate between you, the consumer, and the manufacturer of the product. Several of these service providers can be replaced by IAs.

Travel agents already face major consequences from on-line travel agencies. Cyber travel agents will need to take into account hundreds of variables, and have access to a worldwide travel, news, and weather data network. They will have to take into account the tradeoffs between transportation alternatives, seasonal factors, discounts offered, visa requirements, special customer needs and interests, cost limitations, strengths and weaknesses of various travel destinations, and a horde of other factors. We may never be able to provide a "general" travel agent, but a cadre of specialists may work nearly as well, with a little human judgment thrown into the equation.

Agents are distributors, usually of information and expertise, but "you hardly need a real estate agent if you can call up on the screen a fully indexed and illustrated list of all your city's homes for sale" (Case and Useem, 1996). If the typical real-estate buy can configure an agent with the specifications for a desired home (such as location, price, and so on), the agent can go and get all of the properties that match these specifications. That is what a human agent does now by interacting with, typically, the Multiple Listing Service (MLS). But why have the human agent interact with MLS when a cyberagent could do just as well, but more efficiently, and certainly with less cost. Of course, you will have to drive yourself to the property and arrange to get inside for a look, that is, until C3PO happens upon the scene.

Professionals such as loan officers, whose services can be reduced largely to step-by-step procedures, will be impacted by work-flow agents, assistant agents, and agent-based on-line professional services.

Even professionals such as doctors and attorneys will see their careers changed. On-line agents can provide medical information to medical professionals, including the latest advances. These on-line agents could assist with diagnoses and treatment recommendations. Eventually, medically intelligent agents may diagnose and triage patients, referring them to medical professionals as warranted. Agents can also monitor patients constantly—in the hospital or in home-based care systems—and report to nurses on duty. Agents will perform natural-language queries for legal professionals such as attorneys and judges, raising the timeliness and thoroughness of legal research to new levels. On-line legal agents will someday provide direct counsel to clients, and refer them to legal professionals as needed.

Why do we need news reporters? If you need to do anything from monitoring and filtering news sources now we go searching the Internet for information. In the future it will come to us. You will be offered (perhaps on a weekly basis) the trial use of agent-based information filtering services, such as creating a personalized weekly magazine with articles of interest gleaned (and translated) from news sources worldwide.

When a medical diagnosis has to be performed, agents will probably be available to do the job. On-line agent markets will offer thousands of kinds of agents or software robots. These will be everything from vacation planners to auto-parts buyers, customizable to your specific needs. Many of these will probably be offered on a free-trial basis. For example, they will execute in a limited form or for a limited time. Then the user can choose whether or not to purchase the agent and the accompanying agent service. Of course, purchases or subscriptions to agents will be facilitated via electronic money, which will certainly be available from a variety of companies, including American Express, Visa, and MasterCard.

In the realm of Hollywood, studios are already turning to digital gurus to make ever more accurate renditions not only of backdrops, special effects, and impossible creatures, but of human actors and actresses as well. Soon the first feature film will be produced with a cast assembled by *and in* a computer—and we're not talking animation here. Will these digital "Gibsons, Stallones, and Closes" *have* a digital or a real "agent?" Talk about recursion.

9.15 The Global Desktop

When the network computer becomes a reality (the network *is* the computer), the important factor will not be how powerful your workstation is, but how well you connect to the Internet. The whole operating system will be built around providing net access, and you won't have just a dozen or so server choices—you'll have thousands. Agents could play a key role here, since your system configuration could change daily and something needs to keep track of what's available on all servers, how loaded each is over time, what the alternative servers are, and so on.

Agents will automatically connect you by resource provided, not simply by IP address. To improve performance, agents can buffer messages and interact only during times when a server is not as busy. In addition, agents would automatically execute applications and data requests in the most advantageous place—on the server containing the resource (that is, a database) if possible. In addition, other sites will request services and data from yours, and your agents will act as a buffering and filtering system for sharing information. Instead of drive letters such as "C:," you'll have resource icons such as "top ten stock quotes," "nationwide weather," and "local news." When you click one of these, agents will take it from there.

9.16 Agents at Home

An IA is something that is delegated work. One day, agents will care for us in our very homes in every imaginable way. For example, by being part of embedded systems and LANs (for example, like LON, CEBus, and so on) they could:

- Tell you when someone has a sale on Nike tennis shoes
- Help you plan, order, and prepare your family's meals
- Reduce 500 TV channels into one personal channel that you control
- Automatically order your office supply items at the best price
- Make coffee in the morning at 7 A.M. on weekdays and at 8 A.M. on weekends

These may sound like rather trivial assistance, considering the many complex problems civilized human beings face, but let's look at the future possibilities IAs hold for helping humans in personal ways.

9.16.1 A day in the life of an agent-enhanced human

Remember Apple's Knowledge Navigator video of several years ago that featured a futuristic terminal in a home office, with a virtual human—a digital butler if you will—on the screen that conversed with the home owner? With the application of appropriate AI technology, infrastructures, and high-bandwidth net access, the scenario envisioned in that video is not too far off.

When you wake up in the morning, your wrist-based agent informs you of your calendar for the day. Instead of a fat newspaper, a stack of magazines, and ten information channels to sift through, you get only the news you want—specific stock quotes, news and features pertaining to topics in which you've expressed an interest, and advertisements for specified types of products. In communication with observing satellites, your car can not only determine its status, but map out the best route for you to drive to work.

At work your agents keep you informed with summaries of all information—taken from thousands of sources in which you've expressed interest—on what your competitors are doing, what your customers are saying, new research applicable to your job, and so on. Agents will schedule meetings for you, outline your daily tasks, handle all your phone and E-mail messages, prioritizing, scheduling, and coordinating your life.

Your agents can also arrange all travel, and find experts to help you in your work, while monitoring problems encountered by those employees who report to you. The agent handles the minutiae and provides up-to-the-second information. Your job is to create the products and services your customers need, and make intelligent decisions about your customers and competitors. Decision support agents are a key here.

Agents will also negotiate with other agents to get the best price based on current information. Agents can purchase everything from downloadable music and videos to groceries and clothing. Agents can analyze the information and entertainment channels to which you have access, keep tabs on your nutritional, medical, and financial conditions, and act as advisors in each of those domains.

9.17 Agents: The Dark Side

Agents have the potential to improve lives. They can keep us informed and entertained as never before. They can help us get organized, be safer and healthier, make better decisions, and be more creative, freeing us from mundane minutiae.

On the other hand, rogue or maliciously programmed agents can make the worst viruses we've seen look pretty tame. To prevent some hacker's agent from coming into your home and fiddling with your thermostat, set-top box, or, heaven forbid, your toaster, the future of computing and communications must provide a range of safeguards to the average homeowner. The agent infrastructure must provide security such that any incoming software entity must register with a security agent. This security agent checks to see whether the incoming entity was requested by the home platform or owner (whether direct or scheduled). It ensures that it contains no viruses and that its executable software behaves itself and does only what it is contracted to do while on the premises maybe by "wrapping" it in a security "blanket."

9.17.1 Future agent security

Users must feel that their data and privacy are secure from malicious agents, if agents are ever to become a popular way of doing business. As explained in Sections 6 and 7, agents are autonomous processes, and have the authority to act on behalf of a user. Agents may also be mobile, executing on more than one platform. Due to these factors, sophisticated precautions must be taken to make sure agents don't run amok.

As humans begin to trust agent technology, agent builders become user-friendly, and the infrastructures to execute agents become ubiquitous. Not only mere viruses but both intentionally malicious agents and wild rogue agents will become security threats. The future will bring not only intelligent servants, but intelligent spies and vandals as well. Infrastructures where agents execute must therefore be controlled and monitored similarly to the ways corporate information systems are today, including data protection and process validation.

While security will need to be stepped up, all of the security logic will need to be contained within the builder (end-user agent configurator or user-friendly agent creation and dispatch environments) and the infrastructure, and be transparent to the end user. The builder or end-user agent configurator must be restrictive enough to disallow illegal agent configurations, yet flexible enough to support creating a wide range of agents. The infrastructure must check the agent for illegal instructions before executing it, and stop the agent from performing any detrimental action it may try to do in spite of the initial verification.

9.17.2 Privacy

In a world of distributed agents, privacy will become an even bigger issue than it is now. If you monitor the news groups to which a user posts, and what he or she says in the news groups, you can put together a lot of details about the user's life. As agents monitor users, they return the raw data for analysis—user address, employment, fax and phone numbers, E-mail address, occupation, marital status, children's names, even credit card numbers.

Sleuth agents, a software analog of a detective, can put together a file on any user. Does he or she access pornographic material? Religion? Politics? In addition, with ever-growing on-line commerce, agents can monitor buying habits and piece together a fairly complete picture of a user, including annual salary, where he or she banks and invests, and so on. Agents with access to an ever-widening pool of information can develop ever more detailed (and therefore possibly one-sided, inaccurate, or downright unflattering) profiles of people.

9.17.3 You are your agent(s) (at least in cyberspace)

In the future people will increasingly rely on agents to represent them in various forums, venues, and contexts. Even now, so-called avatars are representations of users in specific cyberscenarios (such as game playing). Of course, as we ask (via programming, scripting, configuring, and teaching) agents to carry out tasks for us, they will carry a slice, or aspect, of us into cyberspace with them.

Lanier (1996) fears this prospect, because the agent's representation of us is so much less than what we really are. He further posits that people will be disempowered as agents proliferate and make more and more decisions for us. He suggests that agents will allow more programmers to become lazy and to design "quirky" software, because they could always claim that the programs are "agents" and therefore autonomous (autonomous being associated with quirky). Much of this nay-saying stems from Lanier's, and others', downplaying the current significance of advances in AI.

We propose a more balanced outlook on the future of agents. Clearly, although AI has not lived up to the rosiest past prognostications of its most fervent believers, neither is it a dismal failure. In fact, there are many aspects of current AI in practice and in development that point to its use in more and more applications, and that includes IAs. Indeed, we feel that increasing use of AI techniques will only increase the efficacy of IAs. We also do not feel any angst over aspects of people swimming in cyberspace; that has been happening for decades with the telephone, and even for centuries with letters. Do people really think that a letter or two, or a couple of phone calls with another person indicate, with any accuracy or completeness, the persona of the other person? Yes, you might get a certain impression in a first or second contact, as you will by "meeting" another person through his or her agent.

But humans, as social creatures, have always been embedded in situations in which we progressively learn more and more as the situation unfolds. It will be no different in the future in this respect. As people refine their agents and use them more and more to interact with both other agents and other humans, those others will see as much of us as we allow—just as in other interpersonal modes.

Of course, there is the danger that as agents carry more information about us into an increasingly accessible cyberspace, nefarious entities, both human and IA-based, will waylay our cyberselves for any number of purposes. Sellers will try to gain advantage of us by knowing as much as possible about our desires and finances. Cyberrogues will attempt to hack our agents or put them to their own use. This could go far beyond illegally obtaining credit card numbers. Consider all the sensitive data your financial or health advisory agent would have about you. Consider the reasoning power of a typical future agent hacked or surreptitiously reprogrammed. Someone with a grudge could make a royal mess of your life through your agents.

And of course, Big Brother, in the form of the governmental-industrial complex, will request, and probably be granted, access to our agents under the guise of national security or fighting crime. (Witness the current fight over the government's attempt to listen in to our cyberlife via the V-chip and their usurpation of encryption technology.)

As we encode more of ourselves in the bits and bytes of cyberspace, more opportunities for intrusion into our lives will exist. But as Lanier suggests, if we go into this new realm with eyes open and insist on taking part in all aspects of our cyberlives, we can protect our cyberselves and, indeed, even gain more empowerment over those bits and bytes. With users and developers working together, we believe IAs can both enhance and protect both our real lives and our cyberlives.

9.18 Inventing the Future of Agents at MIT: Work at Software Agents Group, MIT Media Laboratory

A lot of what we will see in the future concerning the proliferation of IAs will come from the commercialization of ideas originating within the hallowed halls of academia. We would now like to give you a flavor of some of the agent-related research going on. One place that has a great deal of notoriety in the IA community is the Software Agents Group at MIT's Media Laboratory. The next several subsections are snippets of work on IAs and related topics ongoing there. The lead researchers on each project are listed after the title.*

We are indebted to Professor Pattie Maes and associates for their kind permission, allowing us to include the following list of projects that relate to the future of IA technology.

9.18.1 Modeling intelligent autonomous agents (Professor Pattie Maes)

An autonomous agent is a computational system that inhabits a complex, dynamic environment. The agent can sense and act on its environment, and has a set of goals or motivations that it tries to achieve through these actions. Depending on the type of environment, an agent can take different forms.

Autonomous robots are agents that inhabit the physical world, computer-animated characters inhabit simulated three-dimensional worlds, while software agents inhabit the world of computer networks. The same basic questions

---

*Navigate to http://casr.www.media.mit.edu/groups/casr/maes.html or to http://lcs.www.media.mit.edu/groups/agents/research.html for more information on agent research at MIT.

are studied in these different domains: How does an agent decide what to do so as to progress toward its goals; how does an agent learn from experience; and how can multiple agents collaborate? Our research team develops new techniques and algorithms to address these issues.

9.18.2 Computational model of emotion for autonomous agents (Professor Pattie Maes and Juan David Velasquez)

This project focuses on designing computational models of how emotions are generated and how they affect the behavior of autonomous agents. By drawing on ideas from several different fields, we have developed Cathexis, a distributed computational model, which offers an alternative approach to model the dynamic nature of different affective phenomena, such as emotions and moods, and provides a flexible way of modeling their influence on the behavior of the agent.

The model has been implemented as part of an extensible, OO framework, which provides enough functionality for agent developers to design emotional agents that can be used in a variety of applications, including entertainment (such as synthetic agents for interactive drama, video games, and so on), education, and human-computer interfaces.

9.18.3 Software agents (Professor Pattie Maes)

Software agents provide active, personalized assistance to a person engaged in the use of a particular computer application. Software agents differ from current-day software in that they are (1) proactive—taking the initiative to help the user by making suggestions and/or automating the more mundane tasks the user normally would have to perform; (2) adaptive—learning the user's preferences, habits, and interests as they change over time; and (3) personalized—customizing their assistance according to what they have learned about the user. The project focuses on generic techniques for building interface agents, and has produced agents for a wide variety of applications described in the following subsections.

9.18.4 Agents that reduce information overload (Professor Pattie Maes and Alexandros Moukas)

This project attempts to deal with the problem of information overload. We are building software agents that make personalized suggestions to a user for items the user may want to select (news articles, videos, music, television shows, restaurants, and so forth). The project employs two different techniques: content-based filtering and collaborative filtering. The former technique is used to detect patterns among the items liked or disliked, based on keywords and other features of the items.

The second technique is used to detect patterns among different users and to make recommendations to people, based on others who have shown similar tastes. It essentially automates the process of "word of mouth" to produce an advanced, personalized marketing scheme. In addition, we are exploring the use of techniques from game theory, market models, and artificial life. We have built filtering agents for applications such as news, music recommendations, book recommendations, and recommendations for WWW documents.

9.18.5 Amalthaea—a multiagent system that discovers, monitors, and filters information resources (Professor Pattie Maes and Alexandros Moukas)

The exponential increase of computer systems that are interconnected in on-line networks has resulted in a corresponding increase in the amount of information available on-line and distributed among heterogeneous sources. Users typically have many overlapping interests, and information relevant to those interests can be found in many different sources. Instead of constructing a large, complex agent that will have to solve the whole problem, we are creating a society of smaller and simpler specialized agents that try to solve the problem collectively.

We are introducing an artificial ecosystem of evolving information filtering and discovery agents that cooperate and compete in a marketlike environment. The system adapts to the user's interests, which may change over time, while proactively exploring new domains (in the WWW) that may be of interest to the user.

9.18.6 Yenta—matchmaking agents (Professor Pattie Maes and Leonard Foner)

This project is developing a software agent that finds people who have never met, but share similar interests, and introduces them to each other. Such introductions can automatically form interest groups and coalitions, and can be used to locate someone knowledgeable in a particular area.

Each participating user runs a copy of the agent, and these individual copies find each other as appropriate on the network and begin the introduction process. The project is an experiment in creating a decentralized, fault-tolerant application that handles potentially sensitive information (such as people's mail, their personal files, or lists of their particular interests) in a responsible and privacy-protecting fashion, using cryptographic and other techniques. The eventual goal is ubiquitous deployment across the Internet.

9.18.7 Remembrance agents (Professor Pattie Maes, Bradley Rhodes, and Thad Starner)

People have notoriously bad memories. We forget where we left our keys, people's names, and simple directions. Computers, on the other hand, almost never forget the information they store. This project is creating agents that help augment human memory. These agents log everything a user does and all the information that passes through the user's hands and can help a user remember some information based on the content or the context (episodic memory) of the situation. We are exploring remembrance agents that can run on desktop computers for work-related applications as well as agents running on wearable computers for daily life.

9.18.8 Using simulated evolution to create adaptive systems (Professor Pattie Maes and Michael Johnson)

Inspired by Darwinian natural selection, genetic programming (GP) is an emerging paradigm that uses simulated evolution to breed computer programs that solve a particular problem. One problem with GP is that it only works on small problems. One current research project involves a distributed model for GP that encourages the formation of species of computer programs.

Each species specializes in a small part of a hard problem. Collectively, these species are able to automatically break down a hard problem into smaller subtasks and solve them with little human intervention. These techniques are applied in a range of agent-related applications, including computer vision problems, computer graphics and animation, and software agents for resource allocation and information overload.

9.18.9 Anthropomorphizing software agents (Professor Pattie Maes and Alan Wexelblat)

It is still an open question whether software agents should be anthropomorphized in the interface. This project explores the pros and cons of software agents with faces, and studies how facial expressions and characteristics can be used in an effective way. In particular, we are exploring an experimental approach to test how the faces can be used to convey information about the internal state of an agent and to make the agent more usable, trustworthy, and likable. This work builds on the research of former graduate student Tomoko Koda.

9.18.10 Browsing large information spaces—emergent structure from collective action (Professor Pattie Maes and Alan Wexelblat)

Present technologies provide no way to ease the browsing of a heterogeneous, unstructured collection of information, such as the WWW. We are investigating ways of analyzing the activity and navigation patterns of thousands of users to develop more structure in a prototype system called Footprints.

In addition to inferring structure based on usage patterns, we are also investigating methods to allow users to explicitly add and change structure. Together the methods developed allow users to benefit from the problem solving, searching, and sense making performed by previous users of the same information.

9.18.11 Kasbah—an agent marketplace for buying and selling goods (Professor Pattie Maes, Anthony Chavez, and Robert Guttman)

The goal of the Kasbah system is to help realize a fundamental transformation in the way people transact goods—from requiring constant monitoring and effort, to a system where a software agent does most of the work on the user's behalf. A user wanting to buy or sell a good creates an agent, gives it some strategic direction, and sends it off to the agent marketplace.

The Kasbah agents proactively seek out potential buyers or sellers and negotiate with them on their creators' behalf. Each agent's goal is to make the "best deal" possible, subject to a set of user-specified constraints, such as a desired price, a highest (or lowest) acceptable price, and a date by which to sell (or buy). We are investigating the application of various AI techniques to building agents that are intelligent enough to perform well in a complex, dynamic marketplace. A key feature of Kasbah is that it is open to adding new types of agents, using different selling strategies.

9.18.12 ALIVE—artificial life interactive video environment (Professors Bruce Blumberg, Pattie Maes, and Alex Pentland)

To date the cumbersome nature of the equipment and the limited nature of the interaction have limited the range of applications of virtual environments. ALIVE is a novel system that allows wireless, full-body interaction between a human and a rich graphical world inhabited by autonomous agents. The ALIVE system provides more complex and very different experiences than traditional virtual reality systems.

In particular, we are exploring novel applications in the areas of training and teaching, entertainment, and last, but not least, digital assistants or interface agents. Current ALIVE worlds that the user can experience include a virtual dog with whom the user can play, and video-game creatures with whom the user can interact.

9.18.13 Modeling synthetic characters for games and interactive storytelling (Professor Bruce Blumberg, Professor Pattie Maes, Bradley Rhodes, and Michael Johnson)

Current video games and interactive story systems are limited to extremely narrow, simple computer-controlled opponents and characters. This project applies techniques from a range of disciplines (AI, artificial life, animation, literature, and theater) in developing autonomous, interactive, synthetic characters for virtual environments.

In particular, we explore how to make such characters act and react in lifelike ways, how to make them learn from experience, how to give them personality and emotions, and, in general, how to make them entertaining and engaging to a user. The resulting characters are used in games and interactive storytelling applications—story systems that dynamically adapt to a user's inputs. For example, we have used the ALIVE virtual environment to allow a user to enact one character in a computer-animated story in which the other characters are synthetic.

9.19 Miscellaneous Agent-Related Projects

The following subsections comprise a list of research projects involving agent technology.*

9.19.1 Intelligent browsing agents

"Human Computer Interaction—Intelligent Browsing Agents" is an ARPA-sponsored project at the Open Software Foundation whose goals include making the WWW infrastructure agent-ready and agent-aware and providing an extensible set of agents that access information sources, network, and user knowledge in the service of user-related objectives. (For more information, navigate to http://www.osf.org/ri/QuadCharts/HCI.QuadChart.frame.html.)

9.19.2 Persona project

The Persona project at Microsoft Research is developing the technologies required to produce conversational assistants—lifelike animated characters that interact with a user in a natural spoken dialogue. (For more information, navigate to http://www.research.microsoft.com/research/ui/persona/home.html.)

9.19.3 On-line cooperating agent architecture

ARCHON (architecture for cooperative heterogeneous on-line systems), Europe's largest DAI project, devised a general-purpose architecture, software framework, and methodology used in a number of real-world industrial domains. (For more information, navigate to http://www.elec.qmw.ac.uk/dai/archon/test_1.html.)

9.19.4 Agents and ontologies

KACTUS is a European ESPRIT-iii project aiming at the development of a methodology for the reuse of knowledge about technical systems during their

---

*See also the http://www.cs.umbc.edu/agents/agents.shtml See the "Example Agents" page for a list of agents you can try using via the web.

life cycle (that is, using the same knowledge base for design, diagnosis, operation, maintenance, redesign, instruction, and so on). The way this will be achieved is by giving these knowledge bases an explicit structure (often called an ontology). (For more information, navigate to http://www.swi.psy.uva.nl/projects/Kactus/home.html.)

9.19.5 Agent architecture

The SRI Open Agent Architecture project is developing an open-agent architecture and accompanying user interface for networked desktop and handheld machines. (For more information, navigate to http://www.ai.sri.com/~cheyer/oaa.html.)

9.19.6 Operating system support for agents

The TACOMA project at the University of Tromsø, Norway, and at Cornell focuses on operating-system support for agents and how agents can be used to solve problems traditionally addressed by operating systems. We have implemented prototype systems to support agents using UNIX and Tcl/Tk on top of Horus. (They take an agent to be a process that may migrate through a computer network in order to satisfy requests made by clients.) (For more information, navigate to http://www.cs.uit.no/DOS/Tacoma/.)

9.19.7 Internet search agent

Silk is a new Internet resource discovery client. Silk, Bunyip's new Internet client program, allows you to access and search for Internet information without having to know where to look or how to look. It is based on the notion of a uniform resource agent (URA). (For more information, navigate to http://services.bunyip.com:8000/products/silk/.)

9.19.8 The ARPA Intelligent Integration of Information (I**3) project

The vision for the I**3 program is to provide easy access to information—in the form needed by end users and high-level applications—by extracting, integrating, and abstracting information from the growing morass of available data. (For more information, navigate to http://haifa.isx.com:80/pub/I3/.)

9.19.9 Guardian: a prototype intelligent agent for monitoring intensive-care and other medical patients

This Stanford project involves dynamic run-time configuration of appropriate reasoning methods and knowledge components, context-relevant selective perception, and situation assessment, planning, and plan monitoring in a dynamic environment. It uses the BB1 software architecture.

9.19.10 Intelligent, ethical agents

IWAH is a project researching the design and implementation of intelligent, ethical agents for the WWW at the University of Houston, Clear Lake. (For more information, navigate to http://rbse.jsc.nasa.gov:80/agents/Intelligent WebAgents/Houston.)

9.19.11 Mail agent

MailBot is "an intelligent rule-based message processing agent for Microsoft Mail that can sort and filter messages and automatically respond to E-mail. (For more information, navigate to http://www.polaris.net/~daxtron/mailbot.html.)

9.19.12 Knowledge-based agents

Distributed communicating agents (DCAs) are used in a project called Carnot. MCC's Carnot project uses distributed, knowledge-based communicating agents. (For more information, navigate to http://galaxy.einet.net/MCC/Carnot/DCA.html.)

9.19.13 Agent collaboration languages

The Agent Collaboration Language project is a joint effort of the U.S. Army Construction Engineering Research Laboratories, CMU, Stanford University, MIT, and the University of Illinois at Urbana-Champaign to conduct research into language standards for collaboration between heterogeneous groups of software applications. (For more information, navigate to http://raven.cecer.army.mil/acl/welcome.html.)

9.19.14 Sulla—a user agent for the web

NASA is constructing a prototype user agent with the following characteristics:

- The ability to acquire and retain an interest profile of its user and act upon one or more goals based on that profile
- The ability to act autonomously, pursuing the goals posed to it by its user, irrespective of whether the user is connected to the system where the agent is based
- The ability to apprise its user of progress toward outstanding goals and to present preliminary results
- The ability to access a variety of information sources, both via direct access to those sources [such as HTML documents, FTP files, WAIS (Wide Area Information Server) databases, articles posted to news groups, and so on] and those referenced by service agents
- The ability to act ethically, exemplified by the guidelines proposed in a WWW Fall 1994 paper in particular; moderation in the acquisition of information during the satisfaction of a goal

(For more information, navigate to http://rbse.jsc.nasa.gov/eichmann/www-f94/ethics/ethics.html.)

9.20 The Future of AI = The Future of IAs

Although hundreds if not thousands of books and papers have been written about whether machines will ever be truly intelligent or will posses consciousness, we'd like to comment here briefly. Of course, we can't do all of the arguments and subtle philosophical debates justice here, but since we are talking about the future in this section, let's summarize some arguments and speculate on the future of IAs with respect to intelligence and consciousness.

Whether intelligent machines are possible can be restated in terms of two positions, quite familiar to those following the AI and other cognitive sciences. The first position, referred to as the *weak AI* position, states that machines (and by inference IA "machines" as both hardware and software) can be made to act as if they were intelligent. This line of argument is closely aligned with the concept of the Turing test discussed in Section 2. Clearly, in certain, now-limited domains (but in the future, maybe not so limited), intelligent machines exist. Although a robust, general-purpose intelligent machine has yet to be built, we think it is just a matter of time (perhaps less than 25 years) before such machines are built. Consider the current work on huge knowledge bases (such as Cyc and the rule base embedded in the chess-playing computer, Deep Blue), emulation of brainlike mechanisms in both hardware and software (neural networks, fuzzy logic), and advances in sheer computation complexity and speed. Based on these advances, it is not a stretch to imagine very sophisticated machines individually, or collectively via their interaction, giving rise to a close semblance of what most people would deem truly intelligent.

The second position is called the *strong AI* position, and it professes that intelligent machines have minds in the same sense that we humans have minds. So given that IAs depend on, among other things, AI, and given our (and others') position that truly intelligent machines will be built, the real important questions is: will IAs have minds in the sense that we humans have minds? This is a much deeper and much more significant question.

Philosophically, much discussion about what we, as conscious entities, are, and whether we can create entities "like" ourselves (intelligent and conscious), centers on the connection between mind and brain. Brain is the physical and mind the nonphysical "software" executing within it, in the view known as dualism.* Of course, the notion that mind *is* what the brain does is a counterpoint that does not depend on any nonphysical entity. But given that we can eventually construct, down to the finest detail, simple neurons (wetware or sil-

---

*There are subclasses of dualism, such as property dualism, but the basic notion mentioned is sufficient for our purposes here.

iconware, or some combination thereof), and get those "neurons" to interact, by what basis could we claim we have created a mind?

Bateson (1979) has proposed several mind-defining criteria, which, if satisfied by an entity, would lead us to conclude that that entity has a mind. These criteria are:

- Mind is a collection of interacting components.
- The interaction between the components is triggered by difference—a non-space-time negentropic or entropic occurrence.
- Mentation requires energy.
- Mentation requires complex chains of determination.
- Differences produce transforms according to rules, which themselves are stable but also transformable.
- Transformations produce a hierarchy of logical types that are immanent in mind.

Certainly each of these criteria is satisfied by a collection of real neurons in our real brains. But all of that seems very mechanistic in nature. What of emotion, desires, intentionality? What of our intuitive feeling of the "I" that is me? Is it possible that all of these human-inspired attributes of mind can be built from the same kinds of mechanisms that produce and satisfy Bateson's criteria? Can even the "I" be explained (perhaps explained away) by the argument that it is a convenient fiction produced by the many interacting, competing subagents within the real or artificial brain or mind?

The mind as an emergent property of a certain arrangement of neurons (real or artificial) is a position that finds favor with some proponents of the strong AI camp. Even if we do not discover or understand what consciousness *is,* if some entity has a sufficiently complex arrangement of "parts" and exhibits the requisite behavior, some strong AI'ers would claim such an entity has a mind. This would also comport with Bateson's view of mind.

Seemingly intelligent phenomena emerge from other systems as well. Systems which are in disequilibrium, chaotic, and that have a certain degree of complexity beget a synergistic emergence. Is the resultant "intelligence" aware of its own existence as we are? Not in the systems studied so far. But given an appropriate "awareness" algorithm, perhaps a sufficiently complex collaboration of IAs might be said to be aware, in the sense that parts communicate with other parts, parts sense an environment, and a hierarchy of ever complete knowledge of self and other arises from architectures supporting elaborate communications.

Consider the future of the predominant architectures for intelligent computing entities. There exist many distributed entities, incorporating task- or domain-specific reasoning capabilities and task-or domain-specific knowledge. These entities are arranged in hierarchies of cooperating subunits or functional units, synergistically contributing to yet higher-level entities having more sophisticated and progressively higher-level or global goals. All of these entities are communicating over essentially unlimited bandwidth networks, able to relocate to make efficient use of computational resources. And if some problem is initially too difficult, the individual entities as well as the whole can recruit other resources as needed, or can learn and evolve as necessary to attack the problem with different methods, or, using real-time AI, decide from moment to moment, what aspects of the problem need to be worked on first, maximizing expected utility each instant.

Such a future is certainly within sight, possibly within the next 25 years or so. *If* such a future comes to pass, we feel that the weak AI position will be fulfilled, for the age of truly intelligent machines will be upon us. But we do not feel that our notions of what it means to have a mind or to have consciousness are sufficiently sophisticated (yet) to warrant that these intelligent machines are mental beings in the same sense that we are mental beings. In contrast to AI and computing, the cognitive neurosciences are in their relative infancy. Indeed, our brains alone are composed of not just interacting neurons, but a veritable soup of hormones, neurotransmitters, electrochemicals, and, let's not forget, all the stuff of physics and the atomic and quantum levels of reality. Since we can't say what mind or consciousness is (indeed, we haven't even found the ultimate "stuff" yet—so how can we know that it does or does not contribute to what consciousness is), other than with surface operational (or behavioral) characteristics, we can't possibly say when some artificial entity has one. Of course that won't stop some in the strong AI camp from claiming that their newest creation is indeed both intelligent and a conscious being.

As we dream of a Data as a possible ultimate IA in our future, let's all reflect on what *that* artificial "artificial being" brought to our attention. Data's quest was to be human. In what senses was he human and in what senses was he not human? No doubt he was intelligent. There was controversy over whether he had a mind or consciousness deserving of equal treatment with humans. He admitted he lacked emotion. Maybe the question in the future, about how we will treat IAs, will hinge on whether we want to create a race of indentured servants, or beings in any sense deserving of legal and ethical considerations similar to ourselves.

9.21 Summary

From the current research as presented we can determine that some of the major trends are cooperative agent systems, personalized agents, increased intelligence, and independence.

We will see agents that are trainable, can go anywhere, and do almost anything for their masters. Many will function without masters, as they are spawned by automated systems. Agents will have a wide-reaching base for negotiations, interactions, sharing information, protecting assets, and so on.

Agents will change the way many workers, including professionals, go about their jobs. Agents will act human, or at least will be represented anthropomorphically in some aspects.

Agents will advise us, protect us, get us organized and informed. Agent-development and execution systems may have significantly more AI horsepower that they now do. Agent systems may form societies to support complex interactions.

Agents could also cause big problems to become significantly worse. While some agents try to make systems efficient, others are gumming up the works. Agents could overload resources, causing computing to grind to a halt. Without tight security, agents may also delete files or access confidential information. With an agent infrastructure in place such as Java or Telescript VMs, these agent-run-amok problems could be minimized to an acceptable level.

Data used to mean files of records. Then it meant databases. Now, increasingly, it means distributed information streams. These streams are too dispersed, diverse, and fast-moving to be handled by the old static programming paradigm. Distributed searching and filtering agents will continue to increase in power and ubiquity as the Internet continues its phenomenal growth, and as databases are distributed across multiple servers. Eventually systems of distributed, cooperative agents will become the only reasonable way to deal with complex information webs from enterprisewide intranets to the Internet.

Agents will change the way many people live. As inexpensive networked computers become integrated into homes, businesses, and even automobiles, tens of millions of computer users may come to rely on server-based agents to sort through thousands of information and entertainment channels, shop everything from cybermalls to on-line grocery stores, locate and download books, music, and videos, program household appliances, and perform many other tasks we can't even imagine now.

As IAs exhibit ever more intelligent behaviors such that some of us believe they have minds or are conscious, what sorts of issues will arise? Will agents have legal rights? Can an agent own property, or be liable for damages? How would it pay or earn a living? Could an agent procreate (remember Data and Lol)? Many of these issues are the subject of science fiction. But, as we move into the future, these issues will, in our opinion, be upon us.

Acronyms

ABP	Agent-based programming
ACK	Acknowledge
ACL	Agent communication language
AFIN	Agent-oriented flexible information network
AI	Artificial intelligence
AIAMA	*Artificial Intelligence: A Modern Approach* (Russell and Norvig, 1994)
AIM	Application interface model
ANN	Artificial neural network
ANS	Agent name server
API	Application programming interface
ARPA	Advanced research projects agency
ASDL	Asynchronous digital subscriber line
ASIC	Application-specific integrated circuit
ATM	Asynchronous transfer mode
C/S	Client/server
C-HI	Computer-human interaction
CBO	Coupling between objects
CDR	Common data representation
CGI	Common gateway interface; computer graphics interchange
CMIP	Common mail interface protocol
COBOL	Common business oriented language
COM	Common object model; common object management
CORBA	Common object request broker architecture
CPU	Central processing unit
CSH	Client-server handle
DBMS	Database management aystem
DCA	Distributed communicating agent
DCE	Distributed computing environment
DCOM	Distributed common object model
DDE	Dynamic Data Exchange

DEC	Digital Equipment Corp.
DFS	Distributed file services
DIT	Depth of inheritance tree
DLL	Dynamically linked library
DNA	Deoxyribonucleic acid
DOM	Distributed object management system
DSOM	Distributed system object model
DSP	Digital signal processor
DSS	Decision support system
DST	Distributed Smalltalk
DXF	Data exchange format
EDI	Electronic data interchange
EM	Evolution mechanism
ES	Expert system
FIFO	First in first out
FOPC	First-order predicate calculus
FPGA	Field-programmable gate array
FTP	File transfer protocol
FU	Function units
FUDR	Functional unit description record
GIF	Graphics interchange format
GP	Genetic programming
GIOP	General inter-ORB protocol
GUI	Graphical user interface
HP	Hewlett Packard
HTML	Hypertext markup language
HTTP	Hypertext transfer protocol
I/O	Input-output
IA	Intelligent agent
IAC	Interapplication communications
IBM	International Business Machines
IDL	Interface definition language
IEEE	Institute of Electrical and Electronics Engineers
IDE	Integrated Development Environment
IIOP	Internet interoperable ORB protocol
IN	Information network
IP	Internet protocol

IPC	Interprocess communication
IRC	Internet relay chat
ISO	International Standards Organization
ISV	Independent software vendor
ITX	Interactive transaction
JAAPI	Java agent API
JAT	Java agent template
JDBC	Java database connectivity
JIT	Just in time
JRMI	Java remote method invocation
KB	Knowledge base
KIF	Knowledge interchange format
KQML	Knowledge query and manipulation language
LAN	Local-area network
LCOM	Lack of cohesion in methods
LIFO	Last in first out
LISP	List programming
LON	Local operating network
MAPI	MAIL API
MCC	Microelectronics and Computer Technology Corporation
MEU	Maximum expected utility
MIDL	Microsoft interface definition language
MIPS	Million instructions per second
MIT	Massachusetts Institute of Technology
MS-RPC	Microsoft remote procedure call
MVC	Model-view controller
NEO	Networked objects
NIS	Network information service
NLP	Natural-language processing
NMA	Network management agent
NMS	Network management system
NNU	Neural network utility
NO-OP	No operation
NOC	Number of children
NOT	Number of tramps
ODBMS	Object database management system
ODL	Object definition language

OLAP	On-line analysis and processing
OLE	Object linking and embedding
OMA	Object management architecture
OMG	Object Management Group
OMT	Object modeling technique
OO	Object-oriented
OOA	Object-oriented architecture
OODBMS	Object-oriented database management system
OOP	OO programming
ORB	Object request broker
OS	Operating system
OSF	Open Software Foundation
OSI	Open system interconnection
OTS	Object transaction service
PBD	Programming by demonstration
PDA	Personal digital assistant
PDO	Portable distributed objects
PDF	Portable document format
PDL	Print description language
PDP	Parallel distributed processing
PGP	Pretty good privacy
PICT	Picture
PID	Proportional, integral, derivative
PIM	Personal information management
POSIX	Portable operating system interface
POTS	Plain old telephone service
RAM	Random-access memory
RDBMS	Relational database management system
RFC	Response for a class
ROM	Read-only memory
RP	Remote programming
RPC	Remote procedure call
RTOS	Real-time operating system
SA	Salutation architecture; subsumption architecture
SAAM	Software architecture analysis method
SCO	Santa Cruz Operation
SDR	Service description record

SGI	Silicon Graphics Inc.
SGML	Standard graphics markup language
SLM	Salutation manager
SMTP	Simple mail transport protocol
SN	Semantic network
SNMP	Simple network management protocol
SOM	System object model
SQL	Standard query language
SSL	Secure sockets layer
TCP	Transmission control protocol
TMCF	Task management common facility
UMBC	University of Maryland at Baltimore County
UML	Unified modeling language
UPS	Uninterruptible power supply
URA	Uniform research agent
URL	Universal resource locator
UUID	Universal unique identifier
VB	VisualBasic
VIP	Visual interactive programming
VM	Virtual machine
VOD	Violations of the law of Demeter
VRM	Virtual meeting room
VRML	Virtual reality markup language
WAC	Weighted attributes per class
WAIS	Wide area information server
WAN	Wide-area network
WMC	Weighted methods per class
WWW	World Wide Web

Bibliography

Abowd, G., J. Engelsma, L. Guadagno, and O. Okon: "Architectural Analysis of Object Request Brokers," *Object Mag.*, March 1996, pp. 44–51, 98.

"A Computational Market Model for Distributed Configuration Design," in *Proc. Nat. Conf. on Artificial Intelligence*, AAAI, August 1995, pp. 401–407; revised in *AI EDAM*, vol. 9, 1995, pp. 125–133.

Aleksander, I: *Neural Computing Architectures: The Design of Brain-Like Machines*, MIT Press, Cambridge, MA, 1989.

"A Market-Oriented Programming Environment and Its Application to Distributed Multicommodity Flow Problems," *J. Artificial Intelligence Research*, vol. 1, 1993, pp. 1–23.

Anderson, J. A., and E. Rosenfeld (eds.): *Neurocomputing: Foundations of Research*, MIT Press, Cambridge, MA, 1989.

Appleby and Steward: "Mobile Software Agents for Control in Telecommunications Networks," *Tech. Rep.*, Martlesham-Heath, UK, 1993.

Atkinson, B. and IBM Intelligent Agents Development Group: Web pages at URL: http://www.raleigh.ibm.com/iag/iahome.html, 1996.

Atkinson, B., et al.: "IBM Intelligent Agents," presented at Unicom Seminar on Agent Software, London, UK, May 25, 1995.

Ball, L.: Network Management with Smart Systems, McGraw-Hill, New York, 1994.

Barr, A., and E. Feigenbaum: *The Handbook of Artificial Intelligence*, vol. 1, William Kaufman, Reading, MA, 1981.

Bateson, G.: *Mind and Nature: A Necessary Unity*, Bantam, New York, 1979.

Baumgartner, P., and S. Payr (eds.): *Speaking Minds: Interviews with Twenty Eminent Cognitive Scientists*, Princeton University Press, Princeton, NJ, 1995.

Bentov, I.: *Stalking the Wild Pendulum: On the Mechanics of Consciousness*, Destiny Books, Rochester, VT, 1988.

Bernstein, P. A.: "Middleware: A Model for Distributed System Services," *Commun. ACM*, vol. 39, no. 2, 1996, pp. 86–98.

Berst, J.: "The Software Industry Is Starting to Wise Up," *PC Week*, vol. 11, no. 17, p. 134.

Bohm, D.: *Wholeness and the Implicate Order*, Routledge Kegan Paul, London, UK, 1980.

Booch, G.: *Object-Oriented Analysis and Design with Applications*, 2d ed., Benjamin/Cummings, Redwood City, CA, 1984.

Brando, T.: "Comparing CORBA and DCE," *Object Mag.*, March 1984, pp. 52–57.

Browning, J.: "Agents and Other Animals," *Scientific American*, February 28, 1996.

Bud, T.: *An Introduction to Object-Oriented Programming*, Addison-Wesley, Reading, MA, 1981.

Campbell, J.: *Grammatical Man: Information, Entropy, Language and Life*, Simon & Schuster, New York, 1982.

———: *Improbable Machine: What the Upheavals in AI Research Reveal about How the Mind Really Works*, Simon & Schuster, New York, 1989.

Case, J., and J. Useem: "Six Characters in Search of a Strategy," *Inc.*, March, 1996.

Chen, H., A. Houston, J. Nunamaker, and J. Yen: "Toward Intelligent Meeting Agents," *IEEE Computer*, August, pp. 62–70, 1996.

Churchland, P.: *NeuroPhilosophy: Toward a Unified Science of the Mind/Brain*, MIT Press, Cambridge, MA, 1993.

CiteWeb: "Web Pages for Fuzzy and Neural Research at URL: http://www.mitgmbh.de/mit/cite.html," 1996.

Clearwater, S. H. (ed.): "Market-Oriented Programming: Some Early Lessons. Market-Based Control: A Paradigm for Distributed Resource Allocation," *World Scientific*, 1995.

Coad, P., and J. Nicola: *Object-Oriented Programming,* Prentice-Hall, Englewood Cliffs, NJ, 1993.
"CORBA and NEO," *Byte,* January, 1996, p. 94.
Corcoran, C. "One Good Brew," *InfoWorld,* March 25, 1996, p. 65.
Cox, E.: "Fuzzy Logic: Where It's Been, Where It's Going," in *Hitchhiker's Guide to Artificial Intelligence,* Miller-Freeman, San Francisco, CA, 1995.
Craig, John C., *Visual Basic Workshop Version 3.0,* Microsoft Press, Redmond, WA, 1993.
Cyc KQML Project, Web pages at URL: http://www.cs.umbc.edu/, 1996.
Delbruck, M.: *Mind from Matter,* Blackwell Scientific, Palo Alto, CA, 1986.
Dennett, D.: *Content and Consciousness,* Routledge, London, 1969.
____: *Consciousness Explained,* Little, Brown, Boston, MA, 1996.
Dennett, D. C.: *Brainstorms: Philosophical Essays on Mind and Psychology,* MIT Press, Cambridge, MA, 1986.
Dewdney, A. K.: *Turing Omnibus: 61 Excursions in Computer Science,* Computer Science Press, Rockville, MD.
"Distributed Smalltalk," Web pages for ParcPlace-Digitalk at URL: http://www.parcplace.com/bod_prod.htm, 1996.
Drexler, E. K.: *Nanosystems: Molecular Machinery, Manufacturing and Computation,* Wiley, New York, 1992.
Dunbar, T.: "OpenDoc and OLE," *Byte,* January 1986, p. 90.
Edelman, G. M.: *The Remembered Present: A Biological Theory of Consciousness,* Basic, New York, 1989.
____: *Bright Air Brilliant Fire: On the Matter of the Mind,* Basic, New York, 1992.
Edmonds, E. A., L. Candy, R. Jones, and B. Soufi: "Support for Collaborative Design: Agents and Emergence," *Commun. ACM,* vol. 37, no. 7, 1994, pp. 41–47.
Eldredge, N.: *Unfinished Synthesis: Biological Hierarchies and Modern Evolutionary Thought,* Oxford University Press, New York, 1985.
Eliot, L., "Intelligent Agents are Watching You," *AI Expert,* August 1994, pp. 9–11.
Eskow, D.: "IBM's Worldwide Strategy for the World Wide Web," *NetReady Advisor,* Winter 1996.
Etzioni, O.: "Software Agents," *AI Mag.,* vol. 15, no. 3, 1994, p. 27.
____: and D. Weld: "A Softbot-Based Interface to the Internet," *Commun. ACM,* vol. 37, no. 7, 1994.
Fang, H.-L., P. Ross, and D. Corne: "A Promising Genetic Algorithm Approach to Job-Shop Scheduling, Rescheduling, and Open-Shop Scheduling Problems," in *Proc. ICGA93,* 1993, pp. 375–382.
Farley, Steven R.: "Mobile Agent System Architecture," *JAVA Report,* vol. 2, no. 5, May 1997.
Finin, T., D. McKay, and R. Fritzson (eds.): "An Overview of KQML: A Knowledge Query and Manipulation Language," KQML Advisory Group, March 1992.
Fischler, M. A., and O. Firchein: *Intelligence: The Eye, the Brain and the Computer,* Addison-Wesley, Reading, MA, 1987.
Flanagan, O.: *Consciousness Reconsidered,* MIT Press, Cambridge, MA, 1992.
Fodor, J.: *The Language of Thought,* Thomas Y. Crowell, New York, 1975.
Forsythe, W., and R. M. Goodall: *Digital Control,* McGraw-Hill, New York, 1991.
Franklin, S., and A. Graesser: "Is It an Agent, or just a Program?: A Taxonomy for Autonomous Agents," Institute for Intelligent Systems, University of Memphis, Web pages at URL: http://www.msci.memphis.edu/~franklin/AgentProg.html, 1996.
Frost, R.: "Java Agent Template (JAT)," developed by Robert Frost's group at CrossRoute Software, Inc., 1996.
Fuller, R. B.: *Synergetics: Explorations in the Geometry of Thinking,* Macmillan, New York, 1975.
____: *Synergetics II: Further Explorations in the Geometry of Thinking,* Macmillan, New York, 1979.
Gamma, E., R. Helm, R. Johnson, and J. Vlissides: *Design Patterns; Elements of Reusable Object-Oriented Software,* Addison-Wesley, Reading, MA, 1995.
Gazzaniga, M. S.: *Mind Matters: How the Mind and Brain Interact to Create Our Conscious Lives,* Houghton Mifflin, Boston, 1988.
Genesereth, M. R.: "Interoperability: An Agent-Based Framework," *AI Expert,* March 1995, pp. 34–40.
Genesereth, M. R., and S. P. Ketchpel: "Software Agents," *Commun. ACM,* vol. 37, no. 7, 1994, pp. 48–53.
Gleick, J:. *Chaos: Making a New Science,* Viking, New York, 1987.
Gosling, J.: "Java Complete," *Datamation,* March1, p. 32, 1996.

Graham, I. S.: *HTML Sourcebook; a Complete Guide to HTML,* Wiley, New York, 1995.
Greenwald, J.: "Buying a Car Without the Old Hassles," *TIME Mag.,* vol. 147, no. 12, 1996.
Gregory, R. L.: *The Oxford Companion to the Mind,* Oxford University Press, Oxford, UK, 1987.
Grossberg, S. (ed.): *Neural Networks and Natural Intelligence,* MIT Press, Cambridge, MA, 1988.
Guha, R. V., and D. B. Lenat: "Enabling Agents to Work Together," *Commun. ACM,* vol. 37, no. 7, 1994, pp. 127–142.
Hafez, W. A.: "Autonomous Planning under Uncertainty: Planning Models," *Int. J. General Systems,* 1989, pp. 188–193.
Halfhill, T. R.: "Agents and Avatars," *Byte,* February, 1996.
Harding, E. U.: "Distributed Objects Remain Rare," *Software Mag.,* vol. 13, no. 18, 1993, p. 25.
Haugeland, J. (ed.): *Mind Design: Philosophy, Psychology, Artificial Intelligence,* MIT Press, Cambridge, MA, 1981.
Head, J.: "Adding Real Intelligence to Mutual Management," *LAN Times,* vol. 11, no. 2, January 24, 1994, pp. 30–33.
Heitkoetter, J., and D. Beasley (eds.): Web pages: "The Hitch Hiker's Guide to Evolutionary Computation: A List of Frequently Asked Questions (FAQ)," at URL: http://www.cs.cmu.edu/Web/Groups/AI/html/faqs/ai/genetic/part1/faq.html, 1996.
Henry, J.: "Lost in Hyperspace," *OEM,* November 1995, pp., 21–24.
Hillis, W. D.: "Intelligence as an Emergent Behavior; or, The Songs of Eden," *Daedalus: J. American Academy of Arts and Sciences,* vol. 117, no. 1, Winter 1988, pp. 175–186.
Hofstadter, D.: *Godel, Escher, Bach: An Eternal Golden Braid,* Random House, New York, 1980.
Hofstadter, D., and D. Dennett: *The Mind's Eye: Fantasies and Reflections on Self and Soul,* Bantam, New York, 1982.
Hofstadter, D., and the Fluid Analogies Research Group: *Fluid Concepts and Creative Analogies: Computer Models of the Fundamental Mechanisms of Thought,* Basic, New York, 1995.
Hofstadter, D. R.: *Metamagical Themas: Questing for the Essence of Mind and Pattern,* Basic, New York, 1985.
Honeywell Industrial Automation Division, Web pages at URL: http://www.iac.honeywell.com, 1996.
Hookway, C. (ed.): *Minds, Machines and Evolution: Philosophical Studies,* Cambridge University Press, Cambridge, UK, 1984.
Hopfield and Kohonen
Hortuchi, H.: AT&T Wireless Services, 1996.
Jacobson, I.: *Object-Oriented Software Engineering: A Use Case Driven Approach,* Addison-Wesley, Reading, MA, 1993.
Johnson, J.: "Where Java Fits into the Distributed Object Paradigm," *Object Mag.,* June 1996.
____: R. Skoglund, and J. Wisniewski: *Program Smarter Not Harder,* McGraw-Hill, New York, 1995.
Johnson, R. C.: "Is Cognition Really Compression," *Electron. and Eng. Times,* October 1995, p. 47.
Kantrowitz, M., E. Horstkotte, and C. Joslyn: Web pages: "Answers to Frequently Asked Questions (FAQ) about Fuzzy Logic and Fuzzy Expert Systems," at URL: http://www.cs.cmu.edu/Web/Groups/AI/html/faqs/ai/fuzzy/part1/faq.html), November 1996.
Kauffman, S.: *The Origin of Order: Self-Organization and Selection in Evolution,* Oxford University Press, New York, 1993.
Kautz, H. A., B. Selman, and M. Coen: "Bottom-up Design of Software Agents," *Commun ACM,* vol. 37, no. 7, 1994, pp. 143–147.
Kelly-Bootle, S.: "OpenDoc—The Grand Tour," *Cross Platform Strategies,* Fall, 1995, pp. 16–20.
Kendall, E., et. al.: "The Application of Object-Oriented Analysis to Agent-Based Systems," *J. Object-Oriented Programming,* vol. 9, issue 9, February, 1997, pp. 56–62.
Keynes, M.: "Thin Client Called Threat to Wintel," *Electron. Eng. Times,* September, 1996, p. 18.
King, J. A.: "Intelligent Agents: Bringing Good Things to Life," *AI Expert,* February, 1995, pp. 17–19.
____: "Intelligent Agents: Part 2," *AI Expert,* March 1995, pp. 10–12.
KQML Advisory Group 1996, Web pages at URL: http://www.cs.umbc.edu/kqml, 1996.
Kurzweil, R.: *The Age of Intelligent Machines,* MIT Press, Cambridge, MA, 1992.
Krishnamurti, J.: *The Network of Thought,* Harper & Row, San Francisco, CA, 1982.
Laird, J., A. Newell, and P. Rosenbloom; "SOAR: An Architecture for General Intelligence," *Artificial Intelligence J.,* vol. 23, 1984, pp. 269–294.
Lanier, J.: "My Problem with Agents," *Wired,* November 1996, pp. 157–158.

Larijane, L. C.: *Virtual Reality Primer,* McGraw-Hill, New York, 1994.
Laural, B. (ed.): *Art of Human-Computer Interface Design,* Addison-Wesley, Reading, MA, 1993.
Lee, K.-C., W. H. Mansfield, Jr., and A. P. Sheth: "A Framework for Controlling Cooperative Agents," *IEEE Computer,* vol. 26, no. 7, 1993, pp. 8–16.
Lenat, D. B.: "Artificial Intelligence: A Crucial Storehouse of Commonsense Knowledge Is Now Taking Shape," *Scientific American,* September, 1995, pp. 80–82.
Lenat D. B., and Cycorp: Web pages at URL: http://www.cyc.com, 1996.
Leonard, A.: "Bots Are Hot," *Wired,* April, 1996, pp. 114–117, 166–172.
Levine, R., D. Drang, and B. Edelson: *Comprehensive Guide to Artificial Intelligence and Expert Systems,* McGraw-Hill, New York, 1986.
Levy, S.: *Artificial Life: The Quest for a New Creation,* Pantheon, New York, 1992.
Linthicum, D. S.: "Integration, Not Perspiration," *Byte,* January 1996, pp. 83–96.
Lockwood, M.: *Mind, Brain and the Quantum; the Compound "I,"* Blackwell, Oxford, UK, 1989.
Lovejoy, A. O.: *The Great Chain of Being,* Harper & Row, New York, 1933.
Lucky, R. W.: *Silicon Dreams: Information, Man and Machine,* St. Martin's Press, New York, 1989.
Maes, P.: "Agents that Reduce Work and Information Overload," *Commun. ACM,* vol. 37, no. 7, 1994, pp. 31–40.
___: Web pages for Software Agents Group, MIT Media Laboratory, at URL: http://agents.www.media.mit.edu:80/groups/agents/research.html, 1996.
Majewski, S. D.: Mail message to agents@sun.com mailing list in "Clarifications and Additions for AI: A Modern Approach," Web page at URL: http://www.cs.berkeley.edu/~7Erussell/clarify.html, 1996.
Mayfield, J., T. Finin, R. Narayanaswamy, C. Shah, W. MacCartney, and K. Goolsbey: "The Cycic Friends Network: Getting Cyc Agents to Reason Together," CIKM Conf., 1995.
McKie, S.: "Software Agents: Application Intelligence Goes Undercover," *DBMS,* April, 1995.
Mead, C.: *Analog VLSI and Neural Systems,* Addison-Wesley, Reading, MA, 1989.
Merriam-Webster's Collegiate Dictionary, 10th ed., 1993, p. 22.
Minsky, M.: *The Society of Mind,* Simon & Schuster, New York, 1986.
Minsky, M., and S. Papert: *Perceptrons,* expanded ed., MIT Press, Cambridge, MA, 1998.
Mowbray, T.: "Essentials of Object-Oriented Architecture," *Object Mag.,* September, 1995, pp. 28–32.
Mowbray, T. J., and T. Brando: "The Goal of True Interoperability Has Yet to Be Reached," *Object Mag.,* September-October 1993, pp. 51–54.
Muller, J. P., and Mitsubishi Electric Digital Library Group: "The Design of Intelligent Agents: A Layered Approach," London, UK, 1996.
Murray, D.: "Developing Reactive Software Agents," *AI Expert,* March 1995, pp. 27–29.
Negroponte, N.: *Being Digital,* Alfred A. Knopf, New York, 1995.
Norton, P., S. Holzner, H. Davis, and P. Davis, "Peter Norton's Guide to Visual Basic 4 for Windows 95," 1995, Sams Publishing, a division of Macmillan Publishing, USA, Indianapolis, IN.
Norton, P.: *Visual Basic for Windows Release 3.0,* 1993.
Orfali, R., and D. Harkey: "Client/Server with Distributed Objects," *Byte,* April 1995, pp. 151–161.
Orfali, R., D. Harkey, and J. Edwards: "Client/Server Components: CORBA Meets OpenDoc," *Object Mag.,* May 1995, pp. 55–59.
___: *The Essential Distributed Objects Survival Guide,* Wiley, New York, 1996.
Parsaye, K., M. Chignell, S. Khoshafian, and H. Wong: *Intelligent Databases: Object-Oriented, Deductive Hypermedia Technologies,* Wiley, New York, 1989.
Penrose, R.: *The Emperor's New Mind,* Oxford University Press, Oxford, UK, 1989.
___: *Shadows of the Mind: A Search for the Missing Science of Consciousness,* Oxford University Press, Oxford, UK, 1994.
Petrie, C. J.: "Agent-Based Engineering, the Web, and Intelligence," Stanford Center for Design Research, Web pages at URL: http://cdr.stanford.edu/NextLink/Expert.html, 1996.
Pietsch, P.: *Shufflebrain: The Quest for the Holographic Mind,* Houghton Mifflin Company, Boston, MA, 1981.
Pleas, K.: "OLE's Missing Links," *Byte,* April, 1996, p. 102.
Pohl, I.: *Object-Oriented Programming Using C++,* Benjamin/Cummings, Redwood City, CA, 1993.
Polese, K.: "I Don't Know Why You Say Goodbye, I Say Hello," *Java Rep.*, March/April, 1996.
Popper, K., and Sir J. Eccles: *The Self and Its Brain: An Argument for Interactionism,* Routledge Kegan Paul, London, UK, 1983.

Proc. 3rd Int. Workshop on Agent Theories, Architectures, and Languages, Springer, 1996.

Push-Pull Technologies: Web page at URL: http://www.netscape.com/assist/net_sites/pushpull.html, 1997.

Putnam, H.: *Representation and Reality,* MIT Press, Cambridge, MA, 1988.

Rao, V. B., and H. V. Rao: *C++, Neural Networks and Fuzzy Logic,* Henry Holt, New York, 1993.

Restak, R. M.: *The Modular Brain,* Simon & Schuster, New York, 1994.

Riecken, D.: "M: An Architecture of Integrated Agents," *Commun. ACM,* vol. 37, no. 7, 1994, pp. 107–116.

Rosenblatt, *Principles of Neurodynamics: Perceptions and the Theory of Brain Mechanisms,* Spartan Books, New York, 1962.

Rosenschein, J. S., and G. Zlotkin: *Rules of Encounter; Designing Conventions for Automated Negotiation among Computers,* MIT Press, Cambridge, MA, 1994.

Rosenschein, J. S., and G. Zlotkin: "Designing Conventions for Automated Negotiation," *AI Mag.,* vol. 15, no. 3, 1994, pp. 29–46.

Roy, M., and A. Ewald: "Combining CORBA and OLE Automation," *Object Mag.,* November-December, 1995, pp. 79–80, 85.

Rumbaugh, J., et. al.: *Object-Oriented Modeling and Design,* Prentice-Hall, Englewood Cliffs, NJ, 1996.

Rumelhart, D. E., J. McClelland, and the PDP Research Group: *Parallel Distributed Processing: Explorations in the Microstructure of Cognition,* vols. 1, 2: *Foundations,* MIT Press, Cambridge, MA, 1988.

Russell, S., and P. Norvig: *Artificial Intelligence: A Modern Approach,* Prentice-Hall, Englewood Cliffs, NJ, 1994.

Ryner, J. R.: "Muddle in the Middle," *Byte,* no. 3, 1996, pp. 67–70.

Schwartz, J. T.: "The New Connectionism: Developing Relationships between Neuroscience and Artificial Intelligence," *Daedalus: J. American Academy of Arts and Sciences,* vol. 117, no. 1, Winter, 1988, pp. 123–124.

Searle, J.: *Rediscovery of the Mind,* MIT Press, Cambridge, MA, 1994.

Smith, D. C., A. Cypher, and J. Spohrer: "KIDSIM: Programming Agents without a Programming Language," *Commun. ACM,* vol. 37, no. 7, 1994, p. 55.

Soucek, B., and M. Soucek: *Neural and Massively Parallel Computers,* Wiley, New York, 1988.

Sugawara, K., T. Suganuma, G. Chakraborty, M. Moser, T. Kinoshita, and N. Shiratori: "Agent-Oriented Architecture for Flexible Networks," in *Proc. IEEE 2d Int. Symp. on Autonomous Decentralized Systems,* 1995, pp. 135–141.

Tognazzini, B.: *TOG on Interface,* Addison-Wesley, Reading, MA, 1992.

Valiant, L. G.: *Circuits of the Mind,* Oxford University Press, New York, 1994.

Wah, B., and G. J. Li: "Intelligent Control," in *Proc. Symp. on Proceedings Computers for Artificial Intelligence Applications: Tutorial,* IEEE Computer Society Press, Los Angeles, CA, 1989.

Watson, M.: *Programming Intelligent Agents for the Internet,* McGraw-Hill, New York, 1996.

Wayner, P.: *Agents Unleashed,* AP Professional, Boston, MA, 1995.

Wiener, N.: *Cybernetics: Or Control and Communication in the Animal and the Machine,* MIT Press, Cambridge, MA, 1948.

Weiner, R. S., and L. J. Pinson: *An Introduction to Object-Oriented Programming and C++,* Addison-Wesley, Reading, MA, 1988.

White, J. (ed.): *Frontiers of Consciousness: The Meeting Ground between Inner and Outer Reality,* Julian Press, New York, 1985.

White, J. E.: "Telescript Technology: Mobile Agents," a General Magic White Paper, General Magic, October, 1995.

Whitehead, A. N.: *Process and Reality,* corrected ed., Macmillan, New York, 1929.

Wilber, K.: *The Spectrum of Consciousness,* Theosophical Publishing House, Wheaton, IL, 1977.

Winblad, A. et al.: *Object-Oriented Software,* Addison-Wesley, Reading, MA, 1990.

Wirfs-Brock, R., B. Wilkerson, and L. Weiner: *Designing Object-Oriented Software,* Prentice-Hall, Englewood Cliffs, NJ, 1990.

Woelk, D., et al.: "Uncovering the Next Generation of Active Objects," *Object Mag.,* July–August, 1995.

Wolf, F. A.: *Star Wave: Mind, Consciousness and Quantum Physics,* 1984.

Wolff, G.: Web pages: "Computing as Compression: The SP Theory and the SP System," School of Electronic Engineering and Computer Systems, University of Wales at Bangor, at URL: http://www.sees.bangor.ac.uk/~gerry/sp_summary.html, November, 1996.

Wolfram, D. D., T. J. Dear, and C. S. Galbraith: *Expert Systems for the Technical Professional,* Wiley, New York, 1987.

Wooldridge, M., J. P. Mueller, and N. R. Jennings (eds.): "Intelligent Agents III: Agent Theories, Architectures, and Languages," in *Lecture Notes in AI,* vol. 1193, Springer, New York, 1997.

Yamamoto, G., and D. Chang: "Programming Mobile Agents in Java: A Demonstration of Aglets Workbench," IBM Corp, Tokyo, (as witnessed by Michael D. Knapik at OOPSLA 1996), 1996.

Young, A. M.: *The Reflexive Universe: Evolution of Consciousness,* 1976.

Yourdon, E.: *Decline and Fall of the American Programmer,"* Prentice-Hall, Englewood Cliffs, N J, 1993.

Zadeh, L. A.: "Fuzzy Logic, Neural Networks, and Soft Computing," *Commun. ACM*, vol. 37, no. 3, 1994, p. 78.

____: and J. Kacprzyk: *Fuzzy Logic for the Management of Uncertainty,* Wiley, New York, 1992.

____: "Fuzzy Sets," *Information and Control,* 8, 1965, pp. 338–353.

Index

Abstract data types, 65
Abstractions, in object orientation, 53–54
Acronyms (list), 369–373
ActiveX, 130, 133–135
    developing agents based on, 135–136
    and Java, 135
    technologies in, 134–135
Adler, Darin, 310
Advertising agents, 330
AFIN (*see* Agent-oriented flexible information network)
Agencies, 3
    of human mind, 24–25
Agent Collaboration Language, 364
Agent Communication Language, 164–167
Agent-oriented flexible information network (AFIN), 188–189
Agents:
    and communications, 34–36
    definition of, 2
    goal-driven, 31–32
    of human mind, 24
    knowledge-based, 30–31
    learning by, 33–34
    reasoning by, under uncertainty, 32–33
    [*See also* Intelligent agents (IAs)]
Agentware, 113
Aglets, 283–284
AI (*see* Artificial intelligence)
AIAMA (*see* Artificial Intelligence: A Modern Approach)
ALCELIS, Inc., 106
ALIVE, 361
Annealing, simulated, 30
ANNs (*see* Artificial neural networks)
Anthropomorphic agents, 338
API (*see* Application programming interface)
AppleEvents, 147–148
Applets, 282
Application programming interface (API), 42
Applications of agents, 317–340
    advertising agents, 330
    anthropomorphic agents, 338
    artificial neural networks (ANNs), 112
    assistant agents, 323
    author's assistants, 337–338

Applications of agents (*Cont.*):
    big brother agents, 338–339
    "bots," 337
    commerce agents, 327–330
    communication management agents, 321–322
    computer-aided design assistants, 334
    Cyc, 46–48
    database agents, 320–321, 331
    decision support agents, 336–337
    design engineering agents, 334–335
    e-mail agents, 330–331
    filtering agents, 324–325
    financial agents, 324
    fuzzy systems/techniques, 97–99
    governmental agents, 332–333
    industrial automation/control agents, 331–332
    Java, 279
    M architecture, 160
    medical agents, 333–334
    meeting facilitator agents, 339–340
    military agents, 334
    network agents, 317–320
    research/reporting agents, 325–326
    search agents, 322
    technical assistance agents, 336
    telephony agents, 326–327
    Telescript, 302–310
    virtual communities, 322
    work-flow-automation agents, 323–324
Application-specific integrated circuits (ASICs), 20
Approximate reasoning, 94
Architectures, agent, 151–192
    analyzing, 154–155
    bottom-up prototyping/iteration approach to, 179–181
    characteristics of good, 153–154
    collaborative approach to, 184–187
    communications infrastructure vs., 117, 152–153
    complex, 156
    and dealing with change, 188–192
    and emergence, 187–188
    integration-of-diversification approach to, 157–160

Architectures, agent (*Cont.*):
  interoperability approach to, 161–169
  ITX, 181–184
  moderately complex, 156
  negotiations approach to, 170–179
  simple, 155–156
ARCHON, 362
Artificial Intelligence: A Modern Approach (AIAMA) (Russell & Norvig), 6, 12, 198, 226–227
Artificial intelligence (AI):
  and communications, 34–36
  and computing architecture, 26–29
  development of, 12–15
  future of, 365–367
  and goal-driven agents, 31–32
  human brain as inspiration for, 21–23
  and intelligence, 15–17
  and interpretation, 15
  and knowledge-based agents, 30–31
  and learning, 33–34
  potential impact of, 12
  and reasoning under uncertainty, 32–33
  resources on, 6, 11–12
  and search techniques, 30
Artificial neural networks (ANNs), 21, 23, 107–113
  applications of, 112
  architecture of, 110–111
  brain as model for, 107–108
  incorporation of, into agents, 112–113
  need for, in computational systems, 108–110
  and supervised/unsupervised training, 111–112
ASICs (*see* Application-specific integrated circuits)
Assistant agents, 323
Atkinson, B., 3, 4
AT&T, 310
Authority of mobile agents, 240–241
Author's assistants, 337–338
Autonomous agents, 2, 357–358
Autonomy, Inc., 113
Autonomy of agents, 3
Avatars, 322, 344

BargainFinder, 344
Barr, A., 12, 16
Bateson, G., 227–228, 366
Bernstein, P. A., 116
Big brother agents, 338–339, 347–348
BioComp, 113

Blumberg, Bruce, 361
Bohm, David, 18
"Bots," 337
Bottom-up approach to agent architecture, 179–181
Brain, 365–367
  as distributed system, 28
  as hierarchy, 17–19
  information compression by, 28–29
  as inspiration for AI, 21–23
  as model for ANNs, 107–108
  modular reorganization in, 28
  "standards" in, 24
browsing agents, 362
Bunyip, 363

Cairo, 136
Carnot, 364
CBO (coupling between objects), 69
Chavez, Anthony, 361
Chinese room mind game, 14–15
Classes (object orientation)
  attributes of, 55
  domain-specific, 80–81
  and reuse of agents, 88–90
  RootAgent class, 80
Client/server, 120–121, 204–207
CLIPS/R2, 41
Collaborative agent systems, 16
Collaborative approach to agent architecture, 184–187
COM (*see* common object model)
Commercial agents:
  future trends in, 344–346
  people, finding, 329–330
  products, finding, 328–329
Common object model (COM), 5, 42, 131–132
Common object request broker architecture (CORBA), 5, 128, 137–143, 261–264
Communication management agents, 321–322
Communications:
  and agents, 34–36
  interprocess, 255–258
  object orientation, 64–65
Communications infrastructures, 115–150
  ActiveX, 130
  AppleEvents, 147–148
  architecture vs., 117, 152–153
  choosing, 123–124
  and client/server, 120–121
  CORBA, 137–143

Communications infrastructures (*Cont.*):
　DBMSs, 149
　DCE, 143–146
　evaluating, 121–124
　expert system environments, 150
　functionality of, 117
　groupware, 149–150
　and interoperability "standards," 119–120
　as middleware services, 116
　NEO, 146–147
　OLE, 130–133
　OpenDoc, 125–130
　operating systems, 148–149
　in overall system, 118, 119
　PDO, 147
　scope of, 116–117
Compression, by brain, 28–29
Computer-aided design assistants, 334
Computers *See also* artificial intelligence (AI)
　earliest, 14
　"intelligent," 13
Consciousness, quantum level of, 17–19
　(*See also* Mind)
Constraint agents (databases), 320
Constructors, 56
CORBA (*see* Common object request broker architecture)
Coupling between objects (CBO), 69
Cyc, 43–52
　accessing, 51–52
　applications of, 46–48
　and CycL, 44
　distributed IAs using, 48–51
　inferencing in, 44–45
　interface tools in, 45
　knowledge base in, 44
CycL, 44
Cycorp, Inc., 49–51

Database agents, 320–321, 331, 347
Database management systems (DBMSs), 149
DCAs (*see* Distributed communicating agents)
DCE (*see* Distributed computing environment)
Decision support agents, 336–337
Decision theory, 32, 33
Deduction, 34
DENDRAL, 40
Depth of inheritance tree (DIT), 69
Descartes, R., 28

Design, agent, 193–258
　and behavior of agents, 198–199
　and common agent platforms, 210–212
　and communications between agents, 255–258
　and components of agents, 228–230
　desirable features in, 196–197
　and distributed computing paradigms, 203–210
　and end-user expectations, 197–198
　and end-user programming/configuration, 251–255
　and environmental modeling, 225–228
　and error/exception handling, 248–250
　and execution environments, 200–203
　and goals of agents, 199–200
　human factors in, 212–220
　incorporating goals/planning in, 222–225
　and mobility of agents, 230–248
　requirements analysis for, 194–196
　security considerations in, 250–251
　and software upgrades, 220–221
Design engineering agents, 334–335
Destructors, 56
Development environments, 259–315
　Distributed Smalltalk, 274–278
　Java (*see* Java)
　and modern object-oriented languages, 312–314
　power-user environments, 261–265
　and security, 311
　simple agent builders, 260–265
　Smalltalk (*see* Smalltalk)
　Telescript (*see* Telescript)
DFS. (*see* Distributed file services)
Digital signal processors (DSPs), 20n
Distributed communicating agents (DCAs), 364
Distributed computing environment (DCE), 143–146
Distributed file services (DFS), 144
Distributed Smalltalk, 274–278
　and agent interaction, 277–278
　basic application frameworks in, 275–277
　interactive environment of, 274–275
　open architecture of, 275
DIT (depth of inheritance tree), 69
DLLs (*see* Dynamic link libraries)
DNA, 18, 19
Domains:
　categorization of, in agent architecture, 176–178
　classes specific to, 80–81

DSPs (*see* Digital signal processors)
Dumb agents, 4
Dynamic binding, 65
Dynamic link libraries (DLLs), 106

Edmonds, E. A., 185, 186, 188
E-mail agents, 330–331
Emergence, 38, 187–188
Encapsulation (object orientation), 57, 77–78, 89
End-users, 216–217
  agent programming/configuration by, 251–255
  expectations of, 197–198
Environmental modeling, 225–228
Environments, development (*see* Development environments)
Ethical issues, 348
Etzioni, O., 224–225
European Research in Uncertainty, 101
Evolutionary computing, 101–107
  approaches to, 102–103
  incorporation of, into IAs, 105–107
  need for, in IA systems, 103–105
  scope of, 102
Evolutionary programming, and learning, 34
Evolver, 106
Exception handling:
  and agent design, 248–250
  Smalltalk, 274
Execution environments, 200–203
Expert systems, 16
  "classical," 39–43
  environments in, 150
  [*See also* Cyc; knowledge bases (KBs)]
Explicate order, 18

Feedback, 14
  and learning, 34
Feelings, 27
Feigenbaum, E., 12, 16
Field-programmable gate arrays (FPGAs), 20*n*
Filtering agents, 324–325
Financial agents, 324
Firefly, 338
Flexible Intelligence Group LLC, 106
FlexTool Evolutionary Fuzzy Modeling, 106
Foner, Leonard, 359
FPGAs (*see* Field-programmable gate arrays)
France Telecomm, 310

Franklin, S., 2
Future trends in agents, 341–368
  application infiltration, 346
  avatars, 344
  big brother agents, 347–348
  commercial agents, 344–346
  cooperative information systems, 348–349
  database agents, 347
  developers' role, 349–350
  ethical issues, 348
  examples of, 362–365
  global desktop, 353
  home uses, 353–354
  Internet searching, 346
  military agents, 347
  at MIT Media Laboratory, 357–362
  network management, 343–344
  personal representation, 356–357
  replacement of human roles, 351–353
  security/privacy issues, 354–356
  social issues, 350–351
  and trust in AI by users, 348
Fuzzle for Windows, 100–101
Fuzzy logic, 33, 92, 94, 96–99, 101
Fuzzy sets, 93–96
Fuzzy systems/techniques, 91–101
  applications of, 97–99
  developing, 99–101
  inferencing methods used by, 96–97
  logic of, 96
  need for, in IA systems, 92–94
  and set theory, 94–96
Fuzzy-Box, 101
FuzzyExpert, 100

G2, 150
Game agents, 361–362
General Magic, 5, 310
Genesereth, M. R., 161–168
Genetic algorithms, 102
  and learning, 34
Genetic programming, 360
Gensym, 150
Gestalt theory, 25
Goal-driven agents, 31–32
Goals of agents, 199–200, 222–225, 229
Godel, K., 13*n*, 18
Governmental agents, 332–333
Graesser, A., 2
Granulation, 93
Guardian, 363
Guttman, Robert, 361

Hafez, W. A., 228
Handbook of Artificial Intelligence (Boor and Feigenbaum), 16
Hardware, 20
Harkey, D., 140
Hawking, Stephen, 19
Hebb, D. O., 21–22
Hebbian learning, 21
Hertzfeld, Andy, 310
Heuristics, 30
Hierarchies in intelligent systems, 17–21
  atomic level, 18
  cellular level, 18, 19
  consciousness level, 18–20
  and IA systems, 20–21
  organ level, 18
  quantum level, 17–18
Hill climbing, 30
Hillis, W. D., 26, 27
Holography, 18, 19
Honeywell, 98
HTML (*see* HyperText Markup Language)
HTTP (*see* HyperText Transport Protocol)
Human needs, and agent design, 212–220
HyperText Markup Language (HTML), 208–210
HyperText Transport Protocol (HTTP), 208–209

"I," 19–20, 25, 366
IAs (*see* Intelligent agents)
IBM, 3, 42, 113
IBM/Lotus Notes, 6, 42, 149–150, 261, 321
Implicate order, 18
Incompleteness theorem, 13n, 18
Indigo Software Ltd., 100
Inductive learning, 34
Industrial automation/control agents, 331–332
Inferencing, 30
  in Cyc, 44–45
  in fuzzy systems, 96–97
Information compression, by brain, 28–29
Information overload, 358–359
Information theory, 14
Infrastructures, communications (*see* Communications) infrastructures
Inheritance (object orientation), 57–58, 66
Intelligence, 3
  emergent collective, 38
  leveraging of existing, in agent architecture, 169–170

Intelligent agents (IAs)
  components of, 228–230
  definitions of, 2–4
  "dumb" agents vs., 4
  and hierarchy, 20–21
  simple agents vs., 221–222
Intelligent machines, 13
Intelligent systems, hierarchical organization in, 17–21
Interative improvement algorithms, 30
Internet searching, 346
Internet Softbot, 224–225
Interoperability approach to agent architecture, 161–169
  assisted coordination in, 164–165
  direct communications in, 164–165
  facilitator in, 167–169
  language standardization in, 163–167
  and leveraging of existing intelligence, 169–170
  metaprotocols for, 161–163
Interoperability "standards," 119–120
Interpretation, 15
I**3 program, 363
Itinerant agents (*see* Mobile agents)
ITX architecture, 181–184
IWAH, 364

JAT, 334–335
Java, 6, 278–289
  and ActiveX, 135
  advantages of, 278
  agent applications in, 279
  agent supported facilities in, 288–289
  agent template in, 284–288
  agile agents in, 283–284
  applets/servelets in, 282
  C++ vs., 279
  as OO language, 280–281
  potential of, 312
  Telescript vs., 312, 314
  tools in, 281–282
Johnson, Michael, 360, 361
Junk mail, 5

KACTUS, 362–363
Kasbah, 361
Kautz, H. A., 179–181
KBs (*see* Knowledge bases)
Kelly-Bootle, S., 125–126
KIDSIM, 252–254

KIF (*see* Knowledge interchange format)
King, J. A., 2
Knapik test, 13
Knowledge bases (KBs), 30, 31
Knowledge engineering, 31, 97
Knowledge interchange format (KIF), 256
Knowledge Query and Manipulation Language (KQML), 48, 49, 212, 255–256, 285–288
Knowledge-based agents, 30–31
Knowledge-based systems, 16
KQML (*see* Knowledge Query and Manipulation Language)
Krakatoa, 334
Kreyser, Rick, 346
Krishnamurti, J., 25

Lack of cohesion in methods (LCOM), 70
LAN management agents, 318
Lanier, J., 356
LCOM (lack of cohesion in methods), 70
Learning:
   by agents, 33–34
   hebbian, 21
Lederburg, Joshua, 40
Lenat, D. B., 43, 51
Lexical analysis, 15
LiveObjects, 129–130
Logic, 30–31
Lotus Notes (*see* IBM/Lotus Notes)
Lynch, Kevin, 310

M architecture, 157–160
   components of, 157–159
   features of, 159–160
   sample application of, 160
Maes, P., 218, 357–361
Magic Cap, 5
MailBot, 364
Maintenance (object orientation), 70–71, 73–74
Majewski, S. D., 4
Marshaling (common object model), 131–132
Matchmaking agents, 359
Math Works, Inc., 101
Mathematical reasoning, mechanization of, 13
Matlab, 101
Maximum expected utility (MEU), 32–34, 227
McClelland, J., 21, 22, 23*n*

Medical agents, 333–334, 352–353
Meeting facilitator agents, 339–340
Meetings, agent, 239
Memory management, in Smalltalk, 272–273
Metaprotocols, 161–163
Methods (object orientation), 55–57
MEU (*see* Maximum expected utility)
Middleware services, 116
Military agents, 334, 347
Miltonberger, Tom, 310
Mind, 365–367
   agencies of, 24–25
   and network intelligence, 24
Minsky, M., 23, 23*n*, 24
MIT Media Laboratory, 357–362
Mobile agents, 202–203, 230–248
   authority of, 240–241
   and client/server paradigm, 205–207
   connections between, 239–240
   features of, 232–233
   fixed/stationary agents vs., 231–232
   meetings of, 239
   permits for, 241–242
   phases in life cycle of, 248
   and places, 236–237
   RP paradigm for, 234–236
   RPC paradigm for, 233–234
   software technology for, 243–246
   travel by, 238–239, 243
Modal logic, 31
Modeling:
   environmental, 225–228
   object orientation, 61–63, 76
Model-view controller (MVC), 116
Modico, Inc., 100–101
Modifiers, 56
Modular design, and brain function, 28
Monikers (common object model), 132
Moukas, Alexandros, 358–359
Mountcastle, Vernon, 28
Multitasking, 202–203
   object orientation, 63–64
Multithreaded operating systems, 202–203
MVC (model-view controller), 116
MYCIN, 40

NASA, 364
Negotiations approach to agent architecture, 170–179
   domain categorization in, 176–178
   and game theory, 172
   maximization in, 179

Negotiations approach to agent architecture (Cont.):
  salutation architecture, 172–175
  trade offs in, 171–172
NEOs (see Networked Objects)
NetWare, 318–319
Network agents, 317–320
Network intelligence, 24
Networked Objects (NEOs), 146–147
Networks, future trends in, 343–344
Neural Network Utility (NNU), 113
Neural networks, artificial [see Artificial neural networks (ANNs)]
NeuroGenesis, 113
Newt system, 325
NNU (see Neural Network Utility)
NOC (number of children), 69
Norman, Donald, 23n
Norvig, P., 2–3, 6, 12, 29–36, 226–227
NOT (number of tramps), 70
Number of children (NOC), 69
Number of tramps (NOT), 70

Object database management systems (ODBMSs), 120
Object linking and embedding (OLE), 130–133, 261–264
  and ActiveX, 130, 133–135
  architecture of, 130–131
  and common object model, 131–132
  developing agents based on, 135–137
  and OpenDoc, 127
  structure of, 132–133
Object orientation (OO), 52–90
  abstract data types in, 65
  abstractions in, 53–54
  agent development costs in, 72
  attributes of agents in, 77
  classes in, 55, 78, 80–81
  communications considerations in, 64–65
  and composition, 74–75
  composition of agents in, 76–77, 89
  defining agents in, 76
  design goals for, 60–61
  and design process, 63
  drawbacks of, 67
  and dynamic binding, 65
  dynamic model of, 61–62
  encapsulation in, 57, 77–78, 89
  extensibility of agents in, 74
  flexible agent structuring in, 73
  inheritance in, 57–58, 66

Object orientation (OO), 52–90 (Cont.):
  instantiation in, 65–66, 78, 80, 90
  intrinsic system knowledge in, 75–76
  and Java, 280–281
  and maintenance, 70–71, 73–74
  methods in, 55–57
  metrics for, 68–70
  models/simulations in, 76
  modern OO languages, 312–314
  and multitasking, 63–64
  objects in, 53
  operations of agents in, 77
  persistence in, 60
  polymorphism in, 58–60
  prototyping, 62–63
  relationships among agents in, 81–88
  requirements analysis in, 60
  reuse of agents in, 71–72, 88–90
  RootAgent class in, 80
  states in, 54–55
  static model of, 61
  and team-based development, 67–68
  understandability of agents in, 74
Objects, 53
  coupling between, 69
  networked, 146–147
  portable distributed, 147
  in Telescript, 293–295
OCX, 41
ODBMSs (see Object database management systems)
OLE (see Object linking and embedding)
On-line agents, 352
Ontologies, 31
OO (see Object orientation)
OpenDoc, 5, 125–130
Operating systems (OSs), 20, 148–149, 202–203
Orfali, R., 140
OSs (see Operating systems)

Papert, S., 23
Parallel distributed processing (PDP), 21–22
ParcPlace-Digitalk, 5
PBD (programming by demonstration), 254–255
PDOs (portable distributed objects), 147
PDP (see Parallel distributed processing)
Penrose, Roger, 18, 19
Pentium MMX chip, 20n
Pentland, Alex, 361
Perceptrons, 22–23

Permits (mobile agents), 241–242
Persistence (object orientation), 60
Persona project, 362
Places, 236–237
Planning, 31–32
Polymorphism (object orientation), 58–60
Portable distributed objects (PDOs), 147
Principles of Neurodynamics (Rosenblatt), 22
Privacy issues, 355–356
Proactivity of agents, 4
Probability theory, 32
Problem solving, with search methods, 30
Programming by demonstration (PBD), 254–255
Publish/Subscribe, 147

Quantum mechanics, 17–19

Rational agents, 3
Reactivity of agents, 3
Reasoning, 30
  approximate, 94
Redundancy, 28–29
Remembrance agents, 359–360
Requirements analysis, 194–196
Research/reporting agents, 325–326
Response for a class (RFC), 70
Restak, R. M., 28
RFC (response for a class), 70
Rhodes, Bradley, 359, 361
RNA, 18
Roaming agents, 125
Robots, 36
RootAgent class, 80
Rosenblatt, 22, 23n
Rosenschein, J. S., 172, 176, 178, 179
RP paradigm, 234–236
RPC paradigm, 233–234
Rumbaugh, J., 54–55
Rumelhart, D. E., 21–23
Russell, Bertrand, 13
Russell, S., 2–3, 6, 12, 29–36, 226–227

SA (*see* Salutation architecture)
SAAM (*see* Software architecture analysis method)
Salutation architecture (SA), 172–175
Schramm, Steve, 310
Schwartz, J. T., 22
Search agents, 322

Search methods, problem solving with, 30
Searle, J., 14, 25
Security:
  and agent design, 250–251
  and agent development environments, 311
  distributed environments, 143
  and future trends, 355–356
Selectors, 56
Self, 25
Semantic analysis, 15
Semantic networks, 31
Servelets, 282
Shannon, Claude, 14
Silk, 363
Simulated annealing, 30
SireneF, 101
Smalltalk, 162, 232, 265–274
  advantages of, 265–266
  characteristics of, 266
  components of agents in, 268–269
  development environments for, 267
  on-line auction (example), 269–270
  as OO language, 280–281
  specializing agents in, 270–271
  support for agents in, 271–274
  See also Distributed Smalltalk
SOAR, 50
Social ability of agents, 3
Social issues, 350–351
The Society of Mind (Minsky), 23
Soft computing technologies, 90–113
  artificial neural networks (ANNs), 107–113
  evolutionary computing, 101–107
  fuzzy systems/techniques, 91–101
Softbots, 36
Software, 20
Software agents, 358
Software architecture analysis method (SAAM), 154–155
SP theory, 29
SRI Open Agent Architecture, 363
Standards, 5
  interoperability, 119–120
  language, 163–167
Starner, Thad, 359
States (object orientation), 54–55
Storytelling agents, 362
Subgoals, 31
Sugawara, K., 188
Sulla, 364–365
Sun Microsystems, 146, 147
Superimplicate order, 18

Symbols:
  encoding of, in brain, 24
  manipulation/processing of, 13–15
Syntactic analysis, 15

Tabriz, 5
TACOMA project, 363
Team-based development, object orientation and, 67–68
Technical assistance agents, 336
Telephony agents, 326–327
Telescript, 236–240, 243–247, 289–310, 351
  agent programming in, 298–302
  agents in, 237–238
  authority of agents in, 240–241
  development of, 310
  engine, 245
  Java vs., 312, 314
  language, 243–245
  meetings of agents in, 239
  mobile agent applications using, 302–310
  and mobile agent technology, 290–293
  objects in, 293–295
  places in, 236–237
  programming in, 295–302
  protocols, 245–246
  travel by agents in, 238–239
Temporal logic, 30–31
Training, and artificial neural networks, 111–112
Travel, by mobile agents, 238–239, 243
Travel agents, 351–352
Trust in agents, 348
Turing, Alan, 13
Turing test, 13, 14

UCI Microelectronique, 101
Understanding, 26–27

Understanding (*Cont.*):
  defining, 14–15
UNIX, 148–149
Utility theory, 32

Velasquez, Juan David, 358
Violations of the law of Demeter (VOD), 70
Virtual communities, 322
Virtual meeting rooms (VRMs), 160
Virtual Reality Markup Language (VRML), 208
VOD (violations of the law of Demeter), 70
von Neumann, John, 14
VRML (Virtual Reality Markup Language), 208
VRMs (virtual meeting rooms), 160

WAC (weighted attributes per class), 70
WEBLS, 42
Weighted attributes per class (WAC), 70
weighted methods per class (WMC), 69
Wexelblat, Alan, 360
Wheeler, John, 19
White, Jim, 310
Wiener, N., 14
WMC (weighted methods per class), 69
Wolff, G., 29
Work-flow agents, 220
Work-flow-automation agents, 323–324

XCON, 41
X-Window, 120

Zadeh, L. A., 93, 94
ZeTec GmbH, 101
Zlotkin, G., 172, 176, 178, 179

ABOUT THE AUTHORS

MICHAEL KNAPIK is a principle software engineer at Honeywell, specializing in AI, object technology, and expert systems. He builds custom homes in his spare time and is President of Conscious Systems, a firm researching consciousness, and the brain/mind connection.

JAY JOHNSON is a senior software engineer at IBM. His 15-year career has focused on real-time process control, object technology, database management, and client/server applications. He is the coauthor of *Program Smarter, Not Harder: Get Mission Critical Projects Right the First Time* (McGraw-Hill, 1995).